# Reformed Theology and Evolutionary Theory

Gijsbert van den Brink

WILLIAM B. EERDMANS PUBLISHING COMPANY
GRAND RAPIDS, MICHIGAN

Wm. B. Eerdmans Publishing Co.
4035 Park East Court SE, Grand Rapids, Michigan 49546
www.eerdmans.com

© 2020 Gijsbert van den Brink
All rights reserved
Published 2020
Printed in the United States of America

26 25 24 23 22 21 20    1 2 3 4 5 6 7

ISBN 978-0-8028-7442-9

**Library of Congress Cataloging-in-Publication Data**
Names: Brink, Gijsbert van den, 1963- author.
Title: Reformed theology and evolutionary theory / Gijsbert van den Brink.
Description: Grand Rapids : Wm. B. Eerdmans Publishing Co., 2020. | Includes bibliographical references and index. | Summary: "An intermediate-level book on how evolutionary theory conflict or agrees with important Christian doctrines"— Provided by publisher.
Identifiers: LCCN 2019026837 | ISBN 9780802874429 (paperback)
Subjects: LCSH: Evolution—Religious aspects—Christianity. | Reformed Church—Doctrines.
Classification: LCC BT712 .B75 2020 | DDC 261.5/5—dc23
LC record available at https://lccn.loc.gov/2019026837

Unless otherwise noted, Scripture quotations are from the New Revised Standard Version of the Bible, copyright © 1989 by the Division of Christian Education of the National Council of the Churches of Christ in the U.S.A., and used by permission.

# Contents

| | |
|---|---|
| *Foreword* by Richard J. Mouw | ix |
| *Preface* | xiii |
| *Introduction* | 1 |

### 1. Reformed Theology as a Distinctive Stance — 10

| | |
|---|---|
| 1.1 Introduction | 10 |
| 1.2 Some Common Misunderstandings | 12 |
| 1.3 Endless Plurality? | 18 |
| 1.4 The Reformed Theological Stance | 22 |
| 1.5 Reformed Theology and the Natural World | 26 |
| 1.6 Seeking the Honeycomb in the Lion's Mouth | 31 |

### 2. Evolutionary Theory as a Layered Concept — 33

| | |
|---|---|
| 2.1 Varieties of Evolution: Terminological Clarifications | 33 |
| 2.2 Gradualism | 37 |
| 2.3 Common Descent | 47 |
| 2.4 Universal Natural Selection | 58 |
| 2.5 What If It Is True? | 69 |

### 3. Evolutionary Theory and the Interpretation of Scripture — 71

| | |
|---|---|
| 3.1 The Bible and Modern Science | 71 |

   3.2  The Search for Harmony: Concordism    74
   3.3  Beyond Category Mistakes: Perspectivism    79
   3.4  Where Science and Christianity Overlap: History    87
   3.5  In Search of Cocceians    94

## 4. Animal Suffering and the Goodness of God    99

   4.1  Gradualism and the Fate of Animals    99
   4.2  The Appreciation of Animals in the Bible    101
   4.3  Is There a Problem at All? Neo-Cartesianism    106
   4.4  The Cosmic-Fall Theory: Animal Suffering and Human Sin    110
   4.5  Animal Suffering as Part of God's Plan    119
   4.6  Animal Suffering and the Demonic    123
   4.7  Evaluation    129

## 5. Common Descent and Theological Anthropology    136

   5.1  Introduction    136
   5.2  A Reformed Concern?    137
   5.3  Human Dignity and the Challenge of Evolution    143
   5.4  Animals in the Image of God?    146
   5.5  Human Uniqueness as a Theological Category    150
   5.6  Why We Are Still Special    158

## 6. Evolution and Covenantal Theology    160

   6.1  The Covenant as a Reformed Key Concept    160
   6.2  The Scientific Story of Human Origins    161
   6.3  Genesis 2–3 and the Historical Adam    165
   6.4  The Fall: Biblical and Theological Backgrounds    180
   6.5  The Fall and Original Sin in an Evolutionary Context    187
   6.6  Human Death as the Wages of Sin    196
   6.7  The Scope of the Christian Message of Salvation    199

## 7. Natural Selection and Divine Providence — 204

7.1 Introduction — 204
7.2 High Stakes: Divine Guidance or a Purposeless Universe? — 206
7.3 Evolution, Chance, and Teleology in the Past — 208
7.4 Evolution, Chance, and Teleology Today — 213
7.5 Beyond Compatibility? Convergence and Consonance — 222
7.6 Conclusion: Consonance instead of Conflict — 228

## 8. Morality, the Cognitive Science of Religion, and Revelation — 230

8.1 Biological, Social, and Cultural Evolution — 230
8.2 The Evolution of Morality and Theological Ethics — 236
8.3 The Cognitive Science of Religion — 241
8.4 A Panoply of Theories: Religion as By-Product or Adaptation — 246
8.5 Revelation Debunked? — 252
8.6 Conclusion — 264

## 9. Any Other Business? — 266

*Acknowledgments* — 275
*Bibliography* — 277
*Index of Names* — 309
*Index of Subjects* — 317
*Index of Scripture References* — 326

# Foreword

In 2015 Gijsbert van den Brink was appointed by Amsterdam's Vrije Universiteit to the newly established University Research Chair for Theology and Science. I was pleased when I heard the news of this new assignment. I was an admirer of Professor Van den Brink's scholarly achievements, and I had discussed faith-and-science topics with him on several occasions, so I knew that he clearly had a strong interest in those matters. Up to that point, however, I had associated him with his important work in traditional topics in systematic theology—pneumatology, *imago dei*, providence, and biblical authority—as evidenced in his 2012 *Christelijke dogmatiek. Een inleidung*, coauthored with Cornelis van der Kooi. That book quickly—and surprisingly, for a work in systematic theology in the Netherlands—went through several printings, and then was published in an English translation by Eerdmans five years later, as *Christian Dogmatics: An Introduction* (2017). While that volume covered much fascinating theological territory, it understandably gave no sustained attention to evolutionary thought and other related scientific matters.

When I learned, then, of the release by Van den Brink of a new book on faith and science, I assumed that he would be offering helpful initial insights into an area of thought that he was only beginning to explore with sustained scholarly seriousness. Given that expectation, I was taken aback by the depth and range in this important book. He covers an amazing range of scientific topics in these pages, including suffering and death in the animal kingdom, common human descent, natural selection, the age of the earth, Middle East culture, and cognitive science. His grasp of details in these areas is impressive. And at key points he supplements his treatment of theological and scientific issues by drawing upon some of the technical discussions of scientific methods and theories in recent Anglo-American analytic philosophy.

## FOREWORD

One of the complaints often raised about discussions of faith and science is that the theologians who address these matters frequently have little more than a novice's grasp of the sciences, while the scientists who weigh in on the issues often know very little theology. There have been, of course, some notable exceptions to this pattern, and now this book sets some new standards for evaluating these efforts. Van den Brink has clearly done his homework in both areas, and his past in-depth explorations of theological matters clearly serve him well in this book.

Those who know the author's previous theological writings will not be surprised that he is explicit about his identity as a Reformed scholar. But the non-Reformed should not avoid reading the book for that reason. For one thing, he offers a compelling case for seeing the Reformed approach as a "stance" rather than as a closed system of doctrines. Nor does Van den Brink, in making use of Reformed themes, focus on the more soteriological teachings associated with classic Calvinism: this is not a book that argues for predestination, election, limited atonement, and the like. In detailing the theologically relevant themes for a dialogue between theologians and scientists, he highlights the doctrines of creation, original sin, the image of God, divine providence, human volitions, and the incarnation—all subjects that are surely important for exploring the relevance of a variety of confessional traditions for scientific inquiry.

It is sometimes said that the North American evangelical debates regarding faith and science have been shaped by public controversies that have not figured largely in British and continental European religious life. There is something to be said for that contention. The nineteenth-century Darwinian controversies in major American denominations, the Scopes trial of the early twentieth century, the often passionate anti-intellectualism of frontier revivalism, and recent local political debates about the teaching of scientific "theories" in public schools—these developments certainly have given expression to distinctly American moods and pieties.

However, Van den Brink's book, while written originally for a Dutch readership, also has profound significance for our own North American context. His openness to key elements in evolutionary thought, for example, has been condemned by representatives of his own orthodox wing of the Dutch Reformed churches—folks who in the past have "owned" him as one of their most capable theological defenders—in language not unlike that of many North American "young earth" proponents. And it is precisely the familiarity of that kind of response that makes this book such a valuable resource for those of us who are committed to preserving the basics of theological orthodoxy while wanting to honor the careful scientific study of created reality.

For me personally, reading Gijsbert van den Brink's profound—and, I would insist, groundbreaking—study of faith and science has been an illuminating intellectual adventure. And while it has stretched my thinking, it has also encouraged me in my journey of faith. This splendid book is a gift to all of us who want to be open in new ways to how, in Van den Brink's words, "the unspeakable glory and majesty of God are underlined by the evolutionary history of the world."

Richard J. Mouw

# Preface

It has been a privilege to work on this book with so many supportive colleagues and others encouraging me to write it. I conceived it and began drafting it during a residential year (2010-2011) at the Center of Theological Inquiry (CTI), Princeton, New Jersey, as the Houston Witherspoon Fellow for Theology and Science. I am greatly indebted to the center's then directors William Storrar and Thomas J. Hastings, and to their assistant Jeanie Mathew, for their hospitality and encouragement, as well as for the many opportunities they created for dialogue and interaction. My fellow members at the CTI—Ann Astell, Brian E. Daley, SJ, and Ola Sigurdson—contributed to this dialogue and interaction in all sorts of fruitful ways, as did J. Wentzel van Huyssteen, then professor of theology and science at Princeton Theological Seminary, who showed a keen interest in my work during this year and beyond.

I am grateful to the Templeton World Charity Foundation (TWCF) for allowing me to pursue this project as part of the research program Science beyond Scientism (SBS) at the Abraham Kuyper Center, Vrije Universiteit, Amsterdam. The views expressed in this book do not necessarily reflect the opinions of TWCF, but I probably could not have written this book without their support. I am grateful to all those involved in the SBS program who read and commented on drafts of several chapters or on the whole manuscript: Leon de Bruin, Cornelis van der Kooi, Rik Peels, Jeroen de Ridder, and René van Woudenberg. I learned much from their invaluable comments and recommendations, which helped me improve the manuscript considerably (of course, I am myself responsible for any remaining infelicities).

Other colleagues were so kind as to offer insightful comments on parts of the manuscript. In this way, a draft of chapter 2 was reviewed by the scientists Cees Dekker, René Fransen, and Piet Slootweg, who not only helped me to "get the science right" but also came up with some quite interesting theological ideas. The same goes for Canadian biologists Harry Cook (The King's University, Edmonton) and Jitse M. van der Meer (Redeemer University College, Hamilton), both of whom read the entire manuscript and made many perceptive comments that helped me improve the draft. Dutch philosopher Herman Philipse offered some helpful critical feedback on chapter 8. I am also grateful to the Issachar Fund for generously granting me a five-week writer's retreat (May-June 2015). By hosting me in their apartment in Grand Rapids, they gave a real boost to the writing project. During this stay in Grand Rapids, I also profited greatly from conversations with Rob Barrett (Colossian Forum); Joel Beeke (Puritan Reformed Theological Seminary); Lyle Bierma, Carl Bosma, John W. Cooper, and Mary Vanden Berg (Calvin Theological Seminary); Dan Harlow (Calvin College); Deborah Haarsma and Jim Stump (BioLogos); and Nicholas Wolterstorff.

I also received inspiration in this project from conversations with, among others, Denis Alexander, Ernst Conradie, Hans Van Eyghen, Lydia Jaeger, Ard Louis, Richard Mouw, Jacques Schenderling, Benno van den Toren, Eva van Urk, and Michael Welker. I had the privilege of discussing drafts of most chapters of this book with ministers of the Protestant Church in the Netherlands who were taking a continuing-education course on Christian faith and evolution, as well as with students preparing for ministry in the Restored Reformed Church (a church established by those who refused to join the merger of three Dutch Protestant churches in 2004). I am grateful to these students for pressing me on a number of issues that led me to rephrase parts of the argument. I am also grateful to PhD student Neda Ghatrouie for her help in composing the bibliography.

A Dutch version of this book was published as *En de aarde bracht voort. Christelijk geloof en evolutie* (Utrecht: Boekencentrum, 2017; 5th printing 2018). This Dutch predecessor was revised in a couple of ways. For example, references to Dutch sources and debates were dropped, and the specific focus on the Reformed tradition was added. Whereas in the Netherlands studies on the relation between Christianity and evolution are scarce, in the Anglo-Saxon world they are numerous. Amid the vast and ever-expanding body of literature on evolution and religion, however, a more detailed scholarly contribution on how Reformed theology in particular is affected by evolutionary theory—and how the evolutionary process can be interpreted from a Reformed perspective—was still missing.

One of the ministers who took part in the continuing-education course mentioned above came up with an appropriate metaphor to express his expectations for this book. He compared the field of evolutionary theory and its ideological complexities to a forest crisscrossed by all sorts of hidden pathways. Some of these pathways lead to breathtaking vistas, others end up in dangerous swamps and morasses. What is needed in such a situation, he suggested, is a guide who knows the territory and who tells you: "Look, this is a place to be avoided, but you can walk that trail until the very end, whereas the winding path over there is safe until that specific place." If this book can serve as such a guide in the complex landscape of evolutionary thinking, where scientific and ideological considerations are often entangled in highly intricate ways, its mission will be accomplished.

Amsterdam<br>Vrije Universiteit

# Introduction

> I have all along had a sensitive apprehension that the undiscriminating denunciation of evolution from so many pulpits, periodicals, and seminaries might drive some of our thoughtful young men to infidelity, as they clearly saw development [i.e., evolution] everywhere in nature, and were at the same time told by their advisers that they could not believe in evolution and yet be Christians. I am gratified beyond measure to find that . . . in showing them evolution in the works of God, I showed them that this was not inconsistent with religion, and thus enabled them to follow science and yet retain their faith in the Bible.[1]

This is a book for Christians who want to make up their minds on evolutionary theory as well as for evolutionists who want to make up their minds on Christianity. For both groups, the question whether Christianity and evolutionary theory can go hand in hand is an urgent one. I especially focus on one particular branch of the Christian tradition, namely, the Reformed one—but large parts of my argument are equally relevant to theists from other traditions. As will be made clear in the first chapter, Reformed theology is not characterized by one or more exclusive doctrinal beliefs but by a series of specific commitments and concerns. It highlights specific doctrinal notions (such as divine sovereignty) that do not receive the same emphasis in other parts of the Christian tradition, while it downplays others (such as the authority of eccle-

---

1. James McCosh, *The Religious Aspect of Evolution* (New York: Scribner's Sons, 1890), ix-x. I found this quote in Hans Schwarz, *Vying for Truth—Theology and the Natural Sciences from the 17th Century to the Present* (Göttingen: Vandenhoeck & Ruprecht, 2014), 76-77.

sial traditions). Thus, there is much common ground, and theists from other traditions can easily see that many issues discussed in this book are relevant to their concerns as well. For example, chapters 4–8 analyze problems that also trouble many Catholic, Jewish, and Muslim believers. Moreover, the term "Reformed" is taken in a broad sense here, as comprising all denominations whose roots go back to the sixteenth-century Swiss Reformation associated with the name of John Calvin and others. It encompasses Presbyterians as well as Methodists, considerable groups of Baptists as well as Congregationalists, and many if not most contemporary Christians who self-identify as evangelicals (or did so until the Trump era).

As a result of their special doctrinal emphases, Reformed folks have their own problems with evolutionary theory, and often a bit more of them than, for example, Roman Catholics. But whereas studies on "Christianity and evolution" abound, and quite a few books on Roman Catholicism and evolution are available, much less work has concentrated on the relationship between *Reformed* theology and evolutionary theory.[2] In fact, the very combination of the phrases "Reformed theology" and "evolutionary theory" may strike many as weird. Doesn't "Reformed theology" go hand in hand with such lofty theological notions as "the authority of the Bible," "the knowledge of God," "the covenant," etc.? It surely does, and all these themes will play a role in this book. But Reformed Christians will also have to make up their minds on much more down-to-earth themes such as evolutionary theory. If they don't, their children will no doubt query them after they see dinosaur skeletons in some museum. When telling people about the research project that underlies this book, I sometimes intentionally pronounced its title very slowly so I could watch facial responses to its constituent parts. Some people were visibly glad to hear that I was writing about Reformed theology but looked concerned

---

2. Though some recent studies written by evangelical Protestants tackle the issues from a broadly evangelical perspective, I know of only one book intentionally locating itself within the Reformed tradition: Deborah B. Haarsma and Loren D. Haarsma, *Origins: A Reformed Look at Creation, Design, and Evolution* (Grand Rapids: Faith Alive, 2007). Interestingly, the title of this book's revised second edition was changed to *Origins: Christian Perspectives on Creation, Evolution, and Intelligent Design* (2011). Cf. for the Catholic side, e.g., Louis Caruana, *Darwin and Catholicism: The Past and Present Dynamics of a Cultural Encounter* (London: Bloomsbury T&T Clark, 2009); Stefaan Blancke, "Catholic Responses to Evolution, 1859–2009: Local Influences and Mid-Scale Patterns," *Journal of Religious History* 37 (2013): 353–68; John F. Haught, "Darwin and Catholicism," in *The Cambridge Encyclopedia of Darwin and Evolutionary Thought*, ed. Michael Ruse (Cambridge: Cambridge University Press, 2013), 485–92.

when I added the phrase "evolutionary theory." Others seemed to find the idea of studying Reformed theology dull and unattractive but suddenly became more interested when I mentioned evolutionary theory. Personally, I find the study of both Reformed theology and evolutionary theory highly instructive, exciting, and rewarding—and all the more so the relationship between the two.

Darwinian evolution continues to be a sensitive issue among evangelicals and other orthodox Christians, whether of a Reformed stripe or otherwise. The intellectual and spiritual landscape is highly polarized by severe critics of evolutionary theory on the one hand and assertive advocates of it on the other. Especially in the United States, both of these groups—as well as in-between groups such as the intelligent design movement—have their own highly active organizations, some of which have considerable financial resources to spend. But debates among Christians on the origins of biological diversity and of the human race are often also fierce and emotionally charged in other countries with a relatively large Protestant constituency (such as the Netherlands).

In this book, it is not my goal to defend a particular stance on evolution. Nor will I—apart from the necessary setting of the scene in chapter 2—extensively discuss the scientific credibility of evolutionary theory or its explanatory power inside and outside the realm of contemporary biology. I cannot entirely escape this issue, but I am not a scientist, so this is not where my expertise lies. As an interested outsider to such discussions, I can only observe that during the past 150 years the Darwinian theory of evolution has shown a remarkable staying power. Critics have often predicted its imminent demise, but, although many questions are still unanswered (cf. §2.4), its epistemic status within the natural sciences has become more and more secure—not least as a result of the so-called genetic revolution.[3] Also, I observe that many Christians who have been trained and are active as scientists accept evolution as unreservedly as their non-Christian colleagues. Apparently, they became convinced that evolutionary theory is the most plausible candidate for explaining the astonishing diversity of life on earth, the varieties, similarities, and interdependence of its manifold species, as well as many other biological and even cultural phenomena. Perhaps some have been persuaded for the wrong reasons. But for those of us who try to face such data with an open mind, it is simply hard to believe that evolutionary theory is just a conspiracy of atheist intellectuals or a collective error that at some point in the future will finally be unmasked and replaced by the "plain biblical truth" (usually conceived of

---

3. Cf. Gijsbert van den Brink, Jeroen de Ridder, and René van Woudenberg, "The Epistemic Status of Evolutionary Theory," *Theology and Science* 15 (2017): 454–72.

as young-earth creationism). Moreover, we should not be unduly selective in our appreciation of science, by, for example, being dismissive of theories that we don't like while at the same time using the most advanced science-based treatments and medicines when we fall ill.[4]

As Pope John Paul II argued, speaking of evolutionary theory in his 1996 address to the Papal Academy of Sciences: "The convergence, neither sought nor fabricated, of the results of work that was conducted independently is in itself a significant argument in favor of this theory."[5] For reasons such as these (let me come out up front with this), I am inclined to accept the theory of evolution. I do not *embrace* evolutionary theory as if I were emotionally attached to it, but I *accept* it as the most plausible scientific theory to date to explain the earth's biodiversity, just as I accept gravitational theory because it neatly explains other natural phenomena.[6] As a child, I was raised in an atmosphere of latent creationism. The first chapters of Genesis were intuitively read as more or less historical records of the beginning of the universe and of life on earth. The twentieth-century culture wars on evolution in the United States had influenced the Dutch scene as well, so in the 1970s and 1980s it was pretty clear that orthodox Reformed Christians were opposed to evolution and supported a "literal" reading of Genesis 1-3.[7] Studies in which, however cautiously, a more nuanced view was suggested were suspect.[8] Though I never actively endorsed

---

4. Sometimes a distinction is made between experimental science (the results of which can be replicated) and historical scholarship (which supposedly is much less reliable). Evolutionary theory is then usually seen as an example of the latter category. But historical claims as well (such as that Alexander the Great actually lived, that the earth has seen at least one glacial period, etc.) can be beyond reasonable doubt, as both Christians and others agree.

5. Pope John Paul II, "Message to the Pontifical Academy of Sciences on Evolution," *Origins* 26, no. 22 (1996): 351.

6. Cf. Norman C. Nevin, ed., *Should Christians Embrace Evolution? Biblical and Scientific Responses* (Phillipsburg, NJ: P&R, 2011); the answer to the title question should probably be "no, but still they may feel they should *accept* evolution." Cf. the title of Denis O. Lamoureux's popular work, which makes the distinction clear: *I Love Jesus and I Accept Evolution* (Eugene, OR: Wipf & Stock, 2009).

7. Cf. Abraham Flipse, "The Origins of Creationism in the Netherlands: The Evolution Debate among Twentieth-Century Dutch Neo-Calvinists," *Church History* 81 (2012): 104-47.

8. E.g., books by the Reformed biologist Jan Lever and Old Testament scholar B. J. Oosterhoff. Unlike Oosterhoff's study, some of Lever's work has been translated into English: Jan Lever, *Creation and Evolution* (Grand Rapids: Kregel, 1958); Jan Lever,

young-earth creationism, I was pretty confident that evolutionary theory was off the mark. As late as 2009 I published a book in which this skeptical attitude toward evolutionary theory was still slightly palpable.[9] By then, however, my thinking had already started to change as a result of conversations with a group of Christians who were scientists and had started to write with appreciation on evolution as a sober scientific theory—while decidedly rejecting the evolutionist worldview.[10] I gradually came to consider evolutionary theory as an option that perhaps need not be rejected out of hand by Christians. Today, though I don't think evolutionary theory has been "proven" (such strong language might better be avoided), I assume that some version of it is most probably true.

Of course, being a fallible human being and, as noted above, not a scientist, I may be wrong here. No part of this book, however, stands or falls with my personal take on evolutionary theory. My argument presumes not the truth of evolutionary theory but its *relevance* as an important player in the scientific field. The question I ask in this book is simple: *What if it is true?* What consequences would that have for one's faith and theology? The reason many orthodox Christians are skeptical of evolution is not usually that they are deeply impressed by the force of the scientific arguments in favor of creationism or another alternative theory. Rather, they fear they cannot continue to be orthodox Christian believers—or, perhaps, Christian believers at all—if they accept evolution. Is this indeed the case? What exactly changes in the household of one's Christian faith when one becomes convinced of the evolutionary background of life? As a student, I had a professor who proclaimed: "Accepting evolution is entirely irrelevant to the Christian faith, because if it is true, evolutionary theory just tells us *how* God did the job—how he created life on earth. Surely that's nice to know, but it does not have any repercussions at all for the faith." As will become clear in this book, it is not as simple as that. Nor do I think that the theory of evolution eradicates the entire Christian faith, or that only a watered-down, extremely liberal variety of it can survive. On the contrary, I will defend the thesis that orthodox forms of

---

*Where Are We Headed? A Biologist Talks about Origins, Evolution, and the Future* (Grand Rapids: Eerdmans, 1970). Though rejecting evolutionary theory, Oosterhoff advocated a reading of Gen. 2–3 that highlighted symbolic elements in the text.

9. Gijsbert van den Brink, *Philosophy of Science for Theologians: An Introduction* (Frankfurt am Main: Lang, 2009), e.g., 60, 123, 222–23.

10. Especially the work of the molecular biophysicist Cees Dekker from TU Delft was influential here.

Christianity, and especially of Reformed theology, are also compatible with evolutionary theory. Those who accept, for example, the Nicene Creed and the Chalcedonian Definition as well as the Reformed confessional heritage can come to terms with evolution without having to betray the faith that is most dear to them.

As a theologian, I consider it my task to think through the relevant theological issues with an eye on what might be called realistic *price tagging*. Accordingly, the questions I ask are of this type: "Imagine that evolutionary theory were true, what would that mean for belief $x$, doctrine $y$, and practice $z$?" So, clearly, there is no need for those who (for whatever reason) are skeptical of evolutionary theory to think that this book was written by an "opponent." For even they might wonder whether there isn't a very small chance—let us say of only 1 percent—that they are wrong and evolutionary theory will turn out to be right after all. Such a thought is a nightmare if one also thinks evolutionary theory rules out the main tenets of the Christian faith. For in that case we can't be completely sure (like Job) that our redeemer lives (cf. Job 19:25). So, even if you think there is only a 1 percent chance that evolutionary theory is correct, this book is good news if it succeeds in its argument that evolutionary theory is *not* at odds with the Christian faith. How about those who are convinced that the chance of evolutionary theory being true is absolutely zero? Even they surely have colleagues, friends, or even loved ones who are not so steadfast in their antievolution stance. What does this mean for them and for their faith? What parts of the traditional "depository of faith" can they sincerely and consistently believe, and where do they have to make revisions?

These are also important questions for missionary reasons. How do we communicate the gospel in a science-imbued world? How do we, to paraphrase Paul (1 Cor. 9:19-22), become an evolutionist to the evolutionists? Should we urge those who are considering Christianity to begin by abandoning evolutionary theory, because only then will they be able to accept the gospel message? Or should evolution be a topic we raise with them later, addressing it as a final residue of their nonbaptized thinking that should still be rectified? Or is it possible to fully communicate the gospel to those who believe that we humans, and all living things on earth, have an evolutionary history, and who would consider the idea of jettisoning this belief ludicrous? These are questions that all Christians have to face—and that many non-Christians might *want* to consider. We might even want to give such questions a more positive and constructive twist: How can Christians, taking their cue from the Bible and confessional traditions, make sense of such a bewildering phenomenon as the evolutionary process? How can we interpret this process in a way that makes clear how God is involved? New and old atheists alike are probably right that, when taken on its own, the evolutionary process is incredibly wasteful and senseless. Therefore, if any hope is to be perceived in it, this can only spring from the belief that it is somehow incorporated in God's ways with this world toward his kingdom.

The issue of Christian faith and evolution is of course a vexed and complicated one, given that so many different topics, commitments, and interests are at stake, many of which are deeply intertwined with each other. On top of that, a huge amount of confusion is caused by secular popular science writers who, whether or not in the slipstream of the so-called new atheists, turn the scientific theory of evolution into an atheist view of life (what we might call the ideology of evolution*ism*) and argue that taking evolutionary science seriously leaves us with no other option. It is with such adherents of metaphysical naturalism (the view that no supernatural beings exist) or scientism (the view that only science can tell us what to believe) that Christians have to take issue in the first place, since that is where the real conflict lies.[11] Obviously, the spiritual battle should not be waged between Christians but between those who want to follow Jesus and those who serve the spiritual powers of the present age—of which metaphysical naturalism is not the least influential one.[12]

In trying to tread carefully yet deal adequately with the issues, I have made a couple of decisions. First, I will speak of "evolutionism" when I have in mind the view of life according to which evolution implies atheism and metaphysical naturalism, and I will speak of "evolutionary theory" to refer to the scientific theory without ideological overtones. This seems to me an important distinction, since all too often the two are conflated and evolutionary theory is rejected by Christians because of its association with evolution*ism*. Whether or not we reject evolutionary theory, we should realize that it cannot simply be equated with evolutionism, since evolutionary theory is just that: a scientific *theory* (i.e., loosely described, a consistent and comprehen-

---

11. Cf. Alvin Plantinga, *Where the Conflict Really Lies: Science, Religion, and Naturalism* (Oxford: Oxford University Press, 2011).

12. This is a problem with books like J. P. Moreland et al., *Theistic Evolution: A Scientific, Philosophical, and Theological Critique* (Wheaton, IL: Crossway, 2017), which target other Christians instead of fighting "where the battle rages" (*D. Martin Luthers Werke. Briefwechsel*, 18 vols. [Weimar, 1930–], 3:81–82). For a response to this book, see Deborah Haarsma, "A Flawed Mirror: A Response to the Book 'Theistic Evolution,'" *BioLogos*, April 18, 2018, https://biologos.org/blogs/deborah-haarsma-the-presidents-notebook/a-flawed-mirror-a-response-to-the-book-theistic-evolution.

sive way of explaining observable phenomena), not an ideology. Second, I attempt to make issues more "manageable" on the theological side by particularly focusing on one confessional tradition, namely, the Reformed one. In doing so, although I will from time to time speak of "Christians" in a more general way (especially when what is being said also applies to other Christian traditions), I try to avoid the impression that Christianity is a monolithic entity. And third, in trying to untie the intricate knot of issues involved here, I will discuss the most important doctrinal topics that are possibly affected by evolutionary theory one at a time, rather than dealing with all of them at the same time.

Leaving aside the impressive work of numerous scientist-theologians and many philosophical theologians, contributions by "proper" systematic theologians to the discussion on evolution are fairly scarce. Most studies on faith and evolution are written by scientists who, although they are often impressive theological autodidacts, lack the acumen to discuss the issues with the necessary theological depth. Many theologians, however, refuse to enter the debate. Subscribing to what Ian Barbour has called the "independence model" in the science and religion debate, they argue that the Christian doctrine of creation and biological evolution are totally unrelated notions, so that there is no need (and no possible way) to study their interconnections.[13] Although these theologians are correct that we should not "smother the differences" between science and religion, it is the task of Christian theology to offer guidance to the many Christians who struggle to make up their minds on evolution—as well as to those "seekers" for whom evolution is self-evident but who wonder whether it can be combined with Christianity.

Theologians should not consider their calling too lofty for such down-to-earth matters, nor should they do their work in splendid isolation of the concerns of so many people both in and outside the pews. Instead, as Alister McGrath argues, they should "bring a theological framework to intrinsically challenging areas of science," such as, among other things, the neo-Darwinian theory of evolution. "This is where it 'hurts,' and this is where more theologians need to be."[14] Of course, theologians are by no means the only ones who have a say on these issues, but they have their distinctive part to play. On top of that, my personal motivation in this connection is aptly adumbrated in

---

13. Cf., e.g., Ian G. Barbour, *Religion and Science: Historical and Contemporary Issues* (San Francisco: Harper, 1997), 77-104.

14. Alister McGrath, "Review Conversation [with Willem B. Drees]," *Theology and Science* 8 (2010): 333-41 (339). Drees presses him on this point.

*Introduction*

the quote of Presbyterian philosopher James McCosh (1811-1894) that serves as this chapter's epigraph—though I am happy to include in my audience thoughtful women next to the "thoughtful . . . men" he mentions. It has been my thought over the past couple of years that the work that McCosh and others have undertaken is not yet complete and therefore must be continued. I hope this book will be a modest contribution to the fulfillment of this task.

CHAPTER 1

# Reformed Theology as a Distinctive Stance

> The Reformed tradition does not stem from Calvin alone. Before him there were other Reformed reformers . . . and there were others who worked with him or were indebted to him. . . . Yet, despite all this diversity, which is with us in even greater measure today, there are a number of doctrinal emphases which are especially characteristic of the Reformed tradition.
>
> —I. John Hesselink[1]

## 1.1 Introduction

What does it mean to be Reformed, or Presbyterian? And what is distinctive of Reformed and Presbyterian *theology* in comparison to other stripes of Christian theology?[2] To investigate the relationship between Reformed theology and evolutionary theory, we first need to clarify what Reformed theology actually is—which is the aim of this chapter. Though debates on the nature and limits of Reformed theology continue to flare up from time to time,[3] one may wonder how relevant the issue actually is. It has been pointed out that "the question of Reformed identity is not very important in Reformed eyes because of the Reformed view of revelation: the sole thing that matters for

---

1. I. John Hesselink, *On Being Reformed: Distinctive Characteristics and Common Misunderstandings*, 2nd ed. (New York: Reformed Church Press, 1988), 89.

2. "Presbyterianism" is a common term for that branch of Reformed Protestantism that traces its origins to the British Isles (cf., e.g., *Wikipedia*); I will not repeat "and Presbyterian" all the time but will henceforth use "Reformed" as the more inclusive concept.

3. This happened, for example, on the Internet after the publication of Oliver Crisp's *Deviant Calvinism: Broadening Reformed Theology* (Minneapolis: Fortress, 2014).

Reformed Christians is how God makes Godself known in his revelation, most particularly in the Bible."[4] Interestingly, however, this argument clearly presupposes that there *is* something like a distinct Reformed theological identity, as this is visible at least in a typical emphasis on divine revelation. Indeed, as we will see in this chapter, it is a focus on the Bible that shapes the Reformed theological identity.

The Bible plays a constitutive role in all Christian traditions, though, and therefore it makes sense to inquire into the specific character of the Reformed approach to the Bible, especially if we want to find out what particular problems Reformed Christians might have vis-à-vis evolutionary theory. Obviously the Christian world is divided into a number of major ecclesial families, such as the Roman Catholic, Pentecostal, Lutheran, Anglican, and Eastern Orthodox families. Depending on their doctrinal and cultural backgrounds, these traditions may relate in different ways to the natural sciences in general and to evolutionary theory in particular. For example, resistance toward evolution seems to be more widespread and tenacious among Reformed Christians than it is in some other ecclesial families, such as Roman Catholicism (Catholics have been more flexible ever since evolutionary theory was deemed acceptable under certain conditions in the encyclical *Humani Generis* in 1950).[5]

To discover why Reformed and evangelical Christians in particular still find it difficult to accept evolutionary theory, we have to examine the specific nature of Reformed theology.[6] I will start with some of the most persistent misunderstandings, since unfortunately, as John Hesselink, among others, has pointed out, misunderstandings abound when it comes to the nature of Reformed theology.[7]

---

4. Hendrik M. Vroom, "On Being 'Reformed,'" in *Reformed and Ecumenical: On Being Reformed in Ecumenical Encounters*, ed. Christine Lienemann-Perrin, Hendrik M. Vroom, and Michael Weinrich (Amsterdam: Rodopi, 2000), 153.

5. John F. Haught, "Darwin and Catholicism," in *The Cambridge Encyclopedia of Darwin and Evolutionary Thought*, ed. Michael Ruse (Cambridge: Cambridge University Press, 2013), 487, calls *Humani Generis* a "watershed moment" in this connection.

6. Apart from its theology, the Reformed tradition is also characterized by a specific ethos and by an encompassing worldview (as was famously argued by Abraham Kuyper in his 1898 Stone Lectures); though these are closely interrelated with Reformed theology, my focus will be on Reformed theology as such. For the ethos of Reformed Protestantism, see, for example, Gijsbert van den Brink and Harro M. Höpfl, eds., *Calvinism and the Making of the European Mind* (Leiden: Brill, 2014).

7. Cf. Hesselink, *On Being Reformed*. That such misunderstandings are persistent is evident from the fact that many years later Kenneth Stewart found reason to take

## 1.2 Some Common Misunderstandings

The first prevalent misconception is the assumption that Reformed theology is synonymous with Calvinism or Calvinist theology.[8] In this view, Calvin is seen as the fountainhead of Reformed theology similar to the way Luther is the father of Lutheranism. This view is wrong for at least three reasons. First, Calvin was by no means the sole originator of Reformed theology. In fact, he was not its originator at all, but "one second-generation codifier among others."[9] Among his predecessors and contemporaries were Reformed figures such as Martin Bucer, Heinrich Bullinger, William Farel, John à Lasco, Johannes Oecolampadius, Peter Martyr Vermigli, Pierre Viret, and Huldrych Zwingli. To be sure, Calvin represented a significant voice in this group (much more so than, e.g., Farel), but he offered by no means the single standard by which all truly Reformed theology was to be measured. That role was assigned to him only much later in certain quarters, in the eighteenth and nineteenth centuries.[10]

Second, Calvin's theology was not original enough to be isolated from other representatives of the broader movement, such as those mentioned above. Writing about the near absence of any distinctive doctrines in Calvin, Richard Muller observes acutely:

> This problem [viz., of suggesting that such distinctive doctrines exist in Calvin's work] has been enhanced by the numerous books that present interpretations of such decontextualized constructs as "Calvin's doctrine of predestination," "Calvin's Christology," or "Calvin's doctrine of the Lord's Supper" as if Calvin actually proposed a highly unique doctrine. We need to remind ourselves that the one truly unique theologian who entered Geneva in the

---

a similar approach: Kenneth J. Stewart, *Ten Myths about Calvinism: Recovering the Breadth of the Reformed Tradition* (Downers Grove, IL: InterVarsity Press, 2011).

8. This assumption is even dominant in John Leith's otherwise excellent *Introduction to the Reformed Tradition: A Way of Being the Christian Community* (Louisville: Westminster John Knox, 1981). It is the first of the ten myths identified in Stewart, *Ten Myths about Calvinism* (21–44). See, for another recent critique of it, Crisp, *Deviant Calvinism*, 2, and passim.

9. Richard A. Muller, *Calvin and the Reformed Tradition* (Grand Rapids: Baker Academic, 2012), 68; cf. Willem J. van Asselt, "Calvinism as a Problematic Concept in Historiography," *International Journal of Philosophy and Theology* 74 (2013): 144–50.

10. Cf. Muller, *Calvin and the Reformed Tradition*, 55.

sixteenth century, Michael Servetus, did not exit Geneva alive. Unique or individualized doctrinal formulation was not Calvin's goal.[11]

Third, the very label "Calvinism" is a later construct; it came into existence as a pejorative term, used by those hostile to the Reformed tradition (especially Lutherans, who had a different view of the Lord's Supper).[12] Originally, representatives of the Swiss magisterial Reformation self-identified as adherents of "la religion réformée," forming the reformed, evangelical, evangelical reformed, or reformed Catholic churches—the term "reformed" emerging as the most common denominator.[13] Historically, the ecclesial boundaries of this "religion" were defined by the *Consensus Tigurinus* (1549)—the agreement on the understanding of the Lord's Supper among Calvin, Bullinger, and a number of other Swiss and Rhineland Reformers.

All this is not to say that terms like "Calvinism" and "Calvinist" are to be completely avoided.[14] They may be retained as descriptors of offshoots and advocates of the Protestant Reformations that were particularly inspired by the reformation of the church in Calvin's Geneva.[15] In this way, the Reformation in, for example, Scotland and the Netherlands took on a distinctively

---

11. Muller, *Calvin and the Reformed Tradition*, 52; one wonders, however, whether Muller is slightly overstating his point here—even when we acknowledge that, for example, Calvin's Christology was not unique, or at least not *meant to be* unique, it might still make sense to investigate its special shape within its proper contexts. In private correspondence, Muller has elucidated that he has "no objection to theologically topical books about Calvin's thought—just to books that work through his theology without any recognition that it has a background and context and treat it as if it arose in a vacuum" (email, June 10, 2015).

12. Cf. Gijsbert van den Brink and Harro Höpfl, "Calvin, the Reformed Tradition, and Modern Culture," in Van den Brink and Höpfl, *Calvinism and the Making of the European Mind*, 8.

13. Philip Benedict, *Christ's Churches Purely Reformed: A Social History of Calvinism* (New Haven: Yale University Press, 2002), xxiii.

14. As has been suggested by Muller, *Calvin and the Reformed Tradition*, 48 ("the terms 'Calvinist' and 'Calvinism' . . . ought to be set aside"), and R. Michael Allen, *Reformed Theology* (London: T&T Clark, 2010), 3 ("the term 'Calvinist' ought to be dropped entirely").

15. Moreover, the terms can be retained for practical reasons, as Benedict, *Christ's Churches Purely Reformed*, xxiii, does: "to make myself clear to the non-specialist reader who is more likely to recognize" these terms; Benedict adds to this, however, that in principle the terms should only be used in some very restricted cases. Another

Calvinistic character.[16] The terms are by no means synonymous with "Reformed," however, since "Reformed" refers more broadly to "the theological tradition that started with the sixteenth-century Reformation in Strasbourg, Zürich, and Geneva as an expression of Christian faith of all times and places."[17] In this definition, not only are figures such as Bucer, Zwingli, Bullinger, and others included but also the catholic intention of Reformed theology is acknowledged.

A second widespread misunderstanding is that Reformed theology might be identified by one core idea, basic belief, or, to use a German term coined by Swiss theologian Alexander Schweizer (1808-1888), *Centraldogma*. Schweizer construed both Lutheran and Reformed theology as hinging on a core idea from which the entire doctrinal scheme could be deduced. For Lutheranism, this was the doctrine of justification, and for Reformed theology, the doctrine of predestination.[18] Being put to the forefront by Calvin, the latter doctrine was further developed into the structuring principle of Reformed theology by post-Reformation Reformed scholastics—and Schweizer evaluated this process positively.[19] Leaving aside the question whether Schweizer was right about Lutheranism, and without disqualifying his immense scholarship, we have to say that he was wrong about Reformed theology. Clearly, his attempt to understand Reformed theology from the perspective of one unifying idea must be seen as anachronistic, reading contemporary concerns back into the tradition. Both Calvin's theology itself and that of his theological heirs—not to speak of the theology of Reformed reformers working alongside Calvin or

---

advocate of retaining the terms is Richard Mouw, *Calvinism at the Las Vegas Airport* (Grand Rapids: Zondervan, 2004), 18-21.

16. Cf. John H. Leith, "The Ethos of the Reformed Tradition," in *Major Themes in the Reformed Tradition*, ed. Donald K. McKim (Grand Rapids: Eerdmans, 1992), 9, 17; Herman Selderhuis, ed., *Handbook of Dutch Church History* (Göttingen: Vandenhoeck & Ruprecht, 2015), 171-84, 202-4.

17. Mission statement of the *Journal of Reformed Theology*, all issues thus far, inside front cover.

18. Alexander Schweizer, *Die protestantischen Centraldogmen in ihrer Entwicklung innerhalb der reformierten Kirche*, 2 vols. (Zürich: Orell, Füslli & Co., 1854-1856).

19. See Willem J. van Asselt and Eef Dekker, eds., *Reformation and Scholasticism: An Ecumenical Enterprise* (Grand Rapids: Baker Academic, 2001), 14-15, 50, 214; Van Asselt, "Calvinism as a Problematic Concept," 145.

prior to him—are much too variegated and insufficiently systematic to be reduced to one central idea.[20]

Still, the central-dogma approach has continued to be enticing. Schweizer bequeathed it to his Dutch admirer J. H. Scholten (1811-1885), and Scholten in turn influenced a student of his at Leiden University who was to become the most influential Dutch theologian ever: Abraham Kuyper. Kuyper's "neo-Calvinism" is largely based on "the view that the doctrine of divine sovereignty and predestination formed the 'central dogma' or the structuring principle of the entire Reformed theological system."[21] From this central principle Calvinism had to be developed into an all-encompassing worldview that could answer the challenges of another worldview: contemporary modernism.[22] So Kuyper selected the notion of God's sovereignty as the governing principle of Reformed theology. In other spatiotemporal locations, other doctrines have been proposed as the structural principle of Reformed theology, such as the concept of the covenant or, most recently, the notion of union with Christ.[23] It is true that these notions may deeply influence the composition or coloring of someone's Reformed theology; in that sense, presumably all three of them were central themes in Calvin's theology. From a historical point of view, it is

---

20. To give one example: Harro M. Höpfl, "Predestination and Political Liberty," in Van den Brink and Höpfl, *Calvinism and the Making of the European Mind*, 155-76, demonstrates that there was no connection at all between the doctrine of predestination and Reformed political theories.

21. Van Asselt, "Calvinism as a Problematic Concept," 145.

22. See esp. Kuyper's *Lectures on Calvinism* (Grand Rapids: Eerdmans, 1931), where in his first lecture Kuyper describes Calvinism as a "life-system" whose "root principle" branches out into every domain of human life; in his second lecture, he identifies this root principle as God's sovereignty—which, by the way, is a broader concept than God's predestination. Cf. James D. Bratt, *Abraham Kuyper: Modern Calvinist, Christian Democrat* (Grand Rapids: Eerdmans, 2013), 262-64.

23. A special Reformed tradition circling around the notion of covenant (and opposed to its "Calvinist" predestinarian strand) was assumed by J. Wayne Baker, *Heinrich Bullinger and the Covenant: The Other Reformed Tradition* (Athens: Ohio University Press, 1980). The notion of the mystical *unio cum Christo* is considered central to Calvin's theology in what has come to be called the New Perspective on Calvin; cf. Charles Partee, "Calvin's Central Dogma Again," *Sixteenth Century Journal* 18 (1987): 191-200; Charles Partee, *The Theology of John Calvin* (Louisville: Westminster John Knox, 2008), 40-43. Cf. the discussion in I. John Hesselink and J. Todd Billings, eds., *Calvin's Theology and Its Reception: Disputes, Developments, and New Possibilities* (Louisville: Westminster John Knox, 2012), 49-94.

not true, however, that any of these (or some other doctrine) has functioned as *the* central dogma of Reformed theology as a whole.

A special instance of the central-dogma theory is to be found in the popular use of the acronym TULIP as a denominator of "truly" Reformed theology. Here, it is not one doctrine that is considered central to Reformed theology but a cluster of closely connected doctrinal distinctions: the so-called five points of Calvinism. Although these points—total depravity, unconditional election, limited atonement, irresistible grace, the perseverance of the saints— pretend to summarize the five Canons of Dordt (1618–1619), they are a later, nineteenth-century conceptualization of its contents.[24] As a shortcut, the TULIP formula does not do full justice to the often more subtle language of the canons, let alone to the views of earlier Reformed theologians.[25] More importantly, the Canons of Dordt were never intended to offer a summary of Reformed theology as a whole; they were only meant to solve one particular problem concerning the nature of God's saving grace that had emerged within the Dutch Reformed church.

The way in which this problem was handled at the Synod of Dordt definitely shows how crucial soteriology was in Reformed thinking, in that the canons aimed at safeguarding the absolutely free, undeserved, and unconditioned nature of God's salvation. Yet, the Dordtian doctrine of grace can by no means be identified with the entire scheme of Reformed theology, and perhaps not even with its "heart." For notions such as infant baptism, a spiritual interpretation of the Lord's Supper, the unity of the covenant, a high view of the Bible, etc., belong to a full-fledged Reformed theology with equal integrity. Moreover, the "five points of Calvinism" are not unique to Reformed theology but can be and have been endorsed within other Christian denominations as well.[26] For similar reasons, it seems that no attempt to identify one central doctrinal notion or cluster of notions as the defining characteristic of Reformed theology will be successful.

24. Muller, *Calvin and the Reformed Tradition*, 58; Ken Stewart, "The Points of Calvinism: Retrospect and Prospect," *Scottish Bulletin of Evangelical Theology* 26 (2008): 187–203. Cf. Gijsbert van den Brink, *Dordt in context. Gereformeerde accenten in katholieke theologie* [The Canons of Dordt in context: Reformed accents in Catholic theology] (Heerenveen, the Netherlands: Groen, 2018), 16–20.

25. Muller, *Calvin and the Reformed Tradition*, 59–60, argues that the acronym (and esp. its *T* and *L*) cannot be traced back to the theological thought of Calvin or other early Reformed Reformers, though it stands in continuity with certain doctrines they held.

26. Cf. Allen, *Reformed Theology*, 3–5, who in turn draws on Richard A. Muller, "How Many Points?" *Calvin Theological Journal* 28 (1993): 427; see also Muller, *Calvin and the Reformed Tradition*, 57–62.

## Reformed Theology as a Distinctive Stance

The third misunderstanding to be noted is that Reformed theology—or, more subtly, what one thinks may "really" count as Reformed theology—is to be equated with one particular strand of thought in it. Again, there are some varieties to be distinguished here, two of which I will briefly mention.

First, coming from the Netherlands, I am familiar with situations in which members of Reformed churches, when they want to make sure a theological view is Reformed, check it against the three "Forms of Unity": the Heidelberg Catechism, the Belgic Confession, and the Canons of Dordt.[27] Often they are unaware that these confessions are only a small part of a much larger corpus of confessional documents that enjoy a similar status in Reformed churches in other geographical regions. Most of these documents display a common theological outlook—yet they are far from identical to each other. Thus, the Westminster Confession is more specific on certain themes than most earlier confessions. For example, unlike the "Forms of Unity," the Westminster Confession specifies that God created the world "in the space of six days" (IV.1). Now, what if someone rejected a tenet included in the Westminster but absent from most other Reformed confessional standards, such as (to anticipate the topic discussed extensively in later chapters of this book) this creation in six days? Would we deny such a person the status of being "truly" Reformed? And what if a person denied the biblical correctness of a doctrinal tenet of the Canons of Dordt, for example, eternal reprobation? It seems that we should not hold an overly parochial view on such matters but rather take into account the variety of early modern Reformed confessional documents as testifying *together*, especially in their shared views, of what it means to be theologically Reformed.

Second, in various quarters of Reformed churches and seminaries it is common to identify Reformed theology with a more or less Barthian theological outlook. In such a reading of Reformed theology, usually Calvin (in a specific christocentric interpretation) and Barth take pride of place as the tradition's main theological heroes. Other voices are neglected, downplayed, or criticized for not being "truly" Reformed. In this way, some studies on the nature of Reformed theology by authors deeply influenced by Barth exhibit attempts to frame Barth's views on, for example, predestination or the order of gospel and law as the only views on these topics that are really, or truly,

---

27. Referring to these particular forms may be fitting in an ecclesiastical context in which they have gained a special confessional status, but, clearly, it is inadequate to use these three documents (or any other selection of Reformed confessions) as a scholarly yardstick for determining what may count as Reformed in the broader international context.

Reformed. In doing so, they tend to obscure the innovative character of Barth's work on these issues. Thus, Barth proposed a fundamental revision of the doctrine of predestination as advocated by both Calvin and the Synod of Dordt. And in changing the order of law and gospel, Barth did not just repeat or intensify an existing Reformed point of view—he really changed something in comparison to at least *some* authoritative earlier Reformed sources (such as the Heidelberg Catechism). One may either acclaim or regret this, but one should not suggest that Barth's views are the only "truly" Reformed ones.[28]

A historical account of Reformed theology should not be reductionist but should respect the tradition's remarkable diversity and plurality, acknowledging similarities and differences from a synchronic point of view as well as continuities and discontinuities from a diachronic perspective. In such a discourse, later expressions of the Reformed theological tradition need not be considered less important or less "pure" than earlier ones, or the other way around.[29] And we should not let our own preferences distort our understanding of Reformed theology in its multifaceted historical expressions.

### 1.3 Endless Plurality?

Thus, in countering the tendencies described in the previous section to reduce the scope of Reformed theology in various ways, we should do justice to the *plurality* of Reformed theological views. As Amy Plantinga Pauw and Serene Jones point out, "Reformed theology has always been polyphonic."[30] This does not mean (and Pauw and Jones do not take it to mean) that "anything goes." Michael Allen has rightly criticized those authors who reduce Reformed the-

---

28. An example of the first shortcut can be found in Matthias Freudenberg, *Reformierte Theologie. Eine Einführung* (Neukirchen-Vluyn: Neukirchener Verlag, 2011), 190–210, where Barth's doctrine of election is portrayed as the apex of preceding Reformed accounts rather than as one Reformed view among others; an example of the second reduction is Eberhard Busch, *Reformiert. Profil einer Konfession* (Zürich: TVZ, 2009), 99–119.

29. It is controversial, however, to what extent modern theologians belonging to a Reformed church, like Friedrich Schleiermacher, should still be considered theologically "Reformed"; but even in his case, there are strong continuities with, for example, Calvin's theological thinking.

30. Amy Plantinga Pauw and Serene Jones, eds., *Feminist and Womanist Essays in Reformed Dogmatics* (Louisville: Westminster John Knox, 2006), xi (they hold that "it is more accurate to speak of Reformed *theologies*").

ology to a revolutionary "habit of mind," as if being Reformed means no more than being fond of change: "Some have usurped the name 'Reformed' for any brand of theology that emphasizes the need for . . . ongoing reform and conceptual revolution. While these deconstructive impulses can bear great affinity with some Reformed criticism of various ideologies, cultural practices, and ecclesial tradition(s), they do not necessarily warrant the title 'Reformed.'"[31] Despite its many voices, there are limits to what might reasonably be labeled Reformed. It seems unlikely that any theological view advocated by a member of a Reformed church should be counted as being truly Reformed. At the very least, to be Reformed means to be *catholic*—that is, to share in the faith of the apostolic church of the first centuries. The very name Reformed reminds us of the Reformed Reformation's leading motive to restore the church to what was considered its original state.

One might even argue that the Reformers did not want to do anything more than that—for example, they did not want to introduce any new doctrines in the church, and they eschewed embracing a particular doctrine or doctrines as "their own" because they wanted to be fully ecclesial instead of sectarian. With the exception of the doctrine of the spiritual presence of Christ in the Lord's Supper (as this was laid down in the *Consensus Tigurinus* of 1549), there is no "typically Reformed" doctrine.[32] That is not to say that Reformed theology has no doctrinal substance, but most of it is none other than the doctrinal heritage of the catholic church. Still, Reformed theology as a matter of fact *did* receive a casting of its own. In the course of time certain characteristics emerged that rightly came to be seen as "typically Reformed."

Attempts are often made to compose a list of such characteristics that, all together, make up what it means to be Reformed. Sometimes such lists are pretty short, sometimes they are longer.[33] On the long lists, one might sup-

---

31. Allen, *Reformed Theology*, 16. For an attempt to define Reformed identity in terms of certain "habits of mind" as opposed to certain doctrinal distinctives, see Brian Gerrish, "Tradition in the Modern World: The Reformed Habit of Mind," in *Toward the Future of Reformed Theology: Tasks, Topics, Traditions*, ed. David Willis and Michael Welker (Grand Rapids: Eerdmans, 1999), 3–20.

32. Even the so-called *extra calvinisticum* as a doctrinal distinction in Christology was shared by patristic and medieval theologians, whereas it eventually came to be rejected by Reformed theologian Karl Barth. See Andrew M. McGinnis, *The Son of God beyond the Flesh: A Historical and Theological Study of the Extra Calvinisticum* (London: Bloomsbury T&T Clark, 2014).

33. See, e.g., Busch, *Reformiert*, 18–31, who distinguishes four Reformed specifics ("Eigentümlichkeiten"); longer lists can be found in Leith, *Introduction*, 96–111, and

pose that one need not exemplify all characteristics mentioned in order to be Reformed, but only a sufficient number of them. Here, Ludwig Wittgenstein's theory of family resemblances seems to offer a helpful analogy. Wittgenstein famously argued that just as family members usually share *some* character traits with each other but not all, so our words used in different contexts (or "language games") usually share some aspects of meaning but not all—and not all the same, as a result of which there is no essential core that applies to all uses of a particular word.[34] Similarly, members of the Reformed theological tradition may be seen as sharing at least some theological accents with one another—but not necessarily the same ones. As long as one subscribes to a sufficient number of theological convictions that are recognizably Reformed, one's thinking may count as Reformed. Of course, there are borderline cases in such an approach, since it can be debated how many "Reformed" theological convictions one needs in order to be "in." The existence of such borderline cases, however, is not a decisive argument against the usefulness of this approach. So-called vague concepts, such as "twilight" (David Hume's example), cannot be demarcated in an unambiguous way, but that does not detract from their usefulness.[35] The case might be similar with "Reformed theology."

Nevertheless, there is another problem with playing the Wittgensteinian card: the analogy of family resemblances does not account for the undeniable *interconnectedness* of the various Reformed theological characteristics. With family resemblances, some family members may have similar noses, others similar ways of laughing or walking, without there being any connection between these various traits (they probably go back to different genes). This is not the case with Reformed theology. Somehow, there is an underlying inner coherence that binds together its various characteristics. To be sure, these connections are never coercive, as if they form a total package; but they are certainly *suggestive*: if one says A, it is quite natural to say B as well. Therefore, it

---

in a very instructive survey of Jean-Jacques Bauswein and Lukas Visscher in their *The Reformed Family Worldwide: A Survey of Reformed Churches, Theological Schools, and International Organizations* (Grand Rapids: Eerdmans, 1999), 26-34. Leith offers seven defining characteristics of Reformed theology, Bauswein and Visscher no fewer than fifteen. See also Donald McKim, *Introducing the Reformed Faith* (Louisville: Westminster John Knox, 2001), 178-80.

34. Ludwig Wittgenstein, *Philosophical Investigations*, trans. G. E. M. Anscombe (Oxford: Basil Blackwell, 1967), §§ 65-71.

35. Cf. Kees van Deemter, *Not Exactly: In Praise of Vagueness* (Oxford: Oxford University Press, 2010).

seems best to identify this distinct character of Reformed theology—that is, of the Reformed modification of the Christian faith—as a *stance*. I am inspired here by the work of philosopher Bas van Fraassen, who, after having concluded that none of the existent definitions of "empiricism" was fully adequate, proposed to consider empiricism as a *stance*: "A philosophical position can consist in a stance (attitude, commitment, approach, a cluster of such—possibly including some propositional attitudes such as beliefs as well). Such a stance can of course be expressed, and may involve or presuppose some beliefs as well, but cannot simply be equated with having beliefs or making assertions about what there is."[36]

In this way, Van Fraassen considers empiricism as a philosophical tradition with different but similar instantiations in various philosophical eras. After having concluded that all attempts to define empiricism in terms of some core belief(s) are inadequate, he appeals to persistent attitudes and approaches, concerns and commitments, inclinations and dispositions that invariably shape its spirit. Empiricism, in other words, is a *stance*. Interestingly, Van Fraassen explicitly compares adopting a philosophical stance (such as empiricism) to adhering to some cause or religion.[37]

How would things look if we described the identity and nature of Reformed theology in terms of a stance? The advantage of this approach is threefold. First, we need not suggest that Reformed theology be identified by some core beliefs, thinkers, or traditions, nor by a set of disparate characteristics that only loosely hang together. Second, we can do justice to the fact that, far from being endlessly plural, Reformed theology does have a recognizable identity of its own, which is to be found in its particular approach to commonly held doctrines and convictions. And third, we need not suggest that these doctrines and convictions are *exclusively* Reformed, as if they cannot be found in other ecclesial traditions as well. Rather, other faith communities may display the same patterns of thought—but often not with the same intensity or commitment or sense of urgency. In this way, Reformed theology may be seen as a specific stance—that is, an *intensification* of some theological doctrines, commitments, and even debates (e.g., on free will) that can also be found, but less emphatically, in other parts of the universal church.[38]

---

36. Bas C. van Fraassen, *The Empirical Stance* (New Haven: Yale University Press, 2002), 47–48.

37. Van Fraassen, *The Empirical Stance*, 61.

38. I owe the concept of intensification to Freudenberg, *Reformierte Theologie*, 15: "For that which Reformed theology and the Reformed church have to say should be primarily understood as an intensification of the common Christian heritage" (*des*

## 1.4 The Reformed Theological Stance

Is it possible to spell out the leading motif or bottom line of the Reformed stance, which is responsible for the underlying coherence of its various emphases and concerns? In theory, many candidates could be considered here, but in my view the most promising one is the famous adage *ecclesia reformata semper reformanda secundum verbum Dei*—"a Reformed church should always be reforming according to the Word of God." In some form or other this motto has been repeated time and again as an appropriate self-description of Reformed believers in widely divergent strands of the Reformed family.[39] Obviously, it is primarily an *ecclesiological* notion, expressing what it means to be church according to the Reformed: to be always reforming oneself in the direction of the Bible. The ramifications of this motto, however, extend far beyond the sphere of ecclesiology. This is especially the case when we include its (often forgotten) final words: *secundum verbum Dei*.

It is here, it seems, that we find the heartbeat of Reformed theology: in the deep-seated wish to let the Word of God and the Word of God alone determine the church's faith and life. Here lies its main commitment. Obviously, this implies a strongly developed doctrine of Scripture, according to which Scripture is the primary locus of authority in the church. It is indeed a typically Reformed concern to let the *entire* Word of God function in this way.[40] Whereas its closest cousin, Lutheran theology, has traditionally been inclined to distinguish a "canon within the canon" by giving priority to those texts that testify to the gospel of God's free grace in the justification of the sinner and marginalizing texts with other emphases, Reformed theology has always felt

---

*gemeinsamen Christseins*); it seems more accurate to say, however, that Reformed theology intensifies *specific aspects or parts or themes* of the common Christian heritage (e.g., the doctrine of grace).

39. The source of the expression is often attributed to the Dutch Pietist Johannes Hoornbeeck, and although that seems false it was in any case very popular in his circles. It was also put to use, however, by Barth and Moltmann, as it was by Kuyper and Bavinck. See Theodor Mahlmann, "'Ecclesia semper reformanda': Eine historische Aufarbeitung. Neue Bearbeitung," in *Hermeneutica Sacra. Studien zur Auslegung der Heiligen Schrift im 16. und 17. Jahrhundert*, ed. Torbjörn Johansson, Robert Kolb, and Johann Anselm Steiger (Berlin: de Gruyter, 2010), 381–442; Fred van Lieburg, "Dynamics of Dutch Calvinism: Early Modern Programs for Further Reformation," in *Calvinism and the Making of the European Mind*, 44–46.

40. Some sixteenth-century Reformed confessions list all sixty-six biblical books identified by Protestants as canonical; cf., e.g., the Belgic Confession, art. 3.

uneasy about this hermeneutics. Next to the *sola Scriptura*, it endorsed the *tota Scriptura*. The Word of God, being itself the final judge, cannot be judged by an internal norm (a "canon within the canon") nor by some external norm—be it an ecclesial tradition, someone's inner voice, or a political authority. This sensitivity for the Word of God as the Christian's sole judge gives Reformed theology its iconoclastic tendency, and, arguably, its preference for the distribution of power among various equally powerful office bearers in church councils and synods over against more hierarchical forms of church order.

Quite a number of mutually related commitments—in fact, most of those that together build up the Reformed stance—seem to follow in one way or another from the *semper reformanda* maxim. Thus, we can understand the Reformed predilection for the literal meaning of biblical texts as opposed to allegorical or symbolic interpretations from this viewpoint. The deep estimation of the Old Testament that is typical for important strands of the Reformed tradition is another case in point. If the Old Testament forms an integral part of the Word of God, it can hardly be considered (as Lutheranism had it) as only the negative foil of the gospel. The Old Testament should be attributed a more constitutive role—which was done in Reformed theology by highlighting the concept of the *covenant*, that is, the one covenant of God with his people that constitutes both Old Testament Israel and the New Testament church consisting of Jews and gentiles. Although this covenant has two different historical periods or "administrations," its essence is one and the same, as John Calvin put it.[41]

From this vantage point, it is also understandable that Reformed theologians accepted *infant baptism*, considering baptism as the New Testament sign of God's covenant with his people. Moreover, the flourishing of covenantal (or "federal") theology in all its diversity within Reformed circles during later centuries can be seen as an elaboration of the same fundamental orientation. Also, the positive attention within the Reformed theological tradition given to the people of Israel—past, present, and future—can be interpreted along these lines.[42] This is not to say that all Reformed theologians shared these commit-

---

41. See Heinrich Bullinger, *De testamento seu foedere Dei unico et aeterno . . . brevis expositio* (Zürich: Froschauer, 1534); John Calvin, *Institutes of the Christian Religion*, ed. John T. McNeill, trans. Ford L. Battles (Philadelphia: Westminster, 1960), 2.10.2; for a succinct survey of Reformed federal theology, see Allen, *Reformed Theology*, 34-53.

42. The Dutch Reformed theologian Hendrikus Berkhof was probably the first who made "Israel" into a locus of its own within the classical dogmatic scheme. Cf. his *Christian Faith: An Introduction to the Study of the Faith*, rev. ed. (Grand Rapids: Eerdmans,

ments, nor is it to say that those from other traditions did not. It is to say that one should not be amazed at encountering a high view of the Old Testament's significance for the Christian church today, an emphasis on the covenant as a leading motif in biblical theology, or a keen interest in God's faithfulness to the Jewish people in Reformed theology.

Another series of typically Reformed accents—or intensifications of catholic affirmations—comes to mind when we realize that in the maxim *ecclesia reformata semper reformanda secundum verbum Dei* the Bible is described as the word *of God*. This does not deny that the Bible was written by humans. The Reformed emphasis was that it is Godself who speaks to us through the Bible, so that it is primarily *his* word. Here we encounter the strong theocentric tendency that is typical of Reformed theology.[43] God's majesty, sovereignty, and honor are the focal point of the Christian faith. The main goal of human life is to know God, because God has created us in order to be glorified in us.[44] Here also a certain predilection for the doctrines of election and predestination becomes understandable, as does Reformed theology's tendency to advocate radical notions of sin and grace and (by extension) an Anselmian doctrine of atonement. Its emphasis on the seriousness and pervasiveness of human sin does not stem from some kind of collective misanthropy but from the wish to avoid all inklings of synergism in soteriology: it is God and God alone to whom we owe our salvation.[45] The ontological distinction between God and humanity is of paramount importance in this regard (thus, many Reformed theologians are critical of the doctrine of *theōsis*, as they fear—rightly or not—a blurring of this distinction).[46] Reformed theology's theocentric commitment can also explain its fear of idolatry, of making God into something that is *not* God by distorting his character—in the form either of physical images,[47] of

---

1986), 225–70. However, the Reformed emphasis on the covenant has also often been interpreted in supersessionist terms and continues to be vulnerable to such interpretations.

43. Cf. McKim, *Introducing*, 179: "Consider the distinctive: Reformed theology lives from stressing *the prior initiative of God and our grateful response*. This is the Reformed faith's inclination, its bent, its proclivity."

44. John Calvin, *Catechism of Geneva* (1545), articles 1 and 2.

45. This is not to deny that, in practice, the Reformed emphasis on the pervasiveness of sin has often been taken as implying a call to self-hate.

46. Cf., e.g., Bruce L. McCormack, "Participation in God, Yes; Deification, No: Two Modern Protestant Responses to an Ancient Question," in *Orthodox and Modern: Studies in the Theology of Karl Barth*, by Bruce L. McCormack (Grand Rapids: Baker Academic, 2016), 235–60.

47. Cf. the typically Reformed division of the Decalogue, which considers the

fixing Christ materially in the elements of the Lord's Supper, or of speculative systems of thought that go beyond the boundaries of Scripture. Instead, we should stick to the way in which God has *revealed* Godself.

Finally, the maxim *ecclesia reformata semper reformanda secundum verbum Dei* evokes the image of an ongoing movement in which ever more domains of life are brought under the dominion of God's will. Although the maxim is ecclesiological in nature, its dynamic quite naturally spills over into the lives of individual Christians. This ongoing movement can be seen concretely in the "typically Reformed" emphasis on sanctification next to justification and on the so-called *tertius usus legis* ("third use of the law," namely, as a "rule of gratitude," next to its civic and pedagogical uses).[48] Reformed theology's emphasis on pneumatology—that is, on the role and work of the Holy Spirit—is crucial in this respect. It is through the Spirit of God, who unites us with Christ, that our lives are being transformed from day to day toward the image of God. But the Spirit does not only transform private lives. Rather, no sphere of life is exempted from the claim of the Word of God and from the transforming power of the Spirit, be it society, culture, philosophy, economics, the arts, politics, or whatever. As Abraham Kuyper famously phrased this sentiment: there is "not a square inch in the whole domain of our human existence over which Christ, who is Sovereign over all, does not cry, Mine!"[49] Once again, not all Reformed theologians would echo Kuyper in such a strong voice, whereas representatives of other traditions might share his commitment—but it is quite natural to find it in the Reformed tradition. During the twentieth century it was found there once again, this time in Karl Barth's strong theological resistance to the Nazification of the German church at the Synod of Barmen in 1934.

This dynamic of always needing to be reformed toward the Word of God implies that in ever-changing conditions the search for implementing God's will is pertinent.[50] It means, among other things, that there can never be one

---

prohibition on making images as a commandment of its own rather than subsuming it under the first commandment not to honor other gods.

48. Cf. the Heidelberg Catechism, Q&A 2, 86, 91–115; cf. Richard A. Muller, *Dictionary of Latin and Greek Theological Terms* (Grand Rapids: Baker Books, 1985), 320–21; Cornelis van der Kooi and Gijsbert van den Brink, *Christian Dogmatics: An Introduction*, trans. Reinder Bruinsma with James D. Bratt (Grand Rapids: Eerdmans, 2017), 686–96.

49. Abraham Kuyper, "Sphere Sovereignty," in *Abraham Kuyper: A Centennial Reader*, ed. James D. Bratt (Grand Rapids: Eerdmans, 1998), 488.

50. Reformed theology has emphasized on many occasions that the will of God is to be obeyed even if this means disobedience toward political rulers. It cannot be upheld, however, that the "right of resistance" to secular authorities (either by the

final, universally binding confession (as is the case in Lutheranism), since in every cultural situation new confessional statements may emerge that should be taken seriously. Reforming to the Word of God is a never-finished business. That also means, of course, that seriously studying the Word of God is of paramount importance. We cannot live by traditional interpretations of Scripture, since these may easily block its powerful living voice. Scripture may always have to say other things to us than we thought. Therefore, we can never be satisfied with short summaries or traditional interpretations of its content. We should continue to study Scripture itself in light of the old and new questions that confront us in our specific context. It is precisely this task that we will undertake in the remainder of this book with regard to a specific concern of our own time: the reception of evolutionary theory. But before broaching that topic, we must make one more preliminary step.

## 1.5 Reformed Theology and the Natural World

Clearly, how we assess evolutionary theory is largely dependent on how much we value contemporary science in general. If we hold a high view of modern science, we will more readily take evolutionary theory seriously than if we distrust the entire scientific endeavor. In the first case, one may even see evolutionary theory as an unassailable stronghold of modern science: let nobody think it is open to debate! In the second case, evolutionary theory is easily discarded or downplayed with an appeal to what is seen as the unreliable character and highly limited scope of scientific theories. Being "just a theory," evolution should not be taken too seriously because scientific theories are by definition shaky and uncertain.

How does Christian theology, especially in its Reformed intensification, value the scientific endeavor? Many contemporary Christians tend to be critical of contemporary science and suspicious of its claims and methods. This seems at least partially due to the deep-seated tendency in Western culture to consider the relationship of science and religion from the perspective of

---

lower magistrates or more generally) is a typically Reformed notion. Although this has often been thought, in fact this notion wandered back and forth from Lutheranism to Calvinism and Roman Catholicism, depending on which confession was denied freedom of religion in a specific situation. Cf. Benedict, *Christ's Churches*, 87–88 (on the Lutherans in Germany), 146–47 (on the Reformed and Catholics in France), 187–88 (on the Reformed in the Netherlands).

the so-called conflict model, according to which science is by its very nature opposed to religion.[51] The two are seen as communicating vessels: the more seriously you take science, the less room will be left for religion in your life—and the other way around. If science and religion are indeed foes of each other, Christians usually do not find it hard to make a choice: science must be wrong in many cases. In this way, in orthodox Reformed circles one can often hear science and faith being pitted against one another by rhetorical questions such as: "Shouldn't we trust the plain Word of God much more than human inquiry?" In such an atmosphere, there is little reason to take evolutionary theory seriously, let alone to carefully weigh its theological consequences.

It is important to observe, however, that "from the beginning it was not so." That is to say: the mainstream of the Christian tradition has been characterized by a *positive* and supportive attitude toward the study of the natural world. The magisterial Reformation that spawned both the Lutheran and the Reformed traditions was open to what counted as established knowledge in its cultural and intellectual environment. Though clearly there have been exceptions, in general the temptation to withdraw into isolated communities and private forms of spirituality was resisted. As we saw above, Reformed theology was characterized by a strong concern about the seriousness of sin and its all-pervasive consequences. Still, Reformed communities maintained a strong focus on society at large, based on the belief that despite its evils the world belongs to God and should therefore be restored toward its proper destination.[52] Throughout the centuries Reformed thinking has been concerned with philosophy, economics, politics, the arts, and the sciences—in particular the study of nature. It was this broader focus that always prevented the Reformed tradition from becoming sectarian. In line with the church catholic, Reformed theology cherished a positive rather than a dismissive and distrustful attitude toward (the study of) the world in which we find ourselves.

---

51. On the history of this conflict model, and the reasons why it has been largely discredited, see, e.g., Thomas Dixon, Geoffrey Cantor, and Stephen Pumfrey, eds., *Science and Religion: New Historical Perspectives* (Cambridge: Cambridge University Press, 2010); Peter Harrison, *The Territories of Science and Religion* (Chicago: University of Chicago Press, 2015). The demise of the conflict thesis was inaugurated by John Hedley Brooke; cf., e.g., his *Science and Religion: Some Historical Perspectives* (Cambridge: Cambridge University Press, 1991).

52. Cf., e.g., the various contributions to Van den Brink and Höpfl, *Calvinism and the Making of the European Mind*.

Let me substantiate this claim by briefly examining a seminal text from the Reformed tradition: the second article of the Belgic Confession (1561)—one of the three "Forms of Unity" mentioned above. The "Belgic" was composed by the Walloonian stained-glass artist and Reformed preacher Guido (or Guy) de Brès (1522-1567), who had studied with John Calvin in Geneva. It was officially adopted as a Reformed confession at the international Synod of Dordt (1618-1619). In its opening articles, De Brès did not immediately jump to the painful differences between Reformed theology and what would come to be called Roman Catholicism, but rather wanted to show how the theology of the Reformers is rooted in the faith of the catholic church of the patristic period. It comes as no surprise, therefore, that article 2 of the Belgic Confession reflects a consensus between Protestant and Roman Catholic theology throughout the centuries. The text runs as follows:

> We know God by two means: First, by the creation, preservation, and government of the universe, since that universe is before our eyes like a beautiful book in which all creatures, great and small, are as letters to make us ponder the invisible things of God: God's eternal power and divinity, as the apostle Paul says in Romans 1:20. . . . Second, God makes himself known to us more clearly by his holy and divine Word, as much as we need in this life, for God's glory and for our salvation.[53]

According to this text, in some way the natural world provides us with knowledge of *God*. In the era in which these words were written, a more forceful recommendation of its investigation could hardly be conceived.[54] It was argued, following up on texts such as article 2, that if nature is indeed a book written by God, then there is every reason to investigate this book closely. And, just as the Reformation (at least officially) put an end to the allegorical method of interpreting the second book—Holy Scripture—because it wanted to stick close to its *sensus literalis*, so it also stimulated staying close to nature when interpreting it rather than looking for allegorical meanings as Augustine and many others had done. Thus, the Reformation fostered an attitude that favored conducting experimental research.[55] Present-day historians of

---

53. Quoted from *Our Faith: Ecumenical Creeds, Reformed Confessions, and Other Resources* (Grand Rapids: Faith Alive, 2013), 26-27.
54. See, e.g., Eric Jorink, *Reading the Book of Nature in the Dutch Golden Age, 1575-1715* (Leiden: Brill, 2010).
55. See Peter Harrison, *The Bible, Protestantism, and the Rise of Natural Science* (Cambridge: Cambridge University Press, 1998), 4-8.

science even consider the metaphor of the two books, operationalized in this way, as a new *paradigm*, in terms of which the changed approach to nature in the seventeenth century can largely be explained. Most seventeenth-century natural philosophers—in an important sense the precursors of modern-day scientists—envisioned their study of nature from this perspective.[56] Their self-understanding was informed by their deeply rooted conviction that nature constituted a source of knowledge of its Maker.

Whereas this belief led some, along Augustinian lines, to view nature as only a collection of illustrations of biblical truths,[57] for others—and they gradually began to form the majority—the two sources of knowledge of God were more independent of each other. In their view, nature deserved to be carefully explored. As a result, they came to investigate nature with a more open mind, curious about the traces of himself that the Creator had left behind, in particular of his power, goodness, and wisdom. Arguably, it was this attitude, articulated in a poetic way in article 2, that gave an important impulse via eighteenth-century physico-theology to the genesis of the contemporary natural sciences.[58] Thus, article 2 was "an extremely important formulation," the implications of which "cannot be stressed too much."[59] It is clear that the impact of the metaphor of the world as a divine text (which also popped up in other sources at that time) casts further doubt on the aforementioned conflict model, which usually describes the rise of modern science as the victory of reason over faith.

The idea of the "book of nature" offers us a better model for understanding the new approach to nature in the seventeenth century, since this is not an

---

56. Jorink, *Reading the Book*, 12-13.

57. Jorink, *Reading the Book*, 22, cites Gisbertus Voetius (52). Another example is Jonathan Edwards, as has been pointed out by Avihu Zakai, *Jonathan Edwards' Philosophy of Nature: The Re-enchantment of the World in the Age of Scientific Reasoning* (London: T&T Clark, 2010), 17-24.

58. See, e.g., Kenneth J. Howell, *God's Two Books: Copernican Cosmology and Biblical Interpretation in Early Modern Science* (Notre Dame: University of Notre Dame Press, 2002). This is not to suggest that scholars of other religious persuasions weren't equally involved in the rise of modern science. Brooke, *Science and Religion*, 109, has pointed out that, for example, many Roman Catholic natural philosophers were as creative and innovative as their Protestant counterparts; cf. Gijsbert van den Brink, "The Reformation, Rationality and the Rise of Modern Science," in *Reformation und Rationalität*, ed. Herman J. Selderhuis and Ernst-Joachim Waschke (Göttingen: Vandenhoeck & Ruprecht, 2015), 198-99, 202-3.

59. Jorink, *Reading the Book*, 20-21.

alien framework imposed on it from a later perspective (such as the modern faith-reason dichotomy) but closely reflects the self-understanding of most early modern natural philosophers. In the middle of all kinds of differences on biblical exegesis, Cartesianism, heliocentrism, and other issues, this was a shared Christian perspective: God reveals himself to us both via the Bible and via that other book—nature. Precisely the book of nature could play a unifying role, since it could be read by everyone. From this perspective it is not strange that during the seventeenth and eighteenth centuries the metaphor of the book of nature became popular not only in the Low Countries but also in England and other European countries.[60] In an unprecedented way, the natural world came to be "read" for what it had to say in and of itself—though still in order to discover the glory, majesty, and wisdom of God.

The use of the metaphor of the two books should thus be understood against the background of the notion of the natural knowledge of God. Whereas it is generally agreed that Roman Catholic theology has always affirmed this notion, it is sometimes suggested that the Protestant Reformers, focusing on issues of sin and grace rather than on nature, abandoned the idea of a natural knowledge of God. On the basis of a representative selection of Reformed sources, Michael Sudduth has shown, however, that "there has been a widely instantiated, deeply entrenched, and historically continuous endorsement of natural theology in Reformed thought."[61] Indeed, the notion of a natural knowledge of God also figures in later parts of the Protestant tradition, for example, in the Westminster Standards.[62]

In conclusion, then, article 2 is part of this larger tradition. By describing the natural world poetically as "a beautiful book" that gives us access to God's majesty and divinity, De Brès intensified an intuition that can be called authentically Christian.[63] By suggesting that we come to know God "more

---

60. For examples from England, see Harrison, *The Bible*, 193-204; on the distribution and diverse use of the two-source theory in the early modern period in all of Europe, see Howell, *God's Two Books*.

61. Michael Sudduth, *The Reformed Objection to Natural Theology* (Burlington, VT: Ashgate, 2009), 40. In Sudduth's terminology, natural theology includes the natural knowledge of God. See more broadly, Wolfhart Pannenberg, *Systematic Theology*, vol. 1 (Grand Rapids: Eerdmans, 1991), 73-74.

62. John V. Fesko, *The Theology of the Westminster Standards: Historical Context and Theological Insights* (Wheaton, IL: Crossway, 2014), 69-71.

63. It can be argued that this goes all the way down to the Bible, where nature is also viewed as a source of (vague) knowledge of God (e.g., Ps. 19; Rom. 1). Cf. James Barr, *Biblical Faith and Natural Theology* (Oxford: Clarendon, 1993).

clearly" through Holy Scripture—that is, no doubt, in a way that reveals to us God's grace, mercy, and love next to his divine majesty—De Brès intensified another notion with deep roots in the Christian tradition, namely, the primacy of Scripture. It is sometimes contended, especially in response to Karl Barth's criticism of natural theology (in which criticism Barth included article 2),[64] that this article should be read back to front: only by using the spectacles of Scripture are we able to read the book of nature in such a way as to draw the right conclusions. However, this clearly goes against the grain of the meaning of the article; surely De Brès did not mean that, after all, Scripture is the first book of God and nature only the second.[65] He argued that the first book offers us some vague and provisional knowledge of God that is then augmented and clarified by the second. Therefore, Reformed Christians who study the natural world with an open mind (instead of taking biblical exegesis as their starting point) are faithful to their confessional tradition in this regard.

### 1.6 Seeking the Honeycomb in the Lion's Mouth

Having sketched what might be considered typical or distinctive of Reformed theology, we are now in a position to discern why coming to terms with evolutionary theory is a special challenge to Reformed Christians. For as we will see more clearly in the next chapter, evolutionary theory elicits serious questions about some of the main commitments and concerns that constitute the "Reformed stance."

The Reformed highlighting of the central role of the Bible, for example, goes hand in hand with a preference for interpreting the Bible—including the Old Testament—as literally as possible, and thus raises the question how evolutionary theory can be brought in line with a "plain" reading of Scripture (see chap. 3). Its theocentric commitment, and especially its concern for safeguarding God's perfect nature, raises questions about the occurrence of so much evolutionary suffering in natural history—is this the work of the Father of Jesus

---

64. Karl Barth, *Church Dogmatics* II/1 (Edinburgh: T&T Clark, 1957), 127. For further analysis, see Gijsbert van den Brink, "As a Beautiful Book: The Natural World in Article 2 of the Belgic Confession," *Westminster Theological Journal* 73 (2011): 273–91 (esp. 283–89).

65. This is argued convincingly in the doctoral dissertation of my former PhD student Bram Kunz, *Als een prachtig boek. Nederlandse Geloofsbelijdenis artikel 2 in de context van de vroegreformatorische theologie* (Zoetermeer, the Netherlands: Boekencentrum, 2013); see the conclusion on 301–2 (or the English summary, 379–83).

Christ (chap. 4)? Next, its covenantal thinking involves a strong focus on the dramatic history of the relationship between God and humanity—a relationship that started with the special creation of humankind "in God's image"; but how special are we humans if we emerged from the animal world (chap. 5)? Further, we noted Reformed theology's tendency to advocate radical notions of sin and grace. God's relationship to humanity took a sharp turn at the fall into sin but was then graciously restored in the history of salvation as culminating in the work of Jesus Christ. Is this narrative still credible given standard accounts of evolution? For example, what happens to the "historical Adam" in such a scenario? What happens to the doctrine of (actual and original) sin, as well as to the sobering view of human death as the consequence of sin? And how about the story of redemption in Christ (chap. 6)? Next, Reformed theology's emphasis on God's sovereignty seems at odds with the pivotal role assigned to randomness and chance by the Darwinian account of evolution (chap. 7). Finally, the extension of Darwinism into notions of *cultural* evolution seems to debunk the Reformed—and more broadly catholic—account of divine revelation as the ultimate source of morality and religion (chap. 8).

It is promising that, due to its dynamic character (as epitomized in the phrase "always reforming"), Reformed theology has never considered itself a static once-and-for-all entity. Reformed Christians have typically been prepared to revisit and rethink hard issues in the light of Scripture rather than taking traditional answers for granted. Moreover, Reformed theology has in its best moments resisted the temptation to withdraw to private forms of pietistic spirituality, focusing instead on the world at large and all that is in it—including the sciences. As it was said by a Reformed author in the seventeenth century: "Wheresoever Truth may be, were it in a Turk or Tatar, it must be cherished . . . let us seek the honeycomb even within the lion's mouth."[66] A serious engagement with what is going on in the realm of the sciences is therefore not at all alien to the Reformed stance. With this in mind, in chapter 2 we will delve more deeply into the specific nature of contemporary evolutionary theory, in order to more firmly grasp the challenges we face. Having outlined the first part of this book's title, we will now explore its second part.

---

66. Johan de Brune (1588-1658), Dutch Calvinist writer and statesman, as quoted (without source reference) by Reyer Hooykaas, *Christian Faith and the Freedom of Science* (London: Tyndale Press, 1957), 12.

CHAPTER 2

# Evolutionary Theory as a Layered Concept

> Evolutionary theory must be one of the greatest paradigm shifts humans have ever made, and in many cases are still making. The conceptual and moral and religious hurdles to evolutionary theory's rapprochement with Christian faith are significant and are still being negotiated as evolutionary theory itself is changing.
>
> —Nicola Hoggard Creegan[1]

### 2.1 Varieties of Evolution: Terminological Clarifications

In what follows we will focus on evolutionary theory as a part of natural science. As indicated in the introduction, it is important to distinguish the scientific theory of evolution carefully from the encompassing worldview that we call evolution*ism*. Evolutionism may be loosely defined as the worldview according to which processes of biological, cultural, and social evolution are sufficient to explain what goes on in the biosphere, so that these processes rule out the existence of God and other supernatural beings.[2] It is, thus, a variety of atheism—and in many cases also of scientism.[3] As an atheistic worldview,

---

1. Nicola Hoggard Creegan, *Animal Suffering and the Problem of Evil* (Oxford: Oxford University Press, 2013), 97.
2. For some of the most well-known defenses of this worldview, see Daniel C. Dennett, *Darwin's Dangerous Idea: Evolution and the Meanings of Life* (New York: Simon & Schuster, 1995), and Richard Dawkins, "Darwin Triumphant: Darwinism as Universal Truth," in *A Devil's Chaplain: Reflections on Hope, Lies, Science, and Love*, by Richard Dawkins (London: Weidenfeld & Nicolson, 2005), 78–90.
3. On scientism, see Jeroen de Ridder, Rik Peels, and René van Woudenberg, eds.,

evolutionism claims either to take over or to make obsolete the traditional roles of religious worldviews. Although at times we will point to this ideological use that is often made of evolutionary theory, in what follows we will focus on evolution as a biological theory. For that is what evolution is: it is not an ideology or an atheistic worldview, nor, for that matter, a mere hypothesis, but a more elaborate scientific *theory*.

There is a lot of confusion about the nature of a scientific theory among the general public, however. In scientific language the concept of a theory is much stronger than in everyday speech (where it may function in sentences such as "My theory is that she has missed the train"). Whereas a hypothesis is usually seen as an assumption with a limited and clearly defined scope that can be tested by following certain procedures, theories often have a much wider scope and a longer trajectory of being elaborated, tested, adjusted, confirmed, etc. They also have a unifying capacity that hypotheses usually lack. Of course, scientific theories can be *false*—like the phlogiston theory in chemistry, which turned out to be false by the end of the eighteenth century. The American Association for the Advancement of Science (AAAS) used to define a scientific theory as "a well-substantiated explanation of some aspect of the natural world, based on a body of facts that have been confirmed through observation and experiment."[4] This definition has been removed from their website, however, presumably because it hardly allows for the possibility of *false* theories.

Still, a theory is not just an incidental try-on or unfounded guess but is usually a group of interrelated hypotheses that turned out to have considerable explanatory power. Therefore, a scientific theory cannot simply be dismissed out of hand as "just a theory." If one is to be rejected, this should be done (as it was with the phlogiston theory) on the basis of overriding evidence. Sometimes, such overriding evidence does not emerge, but, instead, new phenomena are discovered that enrich the theory or can be elegantly explained in

---

*Scientism: Perils and Prospects* (Oxford: Oxford University Press, 2018). Cf. René van Woudenberg, "Limits of Science and the Christian Faith," *Perspectives of Science and Christian Faith* 65, no. 1 (2013): 24–36.

4. Though no longer on its website, www.aaas.org (searched February 6, 2018), quotations of and references to this definition still abound on the Internet. Cf. also an introductory book on evolution written under the auspices of the AAAS: Catherine Baker, *The Evolution Dialogues: Science, Christianity, and the Quest for Understanding* (Washington, DC: American Association for the Advancement of Science, 2006), where a scientific theory is defined as "an explanation of a natural phenomenon well supported by a wide range of evidence from nature" (163).

terms of it. As a result of such processes, the credibility of a theory increases over time.[5] In this way, evolutionary theory has gradually gained increased status in the sciences: it enables us to understand a staggering variety of phenomena in the natural world. In fact, according to many experts, discoveries that have been made in very diverse scientific disciplines make sense only when looked at through the lens of evolutionary theory.[6]

But what *is* evolutionary theory? What exactly does it claim? Unfortunately, that is not a very easy question to answer. Since the word "evolution" just means "development" (which in turn means, essentially, "change over time"), it can be used in different contexts and with different meanings. In the realm of biology, "evolution" originally denoted the idea that life has gradually developed over vast periods of time from the very simplest (mono-cellular) to the most complex forms—for example, there were bacteria before fish, fish before reptiles, reptiles before mammals, etc. In popular uses of the term, this is still the notion that is meant. This definition does not specify, however, how the gradual development of life came about. As is well known, various theories of how the evolutionary process actually works predated Darwin's version. However, it was Darwin's theory, rather than Lamarck's or Richard Owen's or those of others, that came to dominate the field.[7] Therefore, when speaking of evolution in this book, we will usually have in mind Darwinism and its later elaboration, the so-called neo-Darwinian synthesis (or "modern evolutionary synthesis"). The pivotal idea here is that biological adaptation and speciation mainly occur as a result of one specific mechanism: natural selection based on the occurrence of

---

5. For various philosophical views as to what exactly a scientific theory is, see, e.g., Rasmus Grønfeldt Winther, "The Structure of Scientific Theories," *Stanford Encyclopedia of Philosophy*, March 5, 2015, http://plato.stanford.edu/entries/structure-scientific-theories/#SynSemPraVieBas.

6. Note that all this is still not to say that evolutionary theory is proven, or even true—nothing in this book hinges on such epistemologically strong claims. The fact that it is deeply embedded in contemporary science provides sufficient justification for investigating its theological consequences.

7. Nicolaas Rupke has pointed out that the history of evolutionary theories has largely been rewritten by followers of Darwin who wanted to suggest that Darwinism is the only serious alternative to the traditional view that the various species have been created by God *de novo*. See Nicolaas A. Rupke, "Myth 13: That Darwinian Natural Selection Has Been the 'Only Game in Town,'" in *Newton's Apple and Other Myths about Science*, ed. Ronald L. Numbers and Kostas Kampourakis (Cambridge, MA: Harvard University Press, 2015), 103–11. Cf. Rupke's book *Richard Owen: Biology without Darwin* (Chicago: University of Chicago Press, 2009).

phenotypic variation within species. As came to be known in the early twentieth century when classical Mendelian genetics was integrated into Darwin's theory of evolution, this variation has its background in genetic mutations. Thus, here is the most common meaning of terms like "evolution" and "evolutionary theory": they point to the notion that speciation and adaptation are continuous biological processes from the beginning of life on earth until today that are caused by natural selection operating on random genetic mutations.

More precisely, following Fowler and Kuebler, we may distinguish three different levels or layers within evolutionary theory: historical evolution, common descent, and what they call "strong Darwinian evolution."[8] First, by *historical evolution* they mean that there has been a historical sequence of billions of years ("deep time" or "geologic time") during which ever more complex groups and species subsequently appeared on earth, as can be traced from the fossil record. Note that there is no implication here that the various species originated from one another. Some Christians have therefore referred to historical evolution as "progressive creation," suggesting that God created the main biological groups or phyla one after another (each of which then branched into various directions), separated by long periods of time. In what follows, I will use the more neutral term "gradualism" to refer to this first layer.[9] Second, there is the belief that this historical sequence and the shared characteristics of the various forms of life should be explained in terms of the *common descent* of all life from one original source. In this scenario, there is one "tree of life."[10] No theory is implied, however, as to how exactly this tree has developed. A lot of conceivable causal mechanisms might be invoked here. Third, *strong Darwinian evolution* holds that the key for understanding how this process occurred is to be found in the one mechanism of natural selection acting upon random mutations.

---

8. Thomas B. Fowler and Daniel Kuebler, *The Evolution Controversy: A Survey of Competing Theories* (Grand Rapids: Baker Academic, 2007), 28–29. See also the earlier paper by Yale biologist Keith S. Thomson, "Marginalia: The Meanings of Evolution," *American Scientist* 70, no. 5 (1982): 529–31.

9. The term "progressive creation" is not only more theological but also denies common descent, whereas what we need here is a concept that does not yet take a stance with regard to common descent; "historical evolution" is not very helpful either, since it wrongly suggests that the other parts of evolution are ahistorical.

10. In current usage, the metaphor of a "tree of life" is no longer seen as entirely unproblematic, for example, in relation to the so-called prokaryotic life-forms (bacteria and archaea), which are characterized by "horizontal gene transfer." The details of these debates need not bother us here, though.

*Evolutionary Theory as a Layered Concept*

These distinctions are crucial, since as Fowler and Kuebler rightly point out: "Failure to recognize the tiers and their roles has led to completely fallacious claims about the theory of evolution, and is probably the single most important factor contributing to confusion about the subject."[11] Therefore, in what follows, when we use terms like "evolution" and "evolutionary theory," we will usually have in mind the conjunction of these three layers. When focusing specifically on natural selection as the mechanism held to be responsible for speciation and other evolutionary changes, we may speak of (the) Darwinian (theory of) evolution. When we presuppose the incorporation of the genetic revolution into Darwin's theory of evolution, we will use terms like "neo-Darwinian synthesis" and "neo-Darwinism." Since in this view natural selection presupposes the geological timescale and the theory of common descent, neo-Darwinism can be seen as incorporating all three layers: the geological timescale (or "deep time"), common descent, and natural selection.[12] Let us now focus more closely on each of these three layers. How did they become so firmly established in current scientific practice?[13] What possible alternatives do we have for them? And what theological questions and problems do they provoke?

### 2.2 Gradualism

The view that life on earth has been exposed to evolutionary processes for a very long time already could only gain traction after geologists had discovered that our planet and our solar system are much older than was assumed in tra-

---

11. Fowler and Kuebler, *The Evolution Controversy*, 28.
12. The metaphor of "layers" or "levels" is intended to convey the idea that there is a hierarchy at play here: Darwinian natural selection presupposes common descent and deep time, but not the other way around; and common descent presupposes deep time whereas deep time does not presuppose common descent. Note that in theory one can accept natural selection as the main evolutionary mechanism while rejecting common descent and even (as do those who only accept "micro-evolution") deep time. *Darwinian* natural selection, however, presupposes both other claims.
13. In pointing this out, I will assume the perspective of mainstream science (without repeating that all the time). I will not argue for the correctness of this perspective, however, since my goal is more modest, namely, to give the reader an idea of why contemporary scientists think what they think about evolution. In this way, it can become clear why evolutionary theory is so attractive to many of them. For more details on the science parts, summarized for nonspecialists, see, e.g., Denis Alexander, *Creation or Evolution: Do We Have to Choose?* 2nd rev. ed. (Oxford: Monarch Books, 2014), 88–153.

ditional estimations. For the process of evolution can serve as an explanation for the enormous diversity of life on earth only if an astronomical amount of time was involved in it. When in the seventeenth century the discipline of geology emerged, geologists fitted their findings into a time scheme that was derived from an intuitive reading of the Bible: the earth and its rocks were supposed to be only several thousand years old, and the catastrophic worldwide flood in Noah's days explained the fossils that had been found. For example, fossils of fish found in high mountains were there because Noah's flood had lifted the water and its inhabitants to that level.

When in the course of the eighteenth century the scope of geological research rapidly expanded, however, more and more findings emerged that could hardly be squared with this hypothesis. For example, very thick layers of sedimentary rock were discovered and measured. Some of them were miles thick. A single flood of one year could never have deposited enough eroded material to explain such large layers. And in south-central France, volcanic cones were discovered under grasslands. Since there were no reports or legends of volcanoes in that area, they must predate human inhabitation. Haarsma and Haarsma comment as follows: "Upon close inspection geologists were able to map multiple layers of lava flows, showing that the volcanoes in that area had erupted repeatedly, hardening after each eruption and forming additional structures. Evidence also shows significant water erosion taking place between the various volcanic eruptions. This area tells of a longer and more dynamic history than could be fit into a few thousand years, even with a flood."[14]

Especially the work of Charles Lyell (1797-1875), as published in his tripartite *Principles of Geology* (1830-1833), became influential in this connection. Lyell, taking up an older idea of the Scottish naturalist James Hutton (1726-1797), had come to favor "uniformitarianism," the view that the changes in the strata of the earth did not result from sudden catastrophes ("catastrophism") but resulted from very slow and gradual processes that we could still see at work in nature in the present day. But if the processes that alter the earth strata had been uniform over time, then they must have been going on for much longer periods of time than only several millennia! Indeed, it

---

14. Deborah B. Haarsma and Loren D. Haarsma, *Origins: Christian Perspectives on Creation, Evolution, and Intelligent Design* (Grand Rapids: Faith Alive, 2011), 107; cf. more generally, 104-8, and see for more details on the crucial eighteenth century in geology, Davis A. Young and Ralph F. Stearley, *The Bible, Rocks, and Time: Geological Evidence for the Age of the Earth* (Downers Grove, IL: InterVarsity Press, 2008), 71-100.

seemed that a growing number of geological observations were incompatible with catastrophism: the findings could be explained neither by referring to Noah's global flood nor by assuming a larger number of local floods or other catastrophes. As a result of this, around 1840 almost all geologists believed that the earth was at least millions of years old and must have already had a long history before human beings appeared on the scene.

It is important to note that these geologists were not atheists who wanted to disprove or invalidate the Genesis narrative. To the contrary, many were "Bible-believing Christians." If they did not feel required to do justice to their observations, there is little doubt that they would never have abandoned their literal reading of the first chapters of Genesis. Lyell, for instance, was a very religious man, who was opposed to Darwinian evolution because of the role Darwin attributed to natural selection, and also to the view that humans had evolved from the animals.[15] Nor did geologists at the time necessarily have a liberal view of biblical authority. Rather, "many started out with a firm commitment to interpret Genesis as literal accurate history. . . . If the rocks of the earth had been consistent with a young earth and global flood model, these scientists surely would have found it. Instead, [in their experience] the earth itself testified otherwise, over and over again."[16]

Obviously, this raised the question already at this stage—decades before Darwin—of how these observations could be reconciled with the authority of the Bible.[17] Thus, the tensions over the interpretation of the first chapters of Genesis in light of developments in the natural sciences clearly *predated* Darwin's theory of evolution.

We know that Darwin swallowed up Lyell's *Principles*. He had brought its first volume with him during his travel on the *Beagle* (ordering the second

---

15. J. M. I. Klaver, *Geology and Religious Sentiment: The Effect of Geological Discoveries on English Society and Literature between 1829 and 1859* (Leiden: Brill, 1997), xii–xiii; this point is not sufficiently taken into account by Terry Mortenson in his critical portrait of Lyell, *The Great Turning Point: The Church's Catastrophic Mistake on Geology—Before Darwin* (Green Forest, AR: Master Books, 2004).

16. Haarsma and Haarsma, *Origins*, 108.

17. Cf. on this esp. for the British context, Klaver, *Geology and Religious Sentiment*, and the classic of Charles C. Gillespie, *Genesis and Geology: A Study of the Relations of Scientific Thought, Natural Theology, and Social Opinion in Great Britain, 1790–1850* (Cambridge, MA: Harvard University Press, 1996 [1951]). For the strengths and weaknesses of Gillespie's book, see Nicolaas A. Rupke's foreword to the 1996 edition, v–xix; one of the weak points was that Gillespie, following Lyell here, wrongly suggested that the catastrophists were all scientific amateurs at the time.

volume while on his way) and used it to interpret his observations. In particular, Lyell's uniformitarianism brought Darwin to his gradualism, that is, his view that life on earth had developed incrementally from simple to ever more complex forms. Darwin came to envisage the earth as evolution's clock: from the structure and sequence of the strata one could deduce in which order the various species and other taxa had come into existence. Thus, uniformitarianism—including the notion of "deep time" with which it was inextricably linked—functioned as the essential background against which Darwin could develop his theory of evolution.

By the end of the nineteenth century, assessments of the age of the earth had risen to many hundreds of millions of years, and since the development of radiometric dating methods in the twentieth century, it is generally held that the earth and our solar system are no fewer than 4.6 billion years old. And this is only one-third of the time that the universe in which our solar system emerged has existed: approximately 13.8 billion years. The first traces of primitive forms of *life* on earth seem to be 3.5-4.5 billion years old. Although these are no doubt dazzling figures, they are today deeply entrenched in the empirical evidence that has been found—not only by geologists but also, for example, independently from them by astronomers and geneticists. Therefore, as far as I know, there are no atheists or agnostics who challenge these ages. The only people who contest them have *religious* reasons for doing so: Christians and other theists who conclude from their holy scriptures (the Bible or the Qur'an) that the earth and the universe must be much younger. Most "young-earth creationists" calculate the age of the earth at six thousand to ten thousand years. In arguing for this, they usually do not deny that the earth provides very strong indications of a much older age.[18]

One response to this, however, is that the Creator may intentionally have brought about such indications of a very old age, whereas in fact the earth—as well as the universe with all its galaxies—is much younger. So here we have an alternative. Let us have a closer look at this line of thinking.

---

18. Cf. Kenneth D. Keathley and Mark F. Rooker, *40 Questions about Creation and Evolution* (Grand Rapids: Kregel, 2014), 198: "Young-earth creationists admit that they find themselves in a very difficult position. They concede that the empirical case for a young earth is weak."

*Evolutionary Theory as a Layered Concept*

### Omphalism—or Appearance of Age Theory

The idea that the earth may only *seem* to be old goes back to the competent British naturalist Philip Gosse (1810-1888), who was a committed member of the Plymouth Brethren.[19] As a naturalist Gosse was totally aware of the evidence for an old earth, but as a believer in biblical revelation he saw only one way to deal with this: God must have made the earth with everything in and around it *as if* it were many millions of years old. Gosse elaborated this view in his well-researched book *Omphalos*, published in 1857 (two years before Darwin's *Origin*).[20] *Omphalos* being Greek for "navel," Gosse argued that Adam must have been created as an adult man with (like any adult man) a navel. Although this navel had never had a function in his case, every organism bears the traces of a life cycle. Therefore, Adam must have looked much older than he actually was. In the same way, God must have created trees with growth rings and rocks with characteristics suggesting an age much greater than their actual age. In contemporary writings, Gosse's hypothesis is called the "appearance of age theory" or "mature creation argument."[21]

Now it is easy to dismiss Gosse's thoughts out of hand as being absurd. We might better start with an open mind about it, however, not excluding any options prematurely. It might then occur to us that Gosse was clearly right when suggesting that the first human had to be created with characteristics that suggest a certain age—including, presumably, a navel. Similarly, trees should have annual rings right from the beginning, for otherwise they would not be trees as we know them. Such rings would then suggest a much greater age than the trees actually had. So far so good. The appearance of age theory becomes problematic, however, when we start to apply it to the world of fossils. From the Creator's point of view, providing the earth with fossils was superfluous.[22] Fossils were not needed to make the earth a real earth.

---

19. Gosse became especially well known as a result of the beautifully written—but caricatural—biography that his son published about him: Edmund Gosse, *Father and Son* (London: Heinemann, 1907). A good scholarly biography is: Ann Thwaite, *Glimpses of the Wonderful: The Life of Philip Henry Gosse, 1810-1888* (London: Faber & Faber, 2003).

20. Philip H. Gosse, *Omphalos: An Attempt to Untie the Geological Knot* (London: John Van Voorst, 1857); the book was republished by Routledge in 2003, after it had been rescued from oblivion by Stephen J. Gould, *Adam's Navel* (London: Penguin Books, 1995).

21. For a succinct recent discussion, see Keathley and Rooker, *40 Questions*, 217-24.

22. Note that, as Gosse (*Omphalos*, 90-100) already saw, it is hard to combine

Therefore, we can—on the appearance of age theory—only assume that God has scattered them through the strata in order to mislead us. Note that this must have been done in very sophisticated and suggestive ways, through imprints of hundreds of thousands of animals and plants with highly detailed physical structures—which in fact have never been functioning. This is comparable to a situation in which God created the world last week, "complete with history books on library shelves, decayed statues in museums, and false memories in our brains."[23]

It is true that we cannot dismiss this kind of scenario out of hand. After all, whether such a deceiving creator exists does not depend on whether or not we find such a person credible. In his day, Descartes did not find it a priori self-evident to exclude this possibility. It was only because of his belief that God should be perfect that he could shake off the idea of a deceiving deity or an "evil demon" as the creator of the universe.[24] From a Christian perspective, it is indeed precisely this moral perfection of God with which the appearance of age theory seems hard to reconcile. What interest would God have in so massively leading us astray? Isn't God the God of truth? Moreover, if God fools us in nature, how can we know that he does not also do so in Scripture? Some answer that it is not God who leads us astray but we ourselves, since we use wrong assumptions when inferring a certain age. This rejoinder is hardly convincing, however, since it borders on skepticism, making our experience of the created world utterly unreliable. For such reasons, most Christians—including most creationists—reject "omphalism."[25]

In Reformed theology, the "book of nature" is seen as generally reliable, even allowing us to draw some conclusions from creation about the nature of the Creator. Moreover, John Calvin, for one, vehemently opposed ascribing *potentia absoluta* (unbound power) to God—a notion that had led some late medieval nominalist thinkers to wild speculations about the sort of morally

---

the appearance of age theory with flood geology, since the latter presupposes that the strata, in fact, show traces of a "really" young earth. One can only convincingly endorse the mature creation argument if one does so consistently.

23. Haarsma and Haarsma, *Origins*, 113.

24. René Descartes, *Meditations on First Philosophy* (1641); see esp. the first and third meditation.

25. Other reasons are that in this way large parts of the physical universe (e.g., many astronomical data) are rendered illusory (which is more of a gnostic than a Christian view), that the theory is unfalsifiable, that it is just an "epicycle" needed to rescue a literal reading of Gen. 1, etc.

ambivalent things that God, being above the law (*ex lex*), might do.[26] Calvin rejected that as a "chimera" that unduly threatened the certainty of faith. Thus, Reformed Christians arguably have even less reason to embrace the appearance of age theory than others. As Reformed scholar Vern Poythress concludes (after a sympathetic discussion of the theory): "On the basis of the general faithfulness of God, and on the basis of his invitation to explore the world he has created, we have good reason to believe that the apparent ages found in astronomy [and geology] are also real ages."[27]

### Flood Geology

As a result of this, critics of the geological timescale seem to have only two other options: flood geology and agnosticism. As we saw already, flood geology assigns a pivotal role to Noah's flood in explaining the many fossil remains that have been found. Although geologists had abandoned this line of thought already in the eighteenth century when it became clear that it could not be squared with the geological record, it received a boost in the early 1960s from the publication of *The Genesis Flood* by young-earth creationists Whitcomb and Morris.[28] Young-earth creationists today still use flood geology to assign a relatively young age to all sorts of geological phenomena.

From a scientific point of view, flood geology is not compelling, however. Although catastrophes such as sudden huge floods definitely have played a role in the formation of strata and the emergence of fossils, they do not explain all relevant phenomena nearly as well as do more or less uniform processes of gradual erosion and sedimentation over vast periods of time. For example, a deluge cannot elucidate how fossils can be so neatly distributed over the various earth layers. One would expect the older layers to contain especially the traces of older and weaker organisms, who must have been the first to be caught by the water in their flight uphill. But that is not the case. In general, one would expect a much more chaotic pattern than we actually see. Of course, creationists have come up with their responses, but these strike every observer

---

26. Cf. on this Gijsbert van den Brink, *Almighty God: A Study of the Doctrine of Divine Omnipotence* (Kampen: Kok Pharos, 1993), 83-91.

27. Vern S. Poythress, *Redeeming Science: A God-Centered Approach* (Wheaton, IL: Crossway, 2006), 147.

28. John C. Whitcomb and Henry M. Morris, *The Genesis Flood: The Biblical Record and Its Scientific Implications* (Philadelphia: P&R, 1961).

who does not share their specific assumptions about biblical interpretation as artificial and contrived. Reformed geologist Davis Young has especially pointed this out in various publications, showing why from a scientific point of view flood geology is deeply flawed.[29]

### Origins Agnosticism

Those who prefer an agnostic attitude on questions of evolution and human origins (sometimes misleadingly called "cosmological agnostics")[30] recognize this. They see the empirical and theoretical weakness of young-earth creationism, flood geology, and the appearance of age theory, but they don't want to give in to the pressure to accept the geological timescale and gradualism (and the theory of common descent). Instead, they prefer to remain on the sidelines, without putting their stakes on one of the alternatives. But that is unsatisfactory as well. For that means that one recognizes the force of the empirical data but just does not want to adopt the ensuing conclusions. Usually the reasons behind such unwillingness are of a theological nature. For example, doesn't the Bible "teach" a young earth? And didn't death enter the world only after the fall of the first human couple in paradise? Therefore, we have every reason to bring these theological assumptions to the fore and discuss them openly.

Skepticism on questions of origins is often connected with skepticism of the reliability and scope of scientific theories in general. Isn't science, at the end of the day, a human enterprise and therefore by definition fallible? That is definitely true. The history of science is replete with examples of theories that were once broadly shared but turned out to be misguided.[31] We must also take this fallibility of science seriously when it comes to evolutionary theory. Per-

---

29. See esp. the book he coauthored with his colleague Ralph Stearley, *The Bible, Rocks, and Time*, which at the moment counts as the standard work in this field.

30. One is a "cosmological agnostic" when one claims that we cannot know the cause of the big bang; for more precise definitions and distinctions, see Rem B. Edwards, *What Caused the Big Bang?* (Amsterdam: Rodopi, 2001), 23-24.

31. For examples see, e.g., Gijsbert van den Brink, *Philosophy of Science for Theologians: An Introduction* (Frankfurt am Main: Lang, 2009), 25-67; Samuel Arbesman, *The Half-Life of Facts: Why Everything We Know Has an Expiration Date* (New York: Current, 2012). See also (the ongoing debate on) the "pessimistic meta-induction" argument as first put forward by Larry Laudan, "A Confutation of Convergent Realism," *Philosophy of Science* 48, no. 1 (1981): 19-49: from the fact that most scientific theories in past centuries were false, we should conclude that most contemporary theories are false as well.

haps in the future evolutionary theory will be superseded by another theory or paradigm—presumably one that is still more complex and encompassing. Perhaps such a development will even prompt new and different theological questions. We can address the questions and challenges that emerge from the current state of affairs, however, only in order to help each other to deal with them in appropriate ways from the perspective of faith—and perhaps that is precisely what theologians should do.

Moreover, it is highly improbable, to say the least, that future scientific research will come to abandon the notions of deep time and gradualism. We have seen that the community of geologists of the eighteenth century already had good reasons to drop flood geology and other attempts to "save" a young earth. They had no ideological interests in moving in the direction of the geological timescale and the gradual appearance of life on earth—many of them even deeply regretted this (and for a long time refused to go this way) because it brought them in conflict with their reading of the biblical narrative. Still, they ultimately came to the conclusion that there was no other way, and then openly went that way because they did not want to sacrifice their integrity. The conclusion of Young and Stearley is worth quoting in full in this connection:

> It is extremely improbable that future discoveries will lead the geologic community to revive acceptance of a very young Earth. Yes, there have been great revolutions in scientific thought in the past, and we should expect more of them in the future. However, it is futile for proponents of a young Earth to hope for such a revolution that would entail a complete reversal from acceptance of an old Earth that itself resulted from a lengthy scientific revolution. . . . Although some Christians might deny the evidence, turn a blind eye to the evidence or wish that the evidence would just go away because they find it very uncomfortable, and no matter how many Scriptural verses they throw at the rocks, the evidence for Earth's vast antiquity is there—it is diverse, it is voluminous, and it will not vanish.[32]

### Theological Ramifications

What theological questions are raised by accepting the geological timescale and gradualism? I see two of them and will discuss these in chapters 3 and 4. The first issue concerns the doctrine of Scripture. Since the notions of deep

32. Young and Stearley, *Bible, Rocks, and Time*, 475-76.

time and gradualism are at odds with a prima facie reading of the Bible, accepting them prompts us to revisit the doctrine of Scripture and the practice of biblical interpretation. Indeed, it is not just the higher levels of evolutionary theory—common descent and natural selection—that bring the doctrine of Scripture into play. It is sometimes suggested that Reformed Christians (for whom the doctrine of Scripture is of special concern) can easily take deep time on board but should object against common descent and natural selection for scriptural reasons.[33] That strikes me as not true. For the same prima facie reading of the first chapters of Genesis that rules out common descent just as well rules out an earth that is approximately 4.6 billion years old. It is surely possible to stretch the traditional 6,000-10,000 years a bit, since the book of Genesis is clearly not interested in exact figures. But deep time is far too deep to be reconciled with a "literal" reading of the first chapters of Genesis.[34] The genealogies, for example, suggest a much shorter time span between the first humans and the generations we know of by normal historical means than the fossil record allows for. Human-beings-like-us must have been around on earth for much longer than only a couple of thousands of years.

Thus, if one wants to retain a "literal" reading of the Bible, one can only opt for young-earth creationism—there is no other choice. Young-earth creationists are right, it seems to me, when they point out that as soon as one accepts deep time, important theological issues concerning the doctrine of Scripture are raised (though often their language is less cautious), such as whether we need science to tell us how Scripture should be interpreted. However, asking and answering such questions seem unavoidable because of the weight of the evidence for the geological timescale. This is true for the Reformed tradition no less than for other Christian families that have come to take the "book of nature" seriously for theological reasons. In chapter 3, I will therefore discuss the bearing of deep time, gradualism, and evolutionary theory in general on the doctrine of Scripture.

The second issue is of a different nature. It is not directly connected to the notion of deep time but raises its head as soon as we accept gradualism. For if the various forms of life on earth came into existence one after another over vast periods of times (whether through evolutionary processes or by separate

---

33. Cf., e.g., J. van Genderen and W. H. Velema, *Concise Reformed Dogmatics*, trans. G. Bilkes and E. M. van der Maas (Phillipsburg, NJ: P&R, 2008), 272-76.

34. I use scare quotes around "literal" because, as I will point out in chapter 3, it can be questioned whether the intuitive reading of Genesis should indeed be conceived of as literal.

divine interventions), it is hard to avoid the conclusion that *death* must have been around as well during this same time frame. The same goes for extinction and animal suffering, since traces of all these phenomena can be and have been found in the fossil record. In other words: *it is already the first layer of evolutionary theory that confronts us with the so-called problem of evolutionary evil.* This problem becomes especially poignant when we consider natural selection to be the dominant mechanism here. In that case, the struggle for life that characterizes the biosphere today must have been going on for many millions of years already. In this process, the weaker (or, actually, the least adapted) organisms of a species must always have been the most vulnerable ones, their early death enabling the species as a whole to survive by developing toward enhanced fitness. But even if we are not convinced of natural selection as the key mechanism involved here, alternative causes will arguably strike us as equally cruel and wasteful, as far as they are responsible for the same predator-prey patterns that we observe today.

All this may strike us as very troubling. Has this been going on for ages, from the very beginning of life on earth onward? How could a good and gracious God implement such a system? Didn't God tell us that the whole of creation has been made "very good"? And does he not have a special concern for the weak? Indeed, the goodness of God is a nonnegotiable part of the Reformed doctrine of God as much as it is of all other Christian theologies. In chapter 5 we will turn to this problem of evolutionary evil, focusing on the enormous amounts of animal suffering that seem to have been pervasive throughout nature from time immemorial.

We now turn to the second layer of evolutionary theory distinguished above, the notion of the common descent or ancestry of all existing forms of life.

## 2.3 Common Descent

Writing in 1991, prominent evolutionary biologist Ernst Mayr (1904-2005) contended that "there is probably no biologist left today who would question that all organisms now found on earth have descended from a single origin of life."[35] That was—and in our day still is—certainly an exaggeration, if only for the fact that creationists and other groups skeptical of evolutionary theory

---

35. Ernst Mayr, *One Long Argument: Charles Darwin and the Genesis of Modern Evolutionary Thought* (Cambridge, MA: Harvard University Press, 1991), 24.

have their biologists. Yet, it is remarkable that most natural scientists in general and biologists in particular are deeply convinced of the common ancestry of all life on earth. Where does this strong conviction stem from? It seems inspired by the fact that insights on the issue that were gained independently of one another in a whole range of scientific disciplines—such as biochemistry, comparative anatomy and physiology, geology, biogeography, paleontology, and genetics—turned out to confirm and mutually reinforce each other.[36] In a variety of ways, these insights strongly suggested that the diverse forms of life on earth must somehow be profoundly intertwined and related to each other. Let us have a closer look at some of these insights.

### The Fossil Record

When Darwin advanced his theory of evolution (or "descent with modification," as he called it), the fundamental interrelatedness of all forms of life was implied by his theory. However, the evidence for this view was not very strong in his day. Darwin appealed to the archive of fossils—remains of past plants and animals that have been preserved over time by natural processes—and tried to reconstruct lines of descent from it. But this fossil record contained big gaps. Darwin's theory could have been rejected by the scientific community (and as a result almost forgotten) if no new data had emerged that supported it. However, amazing new discoveries were made that turned out to fit exactly in this picture of life as a unified and interconnected whole. For example, an enormous number of additional fossils were found, among which were important in-between forms illustrating the transition from organisms with one type of body plan to organisms with another (e.g., from dinosaurs to birds). At the same time, no fossil evidence was found that spoke against common descent. For example, no fossil remains of contemporary mammals have been found in the Cambrian Period, nor fossil remains of a dinosaur and a human being in one and the same layer of sediment.[37]

---

36. For a short and readable survey of the relevant developments in most of these fields, see Haarsma and Haarsma, *Origins*, 193-204, 231-37.

37. A centerpiece in the case presented by Whitcomb and Morris in their *The Genesis Flood* consisted of pictures showing what seemed to be fossilized remains of dinosaurs next to contemporaneous human footprints in a limestone bed of the Paluxy River, Texas (in the first editions they even claimed that the dinosaur and human remains overlapped each other); their presentation of this material had an

But does the fossil evidence really point to the common ancestry of all life on earth? Aren't there still huge gaps: organisms that we have to assume once existed, because otherwise the transition from one form of life to another cannot be explained, but that have never been found? In that case there could have been multiple lines of descent, which do not all go back to one and the same source of life. A couple of things have to be taken into consideration here, however. First, given the fact that it is only under very special conditions that remains of organisms become fossilized (and almost only those that had harder parts such as nerved leaves and a shell or bones in the case of animals), it is not so strange that many forms of life that once existed have not been found. Obviously, the probability that a dead organism is not only fossilized but also found after millions of years have passed is rather low! Second, as said, many forms of life that were thus far unknown have been found in the fossil archive and continue to be found up to the present day. Among these are so-called transitional fossils that seem to reflect cross-species changes and even changes between more comprehensive classes of species.[38] The existence of some of these species, including the strata in which they were thought to be situated, had been predicted by evolutionary scientists. Third, what is really remarkable is not so much the gaps in the fossil archive but the way in which the archive is ordered: "Dinosaur fossils are found in rock between about 250 million and 65 million years old, but not in rock older or younger. Human fossils are found only in top rock layers. The order of fossils is consistent from location to location, with particular combinations of fossil species consistently found in rock of similar date. Complex life does not occur in the geological record before the oldest, simplest cellular life."[39]

We don't have enough fossils to reconstruct all evolutionary changes that have ever occurred, and we never will. What we do have, however, is enough to give us a general picture of the development of life and to enable scientists to

---

enormous impact on the general public. Later creationists, however, had to admit that the allegedly human footprints were not human at all. Cf. Keathley and Rooker, *40 Questions*, 298–300, and Ronald Numbers, *The Creationists: From Scientific Creationism to Intelligent Design*, expanded ed. (Cambridge, MA: Harvard University Press, 2006), xxx.

38. Cf. Baker, *The Evolution Dialogues*, 63–65. An example of the latter is the so-called *tiktaalik*, a transitional form between fish and vertebrates found in 2006; see "Tiktaalik roseae," University of Chicago, accessed March 29, 2019, http://tiktaalik.uchicago.edu.

39. Baker, *The Evolution Dialogues*, 65.

make a "family tree" showing which species evolved from which other species and when.⁴⁰ All this, especially when coupled with the observation of "homologies" (similar anatomical structures across species) made within comparative anatomy, suggests common descent.

### The Molecular Clock

More confirmation came from biochemical research that led to the discovery of the "molecular clock" of evolution in the early 1960s.⁴¹ It turns out that many biomolecules have a more or less constant changing rate over time, random mutations in their amino-acid and nucleotide sequences emerging in a slow but regular pace. As a result of this, the relative time distances between various lineage splits from a common ancestor can be calculated. Although this method must be used cautiously because particular molecules may in fact vary their rates of change over time, mistakes can be balanced out by examining the differences in various molecules rather than only one.⁴²

Taken on its own, the molecular clock only tells us something about the relative time distances between various lineage splittings; it does not give us any absolute dates. Such concrete dates can be assigned, however, once the molecular clock is calibrated against independent evidence from the fossil record.⁴³ It is possible to calculate the age of many fossils, using a couple of methods (most of them based on radioactivity) to date the age of the strata in which they were found. Remarkably, the various methods in principle yield the same ages, and the fact that it is possible to calculate the relevant time distances in various ways enables scientists to eliminate calculating errors and other mistakes.⁴⁴

---

40. Haarsma and Haarsma, *Origins*, 196.
41. Cf., e.g., Ernst Mayr, *What Evolution Is* (New York: Basic Books, 2001), 37, 288; G. J. Morgan, "Emile Zuckerkandl, Linus Pauling, and the Molecular Evolutionary Clock, 1959-1965," *Journal of the History of Biology* 31 (1998): 155-78, doi:10.1023/A:1004394418084.
42. Mayr, *What Evolution Is*, 37.
43. Cf., e.g., Donald R. Prothero, *Evolution: What the Fossils Say and Why It Matters* (Cambridge: Cambridge University Press, 2007), 96-99.
44. Baker, *The Evolution Dialogues*, 66; Ian Tattersall, *Paleontology: A Brief History* (West Conshohocken, PA: Templeton, 2010), 14-15.

*Evolutionary Theory as a Layered Concept*

### The Genetic Revolution

The most dramatic confirmation of the theory of common descent, however, came from modern genetics. During the first half of the twentieth century, scientists tried to establish which molecules were responsible for the patterns of inheritance that the Augustinian monk Gregor Mendel (1822-1884) had established in his experiments on pea plants, patterns that turned out to be more generally observable. Mendel had shown that somehow physically discrete units must be at work here—units that from 1909 onward were called "genes," a name coined by the Dutch biologist Hugo de Vries. The search for these units led to the discovery of chromosomes, genes, and finally (in 1953) the chemical structure of DNA. It now became clear where the source of variation within populations, which Darwin had tried to track down in vain, was situated: the molecular basis for evolutionary changes was to be found in these DNA molecules. Later research brought to light that generally speaking the genomes—the entire DNA—of species and other taxa have more in common with each other when they are more closely related in the "family tree" of life. For example, the genomes of various trout species have more in common with each other than with those of other fish, and the genomes of fish are more similar to each other than to those of mammals, etc. Especially during the last decades, our knowledge of the genomes of humans, animals, and plants has rapidly increased, and it is still increasing.[45]

The human genome, encoded in approximately three billion "letters," was successfully sequenced in 2003, in a project led by world-renowned geneticist Francis Collins.[46] After animals had been sequenced, comparisons of the human genome with animal genomes showed insightful patterns of similarities and differences. For example, genetically we are very close cousins of the chimpanzees.[47] There is a very strong suggestion here, of course, that life-forms

---

45. Cf., e.g., Francisco J. Ayala, "The Evolution of Life: An Overview," in *God and Evolution: A Reader*, ed. Mary Kathleen Cunningham (London: Routledge, 2007), 64-67.

46. Collins, an evangelical Christian, wrote a seminal study on Christian faith and evolutionary theory in which he defends "theistic evolution," that is (roughly speaking), the view that God somehow guides the evolutionary process. See Francis S. Collins, *The Language of God: A Scientist Presents Evidence for Belief* (New York: Free Press, 2006), esp. chap. 10.

47. For a survey of the rise and development of molecular biology, see, e.g., Francisco Ayala, *Darwin's Gift to Science and Religion* (Washington, DC: Joseph Henry, 2007), 117-35.

with the highest percentage of similar genes diverged relatively recently from a common ancestor, whereas life-forms with fewer similar genes have common ancestors longer ago.[48] Thus, the genetic revolution made it possible to reconstruct a tree of life without any recourse to fossils. Christian biochemist Denis Alexander therefore concludes from all this that "modern genetics has established our common inheritance with the apes beyond any reasonable doubt."[49]

Do we indeed know for sure that the human species originated from other primates? After all, it seems that "*Homo sapiens* is not simply an improved version of its ancestors—it's a new concept."[50] Even the leading neo-Darwinist Ernst Mayr, though accepting the descent of the human species from primate ancestors, concedes that it is "puzzling" why humans and the African apes "are so relatively different in morphology and brain development."[51] Although scientists cannot (yet?) reconstruct how the transition from other primates to *Homo sapiens* has occurred, we might keep in mind the phenomenon of *emergence* here.[52] Emergence is the converging of two or more unrelated natural processes that coalesce in such a way that suddenly an enormous leap forward is made in evolutionary history. Perhaps the enormous differences in brain development, scope of consciousness, etc., between humans and other primates can be explained in this way. In any case, both the genetic and the fossil archive strongly suggest that we humans as well as other species are part of a comprehensive family. Or is an alternative view possible?

*Common Descent or Common Function?*

The notion of a common biological descent of all species continues to be contested, especially by those—usually Christians or adherents of other religions—who are committed to the idea of "special creation." According to this view, the

48. Haarsma and Haarsma, *Origins*, 202.

49. Alexander, *Creation or Evolution*, 237; according to Alexander, our shared inheritance with the apes is "one of the most certain conclusions of contemporary biology" (234).

50. Ian Tattersall, *Becoming Human: Evolution and Human Uniqueness* (New York: Harcourt Brace, 1998), 188; cf. Collins, *Language of God*, 200, where it is claimed that "humans are . . . unique in ways that defy evolutionary explanation."

51. Mayr, *One Long Argument*, 25.

52. For a succinct introduction, see Harry Cook, "Emergence: A Biologist's Look at Complexity in Nature," *Perspectives on Science and Christian Faith* 65 (2013): 233–41, and the literature mentioned there.

various phyla came into existence separately from each other by divine fiat: God created them on a one-by-one basis. Some propose that we substitute a theory of common function for the theory of common descent. Couldn't the Creator have endowed species with similar genes in order to ensure that they had similar bodily structures and functions? In that case, it is not at all strange that humans and chimpanzees share a much larger part of their genome than, say, humans and mice.

There are two problems with this view. First, there is not a direct relation between genes and bodily functions. The genes of flying animals like bats, for example, are more similar to those of mice and rats than to those of birds. And second, the theory of common function can hardly account for the so-called pseudogenes that are also part of genomes. Pseudogenes are stretches of DNA that are similar to ordinary genes except that they have defects that make them dysfunctional and unable to make any useful proteins. The blueprint is there but has undergone certain mutations that stopped the gene from functioning. The Haarsmas offer an illuminating example. For many mammals the gene that allows them to make vitamin C is essential, since they cannot live without vitamin C. Chimps, however, do not need this vitamin C-making gene because they eat a lot of fruit.

> Yet chimps do have a pseudogene for Vitamin C located at the same spot on the genome where most mammals have a functional gene for Vitamin C. The pseudogene has no function, yet it's in their genome. This makes sense if chimps share a common ancestry with other mammals. They inherited the Vitamin C gene from their distant ancestors, but sometime in the more recent past . . . their ancestors had a mutation that turned it into a pseudogene. Because their ancestors were already living on fruit, the loss of the gene's function was not fatal.[53]

Those who defend a theory of common function can account for such pseudogenes only by suggesting that the Creator planted misleading "molecular fossils" in our bodies—in a similar way that some think he planted misleading physical fossils in the rocks. As we saw in §2.2, this suggestion has troubling theological consequences.[54]

---

53. Haarsma and Haarsma, *Origins*, 204.

54. Cf. Alexander, *Creation or Evolution*, 250. Alexander includes an instructive survey of current research on pseudogenes and refers to Graeme Finlay, *Human Evolution: Genes, Genealogies, and Phylogenies* (Cambridge: Cambridge University Press, 2013), for more information.

Another way out for adherents of this theory is to suggest that pseudogenes might have particular biological functions that may still be unknown to us. Indeed, recent research has pointed out that particular pseudogenes have functions other than coding for protein production—which made creationists jump to conclusions: "Pseudogenes are functional, not genomic fossils."[55] Apart from the fact that some of these new functions were newly acquired, however, it seems unwarranted to suggest that all fourteen-thousand-plus pseudogenes that have been detected in the human genome will turn out to be functional after all. And even if they did, this would not imply that the theory of common function is to be preferred, for the other evidence for common descent that we—though admittedly very briefly—highlighted above would still be in place. Especially, it was amazing how precisely the "distances" between various forms of life as calculated from the fossil record were confirmed by the genetic data that became available.

*Old-Earth Creationism*

Those who accept the geological timescale and gradualism while preferring common function to common descent are often called "old-earth creationists."[56] Old-earth creationism (OEC) comes in different varieties, the most important of which are gap creationism, day-age creationism, and progressive creationism. The gap theory inserts an immense period of time between the first two verses of Genesis 1. The first verse of Genesis 1 describes the creation of the universe (including now extinct and fossilized animals on earth) millions of years ago, a creation that was reduced by God to a "formless void" (Gen. 1:2) after it had become spoiled by Satan and his angels—whereas from verse 3 onward God's much more recent creation-in-six-days of all current life forms is described. Day-age interpretations identify the six days of creation in Genesis 1 with geological eras rather than with literal days of twenty-four hours, and attempt to harmonize the chapter with the geological timescale in this way. Progressive creationists hold more generally that creation has been a progressive work of God throughout the centuries. In line with gradualism, they accept the

55. Cf. Jeffrey P. Thomkins, "Pseudogenes Are Functional, Not Genomic Fossils," Institute for Creation Research, June 28, 2013, https://www.icr.org/article/7532.

56. Classic statements of OEC that are still influential include Bernard Ramm, *The Christian View of Science and Scripture* (Grand Rapids: Eerdmans, 1954), and Hugh Ross, *Creation and Time: A Biblical and Scientific Perspective on the Creation Date Controversy* (Colorado Springs: NavPress, 1994).

development of life from simple to more complex forms, but they contend that God must have intervened by occasionally creating new branches of animal life. In the end, God also created humankind in this "interventionist" way.[57]

All these varieties of old-earth creationism accept the geological timescale but reject the theory of common descent. Their adherents usually appeal to the missing links in the fossil record when arguing that God must have occasionally intervened in the process of evolution. But it seems that in this way they turn God into a "God of the gaps," leaving less work for him to be done when more transitional forms are being discovered. Instructive and impressive though the works of some old-earth creationists may be, this is a serious drawback. From a scientific point of view, it can easily turn out to be a slippery slope when more transitional forms (or other, e.g., genetic, pieces of evidence) are found. And from a theological point of view, we may ask whether God is more worthy of our worship when he miraculously intervenes in the evolutionary process from time to time than when he has arranged the process as a whole well in advance. We might just as well follow Charles Kingsley, who famously claimed that a God who "can make all things make themselves" is "much wiser" than a God who could just make all things.[58]

*Theological Ramifications*

Let us therefore suppose that the theory of common descent is true. What are the theological consequences? One conclusion is sometimes drawn from it that in fact does *not* follow. The theory of common descent does not imply that the phenomenon of life can be explained by means of natural processes only, let alone that it implies that the very existence of our universe can be fully explained in immanent ways. For it does not address the most fundamental questions in this connection: Where does life come from? And how did the

---

57. For a succinct introduction and defense, see Robert C. Newman, "Progressive Creationism," in *Three Views on Creation and Evolution*, ed. J. P. Moreland and John Mark Reynolds (Grand Rapids: Zondervan, 1999), 103–33.

58. Charles Kingsley, "The Natural Theology of the Future," in *Westminster Sermons* (London: Macmillan, 1881), xxv; Kingsley referred to John 5:17 in this connection, arguing that God may work through natural selection's continuous scrutinizing of all variations. Shortly before Darwin's *Origin of Species*, Baden Powell had already offered a similar argument in his *Essays on the Spirit of the Inductive Philosophy* (London: Longman, Brown, Green & Longman, 1855), 272.

universe come into existence? The theory of common descent only tells us about what happened *after* life had made its entrance on earth: it developed in such a way that the various species and taxa emerged from one another.

It is important to realize how relevant this observation is. For it implies that, strictly speaking, the theory of common descent cannot be at odds with the Christian doctrine of *creation*. Whereas this doctrine, as expressed for example in the Apostles' Creed ("I believe in God the Father Almighty, Creator of heaven and earth"), affirms the belief that the universe, including all forms of life in it, owes its existence to God, the theory of common descent (like the theory of natural selection) specifies the way in which life on earth may have evolved *after* its inception. Thus, contrary to what the common rhetoric of "creation versus evolution" wants us to believe, there is actually no conflict possible between the two, since the concepts of creation and evolution are aimed at answering different questions. As Reformed theologian Benjamin Warfield already observed, it follows from the very meaning of the terms "creation" and "evolution" that the theory of evolution and the Christian doctrine of creation do not cover common ground.[59] The question is not whether life was the outcome of creation or of evolution, but how life, once created, developed over time—either in a rather static or in a highly dynamic way.[60] Theologically, this is more a matter of providence (or perhaps of "continued creation") than of creation in its proper sense. For this reason, we will not include a chapter on the doctrine of creation in this book.[61]

However, the theory of common descent does pose a challenge to *theological anthropology*—or, as it was formerly called, "the Christian doctrine of man." For if we humans indeed share common ancestry with the apes, can we still maintain that we are created in the image of God whereas apes are not? And how about human dignity? Are we still unique? Clearly, these are issues of serious concern for many Christians. Is it also an area of special concern for

---

59. Benjamin B. Warfield, *Evolution, Scripture, and Science: Selected Writings*, ed. Mark A. Noll and David N. Livingstone (Grand Rapids: Baker Books, 2000), 200-202.

60. Of course, another question is whether the emergence of life from inorganic material is compatible with the Christian doctrine of creation or the Christian faith more generally. This question falls outside the scope of this book, however, since the thesis that life emerged from inorganic material is not part of evolutionary theory.

61. See, for the broader contours of this doctrine, Cornelis van der Kooi and Gijsbert van den Brink, *Christian Dogmatics: An Introduction* (Grand Rapids: Eerdmans, 2017), 200-246. It may be countered that these *broader contours* of the doctrine of creation conflict with evolutionary theory—but on closer inspection, it turns out that usually the doctrine of Scripture is critical, or the goodness of God, or other topics covered in this book.

*Reformed* Christians? I don't think so. Reformed theology has never cherished a particularly high view of the human being. To the contrary, as we saw in chapter 1, more than some other traditions it has emphasized the extent to which we humans are depraved by sin. Reformed theology has at times even displayed a dangerous tendency to pit the honor and glory of God against the indignity and disgrace of the human being, as if we could honor God by debasing ourselves. Yet, the idea that we humans may stem from the higher primates bewilders Reformed Christians as much as it does many others—whereas one could ask if such a "low" origin would not actually be very fitting given our low position before God. According to the creation story in Genesis 1, humans were created by God on the very same sixth day as the land animals—they did not even get a separate day for themselves. The Dutch Reformed theologian Oepke Noordmans (1871-1956), who was influenced by Karl Barth, has even suggested that the human being is "no independent figure" in doctrinal theology. Since according to the creed we believe in the triune God and not in man, we humans should not have any pretensions in Christian dogmatics.[62]

Still, in dogmatic reflection the status of the human being has always had a place of its own. For example, in Christian thought humans were usually situated somewhere between the angels and the beasts. Though the influence of Neoplatonic philosophy should not be underestimated here, the biblical notion of being created in the image of God also played an important role in this connection. And rightly so, because this notion is not just based upon one or two isolated biblical verses but reflects a fundamental theological theme. Therefore, in chapter 5 we will examine to what extent evolutionary theory, and especially the theory of common descent, forces us to give up or reconceive the doctrine of the *imago Dei* and, related to this, the notions of human uniqueness and of dignity. Even though this may not be a specific Reformed concern, it is one many Christians of all traditions share. The Roman Catholic Church, for example, while endorsing evolutionary theory in general, stipulates that the human soul must have been created by a special act of God, thus attempting to uphold both the *imago Dei* and human uniqueness.[63] Should

---

62. O. Noordmans, *Herschepping* [Re-creation] (Zeist, the Netherlands: NCSV, 1934), 47.

63. This view was first put forward by the Roman Catholic anatomist George Mivart (1827-1900), who was later excommunicated because of his criticism of church dogma; eventually it became the pontifical Roman Catholic view, most famously endorsed by Pope John Paul II in a 1996 address to the Pontifical Academy of the Sci-

Christians of other traditions follow suit? Or is this an unhelpful and untenable compromise between science and religion?

Next, according to Reformed federal theology, God made a *covenant* with humanity as soon as the first human being appeared on the scene. It is typical for Reformed theologians to posit that this covenant (either as a "covenant of works" or as a "covenant of grace") was already made with Adam. It was then broken by Adam and Eve in that momentous event that has come to be known as "the Fall" and that led to "original sin" in Adam's posterity. The good news of the gospel, however, is that God did not leave things at that but sent his only begotten Son Jesus Christ in order to sacrifice himself for the sins of humanity and to grant them eternal life through grace alone. What happens to this grand soteriological narrative when we presuppose an evolutionary history of life on earth? Has there ever been a first couple of human beings called Adam and Eve, given the evolutionary insight that new species come with populations rather than with individuals? Next, if there has been such a first couple, did it really all of a sudden fall into sin? And can we still make sense of original sin from the perspective of common descent? Clearly, the neo-Darwinian synthesis forms a challenge to all these notions and concepts. To what extent can they be retained under evolutionary conditions? And what happens in this scenario to the most sensitive of all theological issues: the nature and significance of the saving work of Jesus Christ? It is to these questions that we turn in chapter 6.

## 2.4 Universal Natural Selection

The textbook illustration of how natural selection works is the complicated story of research on the peppered moth in Manchester and other industrialized places in the northwest of England during the second half of the nineteenth century. Before the Industrial Revolution, the vast majority of these moths were light-colored (or, in fact, patchy). Dark-colored specimens were rare. Presumably, they were the result of a genetic mutation (or combination of genetic mutations) that had occurred randomly. By the end of the nineteenth century, however, the dark variety was reported to constitute 95 percent of the total population of peppered moths in Manchester. As a result of emissions by newly built factories, the widespread light-colored lichens that covered the

---

ences. Cf. Paul Jersild, *The Nature of Our Humanity: A Christian Response to Evolution and Biotechnology* (Minneapolis: Fortress, 2009), 24.

trees had died and disappeared, leaving the trees with their own, much darker color. Whereas the light moths were previously "helped" by the blotchy tree barks that served as camouflage, their increased visibility made them a much easier target for preying birds. The black moths, on the other hand, profited from the color change because they became less visible and started to flourish as a result of the spreading pollution. In this way, a change in the ecosystem led to an evolutionary transformation within a relatively short period of time. The results of attempts to test this process experimentally have been contested, but finally the process turned out to be repeatable.[64]

Though by no means the only one, this example of natural selection "in action" is the first one reported. It continues to be one of the clearest cases and rightly deserves its almost iconic place in textbooks of evolutionary biology. It clearly shows the typical roles played by phenotypic variation, inheritance, and adaptation. Since the light-colored peppered moths were less well adapted to their natural environment at the time, the number of them that made it to the stage of reproduction diminished, whereas at the same time the dark-colored moths more often survived to that stage, which allowed them to transmit their hereditary traits to the next generation. Though it is not exactly "the survival of the fittest" (as Herbert Spencer famously but misleadingly dubbed the process of natural selection) at work here, the example definitely shows us the survival of the most *well-adapted* organisms. That this "Darwinian" mechanism plays a role indeed in the development of forms of life is denied by hardly anyone. Even young-earth creationists endorse natural selection to some extent. They restrict its scope, however, to the relatively minor changes that take place *within* species, calling this "microevolution" and carefully distinguishing it from the more encompassing view that natural selection is responsible for the emergence of new species and groups ("macroevolution"). Working with a notion of fixed species and accepting only variation within species, they deny that the entire current biodiversity is the result of natural selection operating on random mutations.

Adherents of the neo-Darwinian account of evolution, on the other hand, argue that the explanatory scope of natural selection extends to the entire spectrum of life on earth, thus pertaining not only to micro- but also to macroevo-

---

64. The experiments were first conducted by Bernard Kettlewell (and Nico Tinbergen) and then extended and finally vindicated by Michael Majerus (who did not live to see his latest research published); cf. L. M. Cook et al., "Selective Bird Predation on the Peppered Moth: The Last Experiment of Michael Majerus," *Biology Letters* 8, no. 4 (November 2, 2011), http://dx.doi.org/10.1098/rsbl.2011.1136.

lution. They even reject the distinction between micro- and macroevolution, because both terms point to one and the same process and species boundaries are often fluid. Let us call this view "universal natural selection." Indeed, contemporary biology has become unthinkable without taking universal natural selection seriously. When the biologist Theodosius Dobzhansky wrote down his famous line: "Nothing makes sense in biology except in the light of evolution," he clearly meant strong Darwinian evolution as based on universal natural selection.[65]

### Scientific Debates on Natural Selection

Meanwhile, this is not to say that, apart from Christians and Muslims who read their holy texts literally, everyone agrees on natural selection as evolution's sole mechanism, or even on how exactly natural selection works. In fact, contemporary Darwinism is far from a monolithic unity. Conor Cunningham may be exaggerating when he compares contemporary Darwinism to the Christian church with its many denominations, but there is definitely a grain of truth in his observation.[66] Important debates are raging on what exactly natural selection operates on: Do "selfish" genes form the unit of selection (as Richard Dawkins argued), or individual organisms and groups (as Darwin thought), or perhaps even entire species (as Stephen J. Gould has suggested)? Further, there is discussion about the *pace* of evolutionary transformations as a result of natural selection. Are these indeed taking place slowly and incrementally? Or are periods of "stasis," in which species are more or less stable, alternated by periods of sudden outbursts in which new species rapidly—relatively speaking, of course—appear on the scene?[67] Other discussions focus on whether natural selection can do the job all on its own or whether other natural mechanisms are involved as well.[68] These debates are often interconnected with each other in complex ways.

---

65. Theodosius Dobzhansky, "Nothing Makes Sense in Biology Except in the Light of Evolution," *American Biology Teacher* 35 (1973): 125-29; Dobzhansky meant Darwinian (macro)evolution here.

66. Conor Cunningham, *Darwin's Pious Idea: Why the Ultra-Darwinists and Creationists Both Get It Wrong* (Grand Rapids: Eerdmans, 2010), xviii.

67. Cf. the classic paper of Niles Eldredge and Stephen J. Gould, "Punctuated Equilibria: An Alternative to Physical Gradualism," in *Models in Paleobiology*, ed. T. J. M. Schopf (San Francisco: Freeman, Cooper, 1972), 82-115.

68. Fowler and Kuebler, *The Evolution Controversy*, 227-327, helpfully introduce the term "meta-Darwinism" to cover those theories that presuppose other natural mechanisms next to natural selection as key to evolutionary change.

Orthodox neo-Darwinians are convinced that natural selection is the only game in town. They point to myriads of phenomena in the natural world that we would hardly be able to understand without ascribing an essential role to natural selection. All kinds of physical traits of both humans and animals that seem to be exactly tailored to the functions they have, for example, suggest a long-term development in which an ever more sophisticated adaptation to their environment took place. Clearly, nature has favored advantageous and functional traits over less advantageous and less functional ones over vast periods of time. The way in which this has happened, however, is not as crystal clear as is sometimes suggested. That is, it has not been demonstrated that the mechanism of natural selection acting upon random mutations is the only and fully sufficient explanation of the rich variety and complexity of life that nature displays.[69]

Skeptics of this thesis have been investigating other natural mechanisms that may be involved in the evolutionary process. Some of these are relatively uncontroversial, such as sexual selection and genetic drift.[70] Others, however, are highly disputed. Fowler and Kuebler distinguish no fewer than eight such disputed theories: punctuated equilibrium, hierarchical selection, exaptation, neutral theory, "evo-devo" explanations, morphogenic fields, self-organization, and endosymbiosis.[71] They could have added the theory of niche construction, which hinges on the ways in which organisms cause changes in their environment that increase their chances of survival—and which has recently received its first theological appropriations.[72] Many of these "meta-Darwinian" theories are intended to complement the theory of natural selection, while others are much more radical and leave only a minor role for it.[73] Some scholars attempt to incorporate such newer theories in the classi-

---

69. Cf. Peter van Inwagen, "A Kind of Darwinism," in *Science and Religion in Dialogue*, vol. 2, ed. Melville Y. Stewart (Malden, MA: Wiley-Blackwell, 2010), 813-24.

70. See for clear and succinct explanations, Baker, *The Evolution Dialogues*, 57-59.

71. Fowler and Kuebler, *The Evolution Controversy*, 279.

72. Cf. F. John Odling-Smee, Kevin N. Laland, and Marcus W. Feldman, *Niche Construction: The Neglected Process in Evolution* (Princeton: Princeton University Press, 2003); Markus Mühling, *Resonances: Neurobiology, Evolution, and Theology; Evolutionary Niche Constructions, the Ecological Brain, and Relational-Narrative Theology* (Göttingen: Vandenhoeck & Ruprecht, 2014), 137-221; Celia Deane-Drummond, *The Wisdom of the Liminal: Evolution and Other Animals in Human Becoming* (Grand Rapids: Eerdmans, 2014), 194-237.

73. A particularly harsh critique has come from self-identifying atheists Jerry Fodor and Massimo Piattelli-Palmarini, *What Darwin Got Wrong* (New York: Picador, 2010);

cal one, aiming at what is called an "extended evolutionary synthesis" (EES) and framing this as a "third way" next to (and beyond) creationism and neo-Darwinism.[74] Interesting though that may be, this is not the place to further discuss or evaluate these developments.[75]

One issue that has drawn quite some attention over time is whether our human rational (or, more precisely, "belief-forming") faculties are generally reliable when they are the result of natural selection operating on random mutations. Darwin himself voiced concerns about the implications of his own theory in this regard. "With me the horrid doubt always arises whether the convictions of man's mind, which has been developed from the mind of the lower animals, are of any value or at all trustworthy. Would anyone trust the convictions of a monkey's mind, if there are convictions in a monkey's mind?"[76] After Darwin, this concern has been repeated and elaborated by theist and atheist thinkers alike. C. S. Lewis ("The relation between stimulus and response is utterly different from that between knowledge and truth") and G. K. Chesterton ("Why should not good logic be as misleading as bad logic? They are both movements of the brain in a bewildered ape") concur with Thomas Nagel ("If we came to believe that our capacities for objective theory were the product of Natural Selection that would warrant serious skepticism about its results") and Jerry Fodor ("There is no . . . Darwinian reason for thinking that we're true believers").[77] The well-known philosopher Alvin

---

as could perhaps be expected, their book was badly received in neo-Darwinian circles. The paperback edition of 2011 includes "Afterword and Reply to the Critics" (165–89).

74. These include Eva Jablonka, Eugene Koonin, Gerd Müller, Denis Noble, James Shapiro, and others. See the Third Way, accessed March 29, 2019, http://www.thethirdwayofevolution.com/.

75. For more information on each of the mentioned theories, see Fowler and Kuebler, *The Evolution Controversy*, 277–324. See also Sy Garte, "New Ideas in Evolutionary Biology: From NDMS to EES," *Perspectives on Science and Christian Faith* 68, no. 1 (2016): 3–11. Garte briefly discusses "concepts of symbiosis, gene duplication, horizontal gene transfer, retrotransposition, epigenetic control networks, niche construction, stress-directed mutations, and large-scale reengineering of the genome in response to environmental stimuli" (3).

76. Francis Darwin, ed., *The Life and Letters of Charles Darwin*, 2 vols. (New York: Basic Books, 1959), 285; the quote is from a letter to William Graham (July 1881).

77. C. S. Lewis, *Miracles* (London: HarperCollins, 2002), 28; G. K. Chesterton, *Orthodoxy* (London: Fontana, 1961 [1909]), 33; Thomas Nagel, *The View from Nowhere* (Oxford: Oxford University Press, 1986), 79; Jerry Fodor, *In Critical Condition* (Cambridge, MA: MIT Press, 1998), 201. Cf. Cunningham, *Darwin's Pious Idea*, 153, 465.

*Evolutionary Theory as a Layered Concept*

Plantinga has even turned this concern into a sophisticated, though highly contested, argument against the sufficiency of natural selection as an explanatory mechanism.[78]

Given the variety of alternative or additional evolution-driving mechanisms that are currently being discussed, we are well advised not to draw any strong conclusions about the scope of natural selection, that is, about the range of phenomena it can actually explain. It seems fair to say that at this moment we just don't know to what extent natural selection is sufficient for explaining the current biological diversity on our planet.[79] However, these ongoing discussions on the status of the neo-Darwinian synthesis and the scope of natural selection as evolution's driving mechanism are *internal debates within the study of evolution*. They should not be taken—as is sometimes done, particularly in Christian circles—as indications that evolutionary theory *as such* is in crisis, or that scientists may not be so sure about evolution after all.[80] Even if next to natural selection one or more other causal mechanisms might be involved, evolutionary theory still stands. For clearly, the other two layers of evolutionary theory discussed above—gradualism and common descent—are logically independent from its top layer: natural selection acting on random mutations. Moreover, even if natural selection cannot account for all biodiversity, it is undisputed that it explains at least part of it. Therefore, we have every reason to take it seriously and examine the theological challenges it raises.

---

78. Plantinga has unfolded and refined this argument in a series of publications, among which are "Against Naturalism," in Alvin Plantinga and Michael Tooley, *Knowledge of God* (Oxford: Oxford University Press, 2008), 1-69, and *Where the Conflict Really Lies: Science, Religion, and Naturalism* (Oxford: Oxford University Press, 2011), 307-50. Plantinga argues that within a theistic framework one does not have this problem, since here we can posit that God created us in such a way that our cognitive faculties are generally reliable.

79. Cf. Van Inwagen, "A Kind of Darwinism," 819; Van Inwagen therefore adheres to what he calls "weak Darwinism": the view that natural selection, though important, is not the only mechanism responsible for the diversity, complexity, and apparent teleology in the biological world.

80. A case in point here is the way in which Gould and Eldredge's theory of punctuated equilibria has been seized upon to discredit evolutionary theory as a whole; Gould and Eldredge remained committed evolutionists and complained about the "quote-mining" practices used by some to suggest otherwise. See, e.g., "The Quote Mine Project," The TalkOrigins Archive, accessed March 29, 2019, http://www.talkorigins.org/faqs/quotes/mine/part3.html.

Suppose, however, that in the end Darwinian evolution turns out to be misguided. That is, suppose that it will become clear that natural selection working incrementally over large periods of time is not the key mechanism that explains biological processes of adaptation and speciation, but that other—perhaps thus far unknown—mechanisms are equally or even more decisive. Would that make much of a difference from the perspective of Christian theology? It does not seem so. Perhaps it will turn out that the role of *randomness* is less critical than is suggested by the neo-Darwinian synthesis. In that case, it would be easier to connect the evolutionary process with notions of divine guidance and providence. As we will try to show in chapter 7, however, the Darwinian mechanism of natural selection working on random mutations is not incompatible with the doctrine of divine providence either. Another difference might be that attempts to extend the scope of natural selection to the domain of *cultural evolution* will prove to be a failure: most probably morality, religion, and all other sorts of human belief and behavior can no longer be explained (away) as products of the struggle for life, or for the enhancement of reproductive success. Perhaps, however, such reductionist explanations are not compelling anyway—a topic we will examine in chapter 8. But even if these two theological consequences of evolutionary theory disappear, we still have to deal with the questions raised by gradualism and common descent: How about the interpretation of the Bible, evolutionary evil, human distinctiveness, the historical Adam, the Fall, etc.? In other words, even if Darwinian evolution is a theory in crisis,[81] this does not mean that "old-time religion" is off the hook, since we would then still have every reason to revisit our theologies in light of the immensity of the geological timescale, the gradual appearance of various forms of life on earth over millions of years, and the common ancestry of all species on earth, including humankind.

A more momentous difference would arise if it could be demonstrated that natural mechanisms *of any kind* are insufficient to explain the world's stunning biodiversity. In that case, careful study of the data would almost necessarily lead us to assume a supernatural intelligence behind the biosphere. This is exactly what has been claimed during the past decades by proponents of the so-called intelligent design (ID) movement.[82] To be sure, even if ID advo-

---

81. Cf. Michael Denton, *Evolution: A Theory in Crisis* (Chevy Chase, MD: Adler & Adler, 1986); Denton's title is misleading since what he means is *strong Darwinian* evolution.

82. Most introductions to intelligent design in the mainstream literature are critical if not downright dismissive. For a more sympathetic overview, see, e.g., Keathley and Rooker, *40 Questions*, 387-96.

cates are right, we would still be faced with the problems we just mentioned, since ID offers no alternative to the first two layers of evolutionary theory but only to the Darwinian mechanism of natural selection. Still, it would make the atheistic twist that has often been given to evolutionary theory impossible—which would definitely be a big deal. Let us, therefore, have a closer look at this most outspoken and most well-known alternative to "blind" (i.e., purposeless) Darwinian evolution: the theory of intelligent design.

### Intelligent Design

During the most recent turn of the century, the notion of intelligent design caused vehement debates in the United States and beyond. Both adherents and opponents of ID were often driven by ideological as much as by scientific agendas. In the popular perception, adherents of ID were often lumped together with creationists—which most of them are not. In fact, their views are much more subtle than is often acknowledged.[83] For example, most adherents of ID accept the geological timescale and gradualism, and some also endorse the theory of common descent. What is common to all ID proponents is a critique of the neo-Darwinian thesis that life's diversity can be almost completely explained by "blind" processes of natural selection operating on random mutations. Thus, they share the concerns of others about the explanatory scope of neo-Darwinism as discussed in the previous section. Unlike these others, however, they claim that some phenomena do not lend themselves to a naturalist explanation at all but instead bear the traces of being intelligently designed. They even argue that such traces of intelligent design can be scientifically demonstrated by means of "inference to the best explanation."[84]

Thus, various ID scholars tried to establish that specific phenomena from the domain of cosmic evolution—in particular the big bang and the fine-tuning of the universe—should be considered as products of intelligent design. They did so in ways that their atheist opponents did not always find easy to refute.

---

83. Cf. Del Ratzsch, "There Is a Place for Intelligent Design in the Philosophy of Biology," in *Contemporary Debates in Philosophy of Biology*, ed. Francisco J. Ayala and Robert Arp (Malden, MA: Wiley-Blackwell, 2010), esp. 357–60.

84. For this method, which is used frequently in historical research, cf. Peter Lipton, *Inference to the Best Explanation* (London: Routledge, 1991): if one has multiple competing hypotheses, the one that best explains *all* the relevant evidence is most likely to be true.

More disputed, however, are their attempts to explain aspects of biological evolution by appealing to intelligent design. For example, in response to Michael Behe's argument in favor of "irreducibly complex systems," it has been pointed out that it is not at all impossible to explain such systems and structures—the eyeball and the bacterial flagellum being the most famous examples—as the result of purely natural processes.[85] Later ID advocates developed similar lines of argument about the first living cell (living cells being units as complex and intricately structured as an urban metropolis) and about the "specified complexity" contained in the human DNA (the only known source of such levels of complexity being intelligent agency).[86]

One of the more recent ID explanations, elaborated in an impressively detailed, scholarly way, concerns the so-called Cambrian explosion, that is, the "sudden" outburst of all sorts of complex animal life-forms some 540 million years ago. Stephen Meyer contends that, given this relatively quick emergence of the major phyla (animal groups) without a trace of the intermediate forms that one would expect on the basis of neo-Darwinism, the mechanism of natural selection is unable to adequately explain this explosion of diversity.[87] Meyer does not deny that natural selection has played an important role in the post-Cambrian evolution of life—his claim is much more specific. How should such a claim be assessed? Meyer has already been criticized for unduly reducing the length of the Cambrian explosion to "only" 5 million years, as well as for shoehorning other data to fit his theory.[88] Suppose, however, that these and other scientific criticisms are misguided. Would we then be well advised to adopt his theory? It does not seem so, since for all we know future research may bring to light other natural mechanisms and processes—perhaps

---

85. See Michael Behe, *Darwin's Black Box: The Biochemical Challenge to Evolution* (New York: Free Press, 1996), and cf., e.g., Kenneth R. Miller, "The Flagellum Unspun: The Collapse of 'Irreducible Complexity,'" in *Debating Design: From Darwin to DNA*, ed. William Dembski and Michael Ruse (Cambridge: Cambridge University Press, 2004), 81-97.

86. See Fazale Rana, *The Cell's Design: How Chemistry Reveals the Creator's Artistry* (Grand Rapids: Baker Books, 2008); Stephen Meyer, *Signature in the Cell: DNA and Evidence for Intelligent Design* (New York: HarperOne, 2009).

87. Stephen C. Meyer, *Darwin's Doubt: The Explosive Origin of Animal Life and the Case for Intelligent Design* (New York: HarperOne, 2013). The Cambrian explosion became known to the general public through Stephen J. Gould's *Wonderful Life: The Burgess Shale and Natural History* (New York: Norton, 1989). Cf. Harry Cook, "Burgess Shale and the History of Biology," *Perspectives on Science and the Christian Faith* 47 (1995): 159-63.

88. See, e.g., Martin R. Smith's review in *Science and Christian Belief* 27 (2015): 101-2.

*Evolutionary Theory as a Layered Concept*

thus far unknown—that help us to elegantly explain the Cambrian explosion without having to resort to the notion of a (supernatural) intelligent design.

It seems that here we encounter the main weakness of all ID theories: they share the basic structure of "God of the gaps" arguments, positing God or another supernatural entity to fill the blank spaces in our scientific knowledge.[89] As German theologian Dietrich Bonhoeffer already saw, this is a very risky strategy, since it tends to leave less and less room for God's involvement:

> How wrong it is to use God as a stop-gap for the incompleteness of our knowledge. If in fact the frontiers of knowledge are being pushed further and further back (and that is bound to be the case), then God is being pushed back with them, and is therefore continually in retreat. We are to find God in what we know, not in what we don't know.[90]

To be sure, this does not preclude intelligent design from playing a role in or behind the evolutionary process. Indeed, studying the structure of the living cell, for example, or of other parts of the biological or cosmological world may fill us with awe and praise for God's incomprehensible majesty—irrespective of whether or not we can explain such phenomena by means of natural mechanisms and processes! It is not at all weird or obsolete to assume God's hand behind such phenomena; and by opening our eyes to the wonderful complexity of the natural world, ID scholars may strengthen us in belief that in the end the entire universe is the handiwork of our Creator. The awe-inspiring character of the natural world (which is sometimes downplayed by atheist evolutionists for ideological reasons) helps us to see that the book of nature has been "written" by the same God we come to know much more clearly in the book of Scripture.

However, the possible or actual failure of ID arguments shows that we should not consider God (or intelligent design) as playing a role in the biologi-

---

89. Strictly speaking, ID does not introduce the God of the Christian faith or any other religious entity in scientific discourse—its arguments just conclude the necessity of design and therefore of some unspecified highly intelligent agent (who might in theory even have evil purposes). Cf., e.g., William A. Dembski, *Intelligent Design: The Bridge between Science and Theology* (Downers Grove, IL: InterVarsity Press, 1999), 107: "Intelligent Design presupposes neither a creator nor miracles. Intelligent Design is theologically minimalist." Therefore, the well-known objection that the design of life is bad or cruel is not an argument against ID.

90. Dietrich Bonhoeffer, *Letters and Papers from Prison*, ed. Eberhard Bethge (New York: Simon & Schuster, 1997), 311-12 (letter to Eberhard Bethge, May 29, 1944).

cal realm *on the same level* as natural explanations. For in that case we are using an "argument from ignorance," which is always dangerous. As Denis Alexander points out, when ID advocates propose that a particular biological phenomenon is inexplicable by natural evolutionary mechanisms, "all we need to do is wait for a decade or so, often less, and a coherent evolutionary account begins to emerge."[91] Thus, we should not consider God as being involved in the evolutionary development of life *on the same level* as natural selection, or as its competitor. For that reason, research on intelligent design, instructive though it may be, should not stop us from taking the neo-Darwinian theory of natural selection seriously.

### Theological Ramifications

We are now in a position to ask the question that will belong to the focus of our study: If all biodiversity on earth, or large parts of it (larger than only that covering microevolution), can be explained by processes of natural selection acting on random mutations, does that have any theological consequences? More specifically, are any typically Reformed concerns at stake? As far as I can see, two issues in particular require our attention in this connection.

First of all, it is often argued that theists cannot subscribe to the (neo-)Darwinian theory of natural selection—and conversely, neo-Darwinians cannot endorse theism—because the randomness of the chance mutations on which natural selection operates rules out all teleology, including any notion of divine guidance or providence. Belief in such divine guidance, however, forms an essential tenet of the faith of most theists. Arguably, it is especially dear to Reformed Christians, given their particularly high view of God's sovereignty and sublime majesty. Thus, the tension between randomness (as implied in neo-Darwinism) and providence (as professed in most monotheistic religions but definitely in Reformed theology) poses a challenge to Christian theology. For does the randomness that is the basis of all evolutionary processes not rule out the possibility that God providentially directs the history of life toward his goal? This depends, of course, on what is actually involved in the randomness of the random mutations. If "random" means that any overarching form of guidance and supervision over the development of life is excluded, then there

---

91. Alexander, *Creation or Evolution*, 403; cf. also his critical comment: "Wherever mysteries are to be found, ID is sure to follow" (408).

is indeed a conflict between (neo-)Darwinism and theism.[92] If, on the other hand, "random" in "random mutations" means something else, such as "unpredictable from our human perspective" or "not aimed at the enhancement of the species' fitness," it is less clear whether the notion cannot be reconciled with divine providence.[93] In chapter 7 we will examine the impact of natural selection on the doctrine of divine providence in more detail.

Second, the mechanism of natural selection is also often appealed to in order to explain phenomena that have emerged in *human culture*. The leading idea here is that evolution is not only a biological but also a cultural phenomenon: it is claimed that various forms of human behavior and belief can be as easily explained by natural selection as can the world's biodiversity. Theologically, this is especially interesting when it concerns human characteristics that in the past have been directly ascribed to God's creative work. Though other poignant examples could be discussed here (first and foremost, human morality comes to mind), we will focus on evolutionary explanations of human *religiosity*. In the cognitive science of religion (CSR), attempts are made to explain the rise and spread of religions and religiosity across cultures along purely natural lines. In particular, it is sketched how processes of natural selection may have been critical in this as well as other domains. Although thus far these attempts are tentative and have not (yet?) convinced the scientific community, in chapter 8 we will extend our "what if?" approach to CSR explanations of religion: If one or more of these explanations will be validated in the course of time, what does that mean for Reformed—and, more broadly, Christian—theology, especially for its doctrine of revelation?

## 2.5 What If It Is True?

Reformed Christians and others may find reasons to doubt whether the neo-Darwinian theory of evolution as sketched in the preceding sections can

---

92. This was the argument of, for example, Charles Hodge in his influential *What Is Darwinism?* (New York: Scribner, Armstrong and Co., 1874); see below, §7.3. Herman Bavinck made essentially the same point in his piece "Evolution" (1907), in *Essays on Religion, Science, and Society*, ed. John Bolt (Grand Rapids: Baker Academic, 2008), 105-18 (cf. 109-10).

93. See René van Woudenberg and Joëlle Rothuizen-van der Steen, "Both Random and Guided," *Ratio* 28 (2014): 332-48, doi: 10.1111/rati.12073; cf. Haarsma and Haarsma, *Origins*, 41-43.

indeed explain the full range of current biodiversity. They may even doubt whether any other natural mechanisms can do the job. As long as we have no overwhelming evidence to the contrary, it cannot be ruled out that God may have brought the first living cell into existence through a special act of his miraculous power. Nor is it unreasonable to suggest that God directly intervened in the process when causing the so-called Cambrian explosion about 540 million years ago or when forming the human being out of its biological predecessors (either by infusing a soul or mind or in other ways that help explain the enormous differences with other primates). Even in the face of extremely strong genetic suggestions of common descent, since theories are always underdetermined by the data, it continues to be possible to adopt an alternative, such as the theory of common function.

Nevertheless, since we want to avoid any God-of-the-gaps arguments for reasons explained above, we will not explore these possibilities further here. Instead, we will ask the question: What if it should turn out at some point in the future that there are no "gaps" in the evidence where an appeal to divine intervention might be rightly placed? What, in other words, if standard neo- or meta-Darwinian accounts of evolution are broadly correct? *What would that mean from a (Reformed) theological point of view?* In the introduction I mentioned that I have personally become convinced of evolution, and I can now make that statement more precise. I have become convinced of the geological timescale and gradualism, and I think the theory of common descent is highly plausible, whereas it seems to me that a natural development of life's diversity, mainly through natural selection but also through other mechanisms, is very well possible.[94] Still, not being a scientist, I will refrain from making sweeping statements here. Having seen how many unresolved questions and areas of debate there still are, I think it is fair neither to discredit or downplay the neo-Darwinian account of evolution nor to endorse it without reservations. For the sake of argument, however, in the remainder of this book I will assume that this account is generally true. Would it be possible for a Reformed Christian to accept this, and what would that mean theologically?

---

94. Thus, I am inclined to ascribe a slightly different epistemic status to each of the three layers of evolutionary theory distinguished above. See Gijsbert van den Brink, Jeroen de Ridder, and René van Woudenberg, "The Epistemic Status of Evolutionary Theory," *Theology and Science* 15 (2017): 454–72.

CHAPTER 3

# Evolutionary Theory and the Interpretation of Scripture

> Nature is as truly a revelation of God as the Bible; and we only interpret the Word of God by the Word of God when we interpret the Bible by science.... When the Bible speaks of the foundations, or the pillars of the earth, or of the solid heavens, or of the motion of the sun, do not you and every other sane man interpret this language by the facts of science? For five thousand years the Church understood the Bible to teach that the earth stood still in space, and that the sun and stars revolved around it. Science has demonstrated that this is not true. Shall we go on to interpret the Bible so as to make it teach the falsehood that the sun moves around the earth, or shall we interpret it by science . . . ?
>
> —Charles Hodge[1]

## 3.1 The Bible and Modern Science

Many people think that accepting the data of evolutionary theory is incompatible with a plain reading of the Bible. According to them, anyone who is convinced of evolutionary theory cannot but reject biblical claims to truth and authority, whereas conversely anyone who wants to be a "Bible-believing Christian" can only reject evolutionary theory. Especially for Reformed Christians this is a sensitive issue, since, as we saw in chapter 1, it belongs to the very heart of Reformed theology's identity to go back to the Bible time and again,

---

1. Charles Hodge, "The Bible and Science," *New York Observer*, March 26, 1863, 98–99, as quoted by Mark Noll, *The Scandal of the Evangelical Mind* (Grand Rapids: Eerdmans, 1995), 184; cf. Charles Hodge, *Systematic Theology*, vol. 1 (New York: Scribner's Sons, 1872), 59, 170–71, 573–74.

and to consider it the final arbiter in matters of faith and life. That is why we start our discussion of doctrinal issues elicited by evolutionary theory with the doctrine of Scripture, which is pivotal in so many contemporary debates on evolution. Although I know of no statistics measuring why groups of Christians reject evolutionary theory, its presumed incompatibility with the Bible is no doubt a very prominent reason among Reformed and evangelical Christians. Moreover, this incompatibility is often seen to extend to each of the three layers of evolutionary theory distinguished in chapter 2. Indeed, if we adopt a so-called literal understanding of the Bible, even its first layer—the notion that forms of life appeared progressively on earth over vast periods of time—can hardly be squared with the biblical witness, since we would have to squeeze millions if not billions of years into each of the six days of creation in Genesis 1.

Strictly speaking, it is not just evolutionary theory but first of all the underlying geological timescale that raises serious questions concerning the "literal" interpretation of Scripture. Those who only take on board the first layer of evolutionary theory may escape most of the dilemmas discussed in this book, but they cannot escape the question of biblical hermeneutics. How did previous generations deal with this question when scientific discoveries caused tensions with traditional understandings of the meaning of biblical texts? A sublime example here is the case of heliocentrism versus geocentrism. Nowadays most people assume that when the biblical authors represented the sun as turning around the earth, they did not propose a particular model of how the universe is structured[2] but simply proceeded on the basis of how we see things when we look around us. In his day, the Reformed theologian Gisbertus Voetius (1589-1676), although well acquainted with this approach, rejected it outright. He considered the idea that the Holy Spirit did not convey the factual truth in every biblical detail as absolutely blasphemous. The patriarchs and the prophets, and the entire people of Israel, were certainly not so stupid as to be incapable of understanding the Copernican system. The Holy Spirit could simply have explained it to them, and through them to the first readers of the Bible. Voetius believed that nothing less than the authority of Holy Scripture, and therefore the entire Christian faith, was at stake here.[3] Voetius's disciple

---

2. It is tempting to speak of a *worldview* in this connection, but since that concept often has religious or metaphysical overtones, I will avoid it here. What I have in mind is a "world picture" (Dutch: *wereldbeeld*), as in Eduard Jan Dijksterhuis's classic *The Mechanization of the World Picture: Pythagoras to Newton* (Princeton: Princeton University Press, 1986 [1961]).

3. Gisbertus Voetius, *Thersites heautontimorumenos* (Utrecht, 1635), 256-83 (espe-

Martin Schoock (1614-1669) agreed with him: in the Bible the Holy Spirit may adapt himself to what we can understand, but never in such a way that the Bible "lies with liars and errs with those who are in error."[4]

As far as I know, no one today agrees with Voetius and Schoock. We have become entirely satisfied with the "observer's perspective interpretation" of Joshua 10:12 and similar texts (after all, we *see* the sun moving around the earth) that at first sight suggests a geostatic model of the world (e.g., Eccles. 1:5; Pss. 19:6; 104:5). Therefore, it seems that in retrospect we must say that Voetius *cum suis* jumped to conclusions too rashly, believing that the authority of the Scriptures was at risk while in fact it was not. Their argument that the Christian faith would perish with the acceptance of the heliocentric model led many Reformed Christians of that day astray. It is an intriguing question how their spiritual heirs could ever have become convinced heliocentrists. In whatever way this historical process precisely took place, in hindsight Christians should be grateful that the required change in hermeneutics eventually took root. Otherwise, the Christian witness would still be hopelessly entangled in an obsolete model of how the universe is built up.

Does not something similar happen when, with an appeal to the Bible, Christians continue to deny any form of "macroevolution"? It seems to me that this is a serious question that contemporary Christian theology cannot ignore. If theology is "the scientific self-examination of the Christian church with respect to the content of its distinctive talk about God," as Karl Barth argued, it should investigate how this talk relates to such dominant contemporary patterns of thought as the theory of evolution.[5] But is it possible to bring this theory into harmony with a truthful reading of the biblical witness? Would this not imply that we turn the Bible into a ventriloquist, or at least that we mix up faith and science in such a way that no justice is done to either of them?

---

cially 271, 281, 283); cf. Rienk Vermij, *The Calvinist Copernicans: The Reception of the New Astronomy in the Dutch Republic, 1575-1750* (Amsterdam: KNAW, 2002), 249-50 (cf. 162-64), and Rienk Vermij, "The Debate on the Motion of the Earth in the Dutch Republic in the 1650s," in *Nature and Scripture in the Abrahamic Religions: Up to 1700*, vol. 2, ed. Jitse M. van der Meer and Scott Mandelbrote (Leiden: Brill, 2008), 605-25.

4. Martinus Schoock, *De scepticismo pars prior* (Groningen: H. Lussinck, 1652), 406, as quoted in Vermij, *The Calvinist Copernicans*, 251.

5. Karl Barth, *Church Dogmatics* I/1 (Edinburgh: T&T Clark, 1956), 3. In his doctrine of creation, however, Barth himself intentionally avoided a discussion of the questions posed by science; see the famous preface of *Church Dogmatics* III/1 (Edinburgh: T&T Clark, 1958), ix-x. Barth rightly sensed, though, that future dogmatic theologians would not be satisfied with this decision (x).

In this chapter I will not discuss the exegesis of individual texts or passages from the Bible in detail but will focus on the underlying hermeneutical problem: How should Christians (for whom the Bible is authoritative) interpret biblical texts that seem to be at odds with what we know from science—and how should we approach scientific claims that seem to be at odds with biblical utterances? First, we will examine some traditional reading strategies that can be subsumed under the label "concordism" (§3.2). Then I will argue that an alternative hermeneutical approach, which I call "perspectivism," is more promising (§3.3). This approach, however, does leave one important question unresolved: how to deal with constitutive historical claims in the biblical narrative (§3.4). I end up with a conclusion (§3.5).

## 3.2 The Search for Harmony: Concordism

A rather obvious way of handling the tensions is to try to attune our exegesis of the relevant Bible passages to the broadly accepted aspects of evolutionary biology. This approach starts with the presupposition that at a fundamental level what the Bible tells us and what scientists have discovered in their work are in harmony with each other, and that it is possible to bring this to light; to put this in terms of the metaphor that we discussed in the first chapter: the two books of God—nature and Scripture—do not contradict but rather confirm each other. The theological challenge, however, is to exhibit this underlying harmony between the Bible and science, to make it visible. To the extent that this succeeds, the divine authority of the Bible is highlighted, since it has been demonstrated that the Bible is in agreement with scientific discoveries that took place many centuries after it was written. Because of its harmony-searching strategy, we will refer to this approach as "concordism."[6]

Here is a provisional definition of this reading strategy: *Concordism is the hermeneutical view that biblical statements pertaining to the physical world*

---

6. This term was coined to refer more particularly to so-called old-earth creationism by Bernard Ramm, *The Christian View of Science and Scripture* (Grand Rapids: Eerdmans, 1954), 145; the reader should notice that I do not use it in this restrictive sense. The more comprehensive meaning I adopt here can, for example, be found in Stanley Jaki, *Genesis 1 through the Ages* (New York: Thomas More, 1992), 43: "Concordism usually denotes the efforts whereby . . . numerous commentators of Genesis 1 [and 2-3] tried to establish its concordance with cosmogonies taken for the last word in science."

*correspond to scientific facts*. The presupposition behind concordism is that the Bible either overtly displays or covertly implies scientifically correct information about the way in which the natural world came into being and is structured. A prime example of the concordistic approach is the theory of so-called young-earth creationism. Adherents of this view maintain that, according to the book of Genesis, the earth is about six thousand to ten thousand years old. They claim that this observation corresponds to a scientific fact, since unprejudiced scientific research would yield the same age.[7] For obvious reasons, however, on further consideration, doubts may arise concerning the correctness of this assessment. Even young-earth creationists themselves nowadays honestly admit that their view is, "at the moment, implausible on purely scientific grounds."[8] One can imagine that this insight leads some of them to conclude that the earth and the cosmos in fact have a much greater age. Oftentimes, the exegesis of Genesis is then adapted to this newly acquired insight, and a form of "old-earth creationism" is adopted.[9] This first happened when in the nineteenth century the so-called *gap theory* was invented, according to which an enormous time gap occurs between Genesis 1:1 and 1:2. The billions of years that have passed since the creation of the cosmos are then located between these two verses.

---

7. For a recent elaboration and defense of young-earth creationism, see Terry Mortenson and Thane H. Ury, eds., *Coming to Grips with Genesis: Biblical Authority and the Age of the Earth* (Green Forest, AR: Master Books, 2008). The best-known present-day representative of this view is perhaps Ken Ham, an Australian who moved to the United States (see "Ken Ham," *Wikipedia*, last edited March 29, 2019, http://en.wikipedia.org/wiki/Ken_Ham).

8. Paul Nelson and John Mark Reynolds, "Young Earth Creationism," in *Three Views on Creation and Evolution*, ed. J. P. Moreland and John Mark Reynolds (Grand Rapids: Zondervan, 1999), 51; cf. Kenneth D. Keathley and Mark F. Rooker, *40 Questions about Creation and Evolution* (Grand Rapids: Kregel, 2014), 195–99. The only reason they continue to endorse the theory that the earth is relatively young is their reading of the Bible.

9. This is not to suggest that old-earth creationism is historically derived from young-earth creationism; in fact, the process was the other way round: young-earth creationism arose as a reaction to old-earth creationism. Old-earth creationism, however, emerged as an adaptation of naïve readings of Gen. 1 that were more openly articulated and defended in later young-earth creationism. On the history of creationism, see Ronald Numbers's landmark study *The Creationists: From Scientific Creationism to Intelligent Design*, expanded ed. (Cambridge, MA: Harvard University Press, 2006).

While the gap theory allows for an exponentially greater age of the earth than the traditional six thousand to ten thousand years, it is still at odds with gradualism: the idea, emerging from the fossil record, that biological groups appeared on earth one after another over long periods of time. Therefore, when people become convinced of gradualism, they tend to adopt another form of old-earth creationism, according to which the "days" of Genesis 1 are interpreted as geological periods of many millions of years.[10] In this view the theory of common descent is still rejected; at the beginning of each biological group or main species is God's creative word. However, the progressive unfolding of God's creative work over long periods of time is accepted. The order in which the main species appeared on earth is considered to mirror the sequence of God's creative acts as recorded in Genesis 1. The separate creation of humans is relatively recent, in accordance with the time period one arrives at when adding up the ages in the genealogies of the first chapters of Genesis. It is acknowledged that death must already have been present on earth before the human fall into sin, in the realms of plants and animals, as is clear from the fossil record. The flood may be seen as a regional rather than a global event, but, except for Noah and his family, all human beings living at the time perished in the water. In this way, data of contemporary science and of traditional biblical exegesis are connected to each other in what is supposed to be a more or less coherent story.

From time to time, however, an old-earth creationist will begin to suspect that there has been an evolutionary development of life on earth that passed the boundaries between the main groups (for instance, fish and amphibians, or reptiles and mammals). This would allow her to better understand the many different intermediate forms that have been found in the fossil material as well as the genetic similarities between all forms of life. If this happens, it may be expected that, after some further consideration, she will come to accept "macroevolution." Most probably, however, she will start to make an exception for the human species, in view of its unique place in Genesis and elsewhere

---

10. The best-known proponent of this view in the USA is probably the astronomer Hugh Ross, with, among many other publications: *The Fingerprint of God: Recent Scientific Discoveries Reveal the Unmistakable Identity of the Creator* (New Kensington, PA: Whitaker House, 1989) and *The Genesis Question: Scientific Advances and the Accuracy of Genesis* (Colorado Springs: NavPress, 1998). For a more concise defense, see Robert C. Newman, "Progressive Creationism," in Moreland and Reynolds, *Three Views on Creation and Evolution*, 103-33; his conclusion is noteworthy: "It seems, then, that harmonization should be our ultimate strategy" (131).

## Evolutionary Theory and the Interpretation of Scripture

in the Bible. This means that she has moved in the direction of what is being called—with a rather poor term—theistic evolution. But she does not yet fully endorse this view, since she continues to accept the special creation of the human species.[11] She will argue that, if correctly understood, the story of the Bible still largely corresponds to the story of the natural sciences: both suggest that the various species and groups developed from each other—or at least followed up on each other—in a process of millions of years, ending up with the appearance of the human being. And although there is no scientific evidence for the special creation of the human being, there is no counterevidence either, so that, for all we know, this part of the story as well may refer to an actual fact.

However, probably it will not be long before it dawns on this person that the intermediate forms between reptiles and mammals do not essentially differ from the other so-called *hominins*—beings which, as testified by their fossil remains, were in between the apes and present-day human beings. She may especially be impressed by the enormous degree of genetic similarity between humans and other primates. All this puts pressure on the idea that in the process of creation God made an exception for the human species by creating it *de novo*. For clearly, in the empirical data nothing can be detected that supports such a special position. Yet, the Bible depicts the human being as somehow unique as compared to all other creatures. This gives rise to the idea that God may have allowed the human body to evolve from earlier forms of life but added a *soul* to this body in a special creative act. Although other parts of the Genesis story can no longer be read concordistically on this view, here still is a clear correspondence between the special status of the human being (which we can observe empirically) and the biblical picture of God breathing the breath of life in the human being (Gen. 2:7).[12]

This view has been the official Roman Catholic position ever since several popes pronounced that "if the body takes its origin from pre-existent

---

11. By special creation is meant here what is sometimes called *creatio de novo*: the immediate instantiation of a new species "out of the blue" rather than through evolutionary processes. For an instructive overview of the different versions of both creationism and theistic evolutionism, see Gerald Rau, *Mapping the Origins Debate: Six Models on the Beginning of Everything* (Downers Grove, IL: IVP Academic, 2012), with a particularly helpful table on p. 41.

12. The idea that God, in an evolutionary context, made Adam into a spiritual being through a special creative act can be found in Bruce Waltke, *An Old Testament Theology: An Exegetical, Canonical, and Thematic Approach* (Grand Rapids: Zondervan, 2007), 203.

living matter, the spiritual soul is immediately created by God."[13] But how convincing is such a dualism between body and soul, both from a scientific and from a biblical perspective? Both the Bible and present-day philosophical anthropology rather seem to describe the human being in holistic terms as a deeply embodied psychosomatic unity. Moreover, could not the rich spiritual aspect of human existence also have emerged along natural lines? When this realization takes hold, the moment is near that "the penny drops" and concordistic attempts to harmonize (i.e., to make correspond) biblical exegesis and scientific data concerning the origin and development of life give way to a different hermeneutics—one that concedes that both the Bible and science have distinct roles to play that should not be conflated. Most probably, this will lead the person in question to adopt a variety of theistic evolution, that is, roughly speaking, the view that the Bible tells us *that* God created life on earth whereas science informs us about *how* God did so. Scientific facts are now no longer pressed into a biblical framework—nor is an exegetical framework imposed on the scientific data.

This brief survey may suffice to clarify which objections can be raised against concordism. First, one has to bend the scientific data and the exegetical findings toward each other in ways that do not do real justice to either side. Both the data of evolutionary theory and the biblical texts must be interpreted in a strained way—which outsiders (e.g., atheists) will regard as forming a complicated brainteaser and experts on both sides will think of as arbitrary and far from convincing. Some biblical passages, for instance, are taken in a literal sense while other passages are read symbolically (or as mythical, or as divine accommodations to a limited human understanding), and often the criteria that are used remain quite murky. Conversely, some aspects of the newer cosmological and biological theories of origin are taken seriously while others are ignored or explained away. Of course, we do not always find people developing from young-earth creationism through old-earth creationism to theistic evolution, but still, it is a fairly common pattern among Christians who gradually become more familiar with the methods of science and the evidence for Darwinian evolution.

Second, as soon as the artificial character of a particular harmonization becomes manifest within the circles where it was developed, the temptation arises to make minor adjustments in order to find a more solid position. This may temporarily reduce one's cognitive dissonance, but as time goes by, the

13. John Paul II, "Message to the Pontifical Academy of Sciences on Evolution," *Origins* 26, no. 22 (1996): 351 (rephrasing a point made earlier by Pope Pius XII).

new position proves to be equally untenable. Thus, as a matter of fact, one is hopping from one ice floe to another, having to recant earlier positions every now and then and ending on very thin ice. This explains why some young-earth creationists are very determined to stay with their position, since they fear that ultimately the core of the Christian faith may be at risk as soon as they start moving even a tiny little bit.[14] Yet, young-earth creationism itself is also based on the concordistic presupposition that the Bible reflects the main contours of a scientifically reliable account of the origin of life on earth. It seems to me that it is precisely this assumption that almost unavoidably leads to the process of ice-floe-hopping. If the Bible does indeed provide us with, or implies, correct scientific knowledge, we will continuously have to adapt our readings in order to harmonize them with current scientific knowledge, making exegetical moves that look more and more spurious.

Now obviously, nothing is wrong with having to change our minds every now and then on the basis of newly acquired information. On the contrary, it is a sign of mental health to be able to undertake such processes of intellectual change. It is different, however, when we know in advance that a position we adopt will almost certainly be wrong, since new information will most probably make it obsolete. This makes it quite understandable that today many Christian natural scientists, theologians, and others plead for a different, nonconcordistic, reading of biblical texts that touch on questions of origin.

### 3.3 Beyond Category Mistakes: Perspectivism

The question may be posed whether the presupposition behind concordism in its various forms is valid: Does the Bible indeed provide—or, in some hidden manner, imply—correct scientific knowledge about how (life on) planet earth developed? Was it the intention of the Bible writers—or of the Holy Spirit—to give us that kind of information? Today it has almost become a cliché to state that "the Bible is not a manual for natural science." Nonetheless, with many

---

14. I was made aware of this in personal correspondence with young-earth creationist Terry Mortenson (2009). See, for similar concerns, Nigel M. de S. Cameron, *Evolution and the Authority of the Bible* (Exeter, UK: Attic, 1983), and Wayne Grudem's foreword to *Should Christians Embrace Evolution? Biblical and Scientific Responses*, ed. Norman C. Nevin (Nottingham: Inter-Varsity Press, 2009), 10: "Belief in evolution erodes the foundations." See also *Theistic Evolution: A Scientific, Philosophical, and Theological Critique*, ed. J. P. Moreland et al. (Wheaton, IL: Crossway, 2017).

the idea that the Bible is such a manual is still fully alive. And it is not true that it has never been claimed or promoted by anyone.[15] At the same time, it is extremely dubious whether the biblical authors, even in places like Genesis 1–3, were really engaged with the historical and scientific questions of origin that so heavily concern us today. In any case, we should not search for answers to questions that the authors of the Bible did not address.[16] We are reminded of this in the letter to the Hebrews, where the writer tells us that *"by faith* we understand that the worlds were prepared by the word of God."[17] If that is so, there is no need for proofs or arguments in this connection. In particular, the Christian belief that God created the universe is not dependent on a detailed correspondence between the Old Testament creation records and the results of scientific research.

This leads us to examine a second approach with regard to biblical insights on the questions of origin, this one based on a different hermeneutical assumption from concordism. This assumption is that the Bible is, when push comes to shove, a fully *theological* book, in the sense that it is primarily focused on the relationship between God, the world, and human beings.[18] To be sure, these human beings are not perceived as disembodied individuals but as deeply embedded in their social, historical, and natural environments. When referring to the natural world as it is investigated by contemporary science, however, the Bible writers intuitively followed the ideas and conventions that

---

15. Gisbertus Voetius, e.g., referred to the Bible as "the book of all sciences"; *Sermoen van de nuttigheydt der Academien ende Scholen mitsgaders der wetenschappen ende consten die in deselve gheleert warden* [Discourse on the benefit of academies and schools as well as the sciences and arts that are taught in them] (Utrecht, 1636), 16. Centuries later, Henry M. Morris, the father of young-earth creationism, stated in his book *Many Infallible Proofs: Practical and Useful Evidences of Christianity* (San Diego: Creation-Life Publishers, 1980), 229: "The Bible is a book of science!"

16. Gordon Wenham, *Genesis 1–15*, Word Biblical Commentary 1 (Milton Keynes, UK: Word, 1991), liii. A highly significant attempt to read Gen. 1–3 in its own ancient Near Eastern context before applying it to our contemporary questions is provided by John H. Walton, *The Lost World of Genesis One: Ancient Cosmology and the Origins Debate* (Downers Grove, IL: IVP Academic, 2009), and John H. Walton, *The Lost World of Adam and Eve: Genesis 2–3 and the Human Origins Debate* (Downers Grove, IL: IVP Academic, 2015); see also chap. 6 below.

17. Heb. 11:3.

18. For a brief elaboration of this view, cf. Cornelis van der Kooi and Gijsbert van den Brink, *Christian Dogmatics: An Introduction* (Grand Rapids: Eerdmans, 2017), 554–61.

were current in their day. Not surprisingly, many of these ideas are no longer plausible. This does not present a problem, however, since it does not in any way diminish the way the Bible leads us to God and confronts us with his message—a message that, according to the Reformed stance, pivots on God's sovereign grace in Christ vis-à-vis human sin and misery. We must concentrate on this message as we read the Bible, for that is where its authority lies; we should not mix up that message with scientific categories, since we are dealing with two different *perspectives* here. Whereas the scientific perspective is focused on all sorts of "facts," the theological perspective focuses on the meaning of life. To be sure, in this theological perspective facts are involved as well, but these are of a unique sort, pertaining to the relationship between God and us.[19] These two perspectives are incongruent, which is to say they should not be merged but should rather carefully be kept separate.[20] Taking the notion of various perspectives as a cue, I will refer to this approach as *perspectivism*.

Let us provisionally define perspectivism as *the hermeneutical view that when the Bible is interpreted, its theological content should be distinguished from the world picture within which this content is embedded*. Whereas the theological content is authoritative, the world picture is circumstantial. This world picture frequently helps us understand the theological meaning of a passage, but the world picture itself is not part of that meaning. For example, when Jesus says that a mustard seed is the smallest of all the seeds (Matt. 13:32), he does not intend to convey a biological truth that we should take to heart, but he wraps up his message in a piece of contemporary conventional wisdom.[21]

---

19. Cf. the opening sentence of Calvin's *Institutes*: "Nearly all the wisdom we possess, that is to say, true and sound wisdom, consists of two parts: the knowledge of God and of ourselves," and these two are "joined by many bonds." John Calvin, *Institutes of the Christian Religion*, ed. John T. McNeill, trans. Ford L. Battles (Philadelphia: Westminster, 1960), 1.1.1.

20. Of course, the two perspectives do not contradict each other, but they do not fit together either, as they do in the various forms of concordism. They relate to each other "as an organ and a vacuum-cleaner," as Karl Barth once wrote in a letter to his niece Christine (meaning by "organ" the musical instrument): "there can be as little question of harmony between them as of contradiction"; Karl Barth, *Letters 1961-1968*, ed. Jürgen Fangmeier and Hinrich Stoevesandt, trans. Geoffrey Bromiley (Grand Rapids: Eerdmans, 1981), 184. The letter is dated February 18, 1965. Barth's view is an example of the "independence model" distinguished by Ian Barbour as one of the models for relating science and religion; cf., e.g., Ian G. Barbour, *Religion and Science: Historical and Contemporary Issues* (New York: HarperOne, 1997), 84-89.

21. We might even surmise that Jesus, being *vere homo* (truly human), actually

More detailed elaborations of perspectivism can be found (1) with biblical scholars who worry that the unique voice of the biblical authors and the integrity of their texts get smothered in our modern creation-versus-evolution discussions, and (2) with natural scientists and theologians—many of them self-identifying as adherents of "theistic evolution"—who feel much impressed by the great amount of material that points toward geological and biological evolution. Let us look more closely at the reasons for perspectivism that are put forward from these points of view by briefly discussing a representative of each of them.

1. Francis Watson, a biblical scholar, resists the common view that, at long last, thanks to Darwin, scientific secularity has triumphed over ancient superstition.[22] This view is based on the assumption that the message of the Bible and that of Darwin are total opposites, and that only one of them can be true. In reality, however, Darwin does not present us with a worldview that allows us "to explain everything" and overthrows the outmoded view of Genesis. Darwin's theory is merely "one possible and partial account of a certain complex development alongside others."[23] Next to it, another story can and should be told—one that allows the biblical texts to speak for themselves. It then becomes clear that the biblical perspective cannot be integrated into the scientific picture but has "its natural habitat within the Christian story of 'salvation.'"[24] To discover this much more important perspective, however, we must abandon the idea that somehow Genesis provides us with correct scientific information—in that respect Darwin has done us a major service by showing us that such information is not to be found in the Genesis narratives.[25]

To support his argument Watson begins with John Calvin's exegesis of Genesis 1:16: "God made the two great lights—the greater light to rule the day

---

believed a mustard seed to be the smallest of all seeds (cf. Mark 13:32 for his human ignorance of certain things). Another example is the grain of wheat that supposedly dies when it falls into the earth (John 12:24; 1 Cor. 15:36); from a biological point of view we now know that grains of wheat do not die before they germinate—but that does not at all detract from the theological meaning and significance of this image.

22. Francis Watson, "Genesis before Darwin: Why Scripture Needed Liberating from Science," in *Reading Genesis after Darwin*, ed. Stephen C. Barton and David Wilkinson (Oxford: Oxford University Press, 2009), 23-38 (24).

23. Watson, "Genesis before Darwin," 24.

24. Watson, "Genesis before Darwin," 35; we should add, of course, that this story was a *Jewish* story before it also became a Christian one.

25. Watson, "Genesis before Darwin," 24, 35-36.

and the lesser light to rule the night—and the stars." It appears that Calvin is not worried that—in spite of what this text suggests—the moon neither gives light nor belongs (together with the sun) to the largest heavenly bodies. In a very down-to-earth way, Calvin comments that Saturn is bigger.[26] But we do not have to ascend into heaven to understand Moses's intentions, for Moses refers to our *earthly experience* of sun and moon as the largest light-giving objects. He wants us to understand that even the light of the moon at night is a gift from God. In our terminology: his message has a theological point, not a scientific one. Watson rightly points out that Calvin does not propose a harmony along the following lines: "And God made two big lights (that is to say, lights that on earth *appear* to be the biggest); the big light to rule the day, and the small one (small in the sense that it derives its light from the sun) to rule the night; and he also made the stars (including the planets, which actually may be bigger than the smallest of these two big lights)." Such a harmony would mix the biblical claims with those of a scientific nature, in such a way that neither of them receives full justice. On such a reading, we would fail to note that Moses intends to make clear that not only the light of the sun but also that of the moon is a gift from God.[27]

Although Watson knows that Calvin was not always fully consistent in this approach, he nonetheless derives from Calvin's detailed exegesis of Genesis 1:16 general guidelines for dealing with questions about the Bible and science: (1) Where the two differ, we must find out whether they perhaps offer different but mutually compatible perspectives of reality, rather than mutually exclusive truth claims. (2) Since the biblical perspective concerns our relationship with God, this perspective is "primary and foundational" as compared to the additional perspectives that science provides; therefore, "the scriptural account should have precedence over the scientific one." (3) The scientific perspective must not be neglected, because it "provokes a more insightful reading" of the biblical text and uncovers the "significance and rationale" of its "fact-like assertions." (4) The difference between the two perspectives must be explained

---

26. Cf. John Calvin, *Commentaries on the First Book of Moses called Genesis* (1554), trans. John King (Grand Rapids: Eerdmans, 1948), 86: "Moses makes two great luminaries; but astronomers prove, by conclusive reasons, that the star of Saturn . . . is greater than the moon." The original Latin text (from 1554) is in G. Baum et al., eds., *Joannis Calvini opera quae supersunt omnia*, vol. 23 (Brunswick, Germany: C. A. Schwetschke, 1882), 22-23. Note that Calvin does not try to cast doubt on the findings of the astronomers, even though they deviate from a plain reading of the Genesis text.

27. Watson, "Genesis before Darwin," 25-26.

and not denied. The integrity of the perspectives of the Scriptures as well as of the sciences is compromised when we try to show that the Bible is confirmed by science, or vice versa. Darwin was right in demonstrating that science had to be liberated from the dominance of the Scriptures, but the Scriptures must also be protected against the dominance of science.[28]

To some degree Watson takes the easy option by focusing on a detail that is doctrinally innocent. He does not make any concrete statements about the implications of letting the two perspectives of Bible and science stand beside each other in matters that seem theologically more important, as, for instance, the origin of the natural world, the historicity (or the lack thereof) of Adam and Eve, the relationship between sin and death, the origin of sin, etc. A significant amount of other literature favoring perspectivism suffers from the same problem, thus ignoring important questions.[29] For example, if the Bible only tells us about God's encounter with human beings, how does the world in which this encounter unfolds relate to God? Is it *God's* world, or is it none of God's business, as Marcion thought? If it is the world of the Father of Jesus Christ, how do we explain its many perils and the widespread suffering of both animals and humans in it? Did human beings at some point break their relationship with God by choosing evil? Or has evil always been an intrinsic part of humanness? Such essential theological questions are too often ignored by perspectivists.

2. This is certainly not the case in an intriguing contribution to the debate on biblical interpretation and evolution by the Canadian biologist, theologian, and "born-again Christian" Denis Lamoureux.[30] Lamoureux wants to

---

28. Watson, "Genesis before Darwin," 27-28.

29. A well-known case in point here is Peter Enns, *The Evolution of Adam: What the Bible Does and Doesn't Say about Human Origins* (Grand Rapids: Brazos, 2012). Enns admits that in this book he "is focused solely on hermeneutical issues . . . and so I make no claim to answer the many intellectual issues that the Christianity/evolution discussion raises" (126).

30. Denis O. Lamoureux, *I Love Jesus and I Accept Evolution* (Eugene, OR: Wipf & Stock, 2009). The phrase "born-again Christian" occurs on the back cover. The title speaks for itself, but note the verbs: I *love* Jesus and I *accept* evolution. Whereas a Christian's relationship to Jesus is existentially charged, his acceptance of evolution is much more sober and down to earth. By the way, in the final sentence of his book Lamoureux adds a typically Reformed twist to the evangelical confession of his love for Jesus: "But more importantly, as the children's Sunday school song has taught me, 'Jesus loves me, this I know, for the Bible tells me so'" (168). For further elaboration of his views, see Lamoureux's earlier five-hundred-page book *Evolutionary Creation: A Christian Approach to Evolution* (Cambridge: Lutterworth, 2008).

endorse the biblical faith without any reservation while also accepting the data of contemporary science in an unqualified way, and he hardly leaves any question about their mutual relationship untouched. Taking science seriously implies for him that, though science is as fallible as any other branch of human endeavor, we should not assume that entire branches of science are rooted in misunderstandings or spiritual deception. Therefore, he also takes evolutionary biology seriously. Lamoureux himself went through a number of phases in his life—among them an atheistic as well as a creationist period. He now labels himself an adherent of "evolutionary creationism" (which term he prefers to "theistic evolutionism"; though this might indeed be considered the better term since the substantive now refers to what is most important—belief in the world as being created—its potential association with other forms of creationism makes it confusing).

Within this stream of thought, however, Lamoureux occupies a radical position. He consistently refuses to consider any concordistic attempts at harmonization and wants the Bible to speak its own language. In Genesis 1 and 2 man is created *de novo* by God; that this happened "from the dust of the ground" (Gen. 2:7) should not be explained in terms of man's evolutionary emergence from the animal world, since clearly that is not what the biblical author had in mind. Genesis 2 and 3 sketch an idyllic picture of worldwide bliss, not a picture of a small garden as an exceptional oasis in an otherwise wild world.[31] This idyllic reality is brutally disrupted by the fall into sin of the first human pair. This led to the death of Adam, Eve, and all their descendants. Genesis does not present this death as some kind of vague "spiritual" death (an existential loneliness because of the rupture in the relationship with God) but simply as physical death. This is how Paul in Romans 5 and 1 Corinthians 15 reads these chapters.[32] As a result of Adam's sin, suffering and death made their entrance into creation. These events did not only affect humans but also made the animal world go awry, causing havoc in all of creation. As a result of this, creation henceforth has to "groan in travail" (see Rom. 8:22).

However, in light of what we now know, this entire presentation must be seen as part of the ancient Eastern *science of the day*;[33] it does not correspond to physical reality but to how physical reality was construed and

---

31. For this view, cf., e.g., Marguerite Shuster, *The Fall and Sin: What We Have Become as Sinners* (Grand Rapids: Eerdmans, 2004), 77.

32. Lamoureux, *I Love Jesus*, 84, 141.

33. Lamoureux, *I Love Jesus*, 144.

imagined at the time. In that sense, it is on a par with the world picture of a three-layered universe (heaven, earth, and netherworld) we meet, for instance, in Genesis 1:6–7, Exodus 20:4, and Philippians 2:10.[34] In his revelation, God went to great lengths to adapt to the ancient Near Eastern picture of the world, prompted by his desire to reach human beings with the message that he is their creator and the creator of the world.[35] Contrary to Voetius, Lamoureux is convinced that the first receivers of God's revelation (Moses, Paul, etc.) would never have been able to understand this message if it had been clothed in the conceptuality of our contemporary model of the universe.[36]

Lamoureux does not mince words when he states what all of this means: Adam never existed, and hence death did not enter the world through him.[37] The vast number of fossils excludes this possibility, for animal remains are found in much older strata of the earth than those in which we find human remains.[38] As a matter of fact, whereas death on earth dates back to the very first manifestations of life, both the image of God and the reality of sin gradually

---

34. In his *Evolution: Scripture and Nature Say Yes!* (Grand Rapids: Zondervan, 2016), Lamoureux provides an extensive description of "ancient science" as it figures in the Bible (28–31, 85–112). It is debatable, however, whether the ancient Near Eastern picture of the world was as uniform, constant, and free from contradictions as is suggested by this type of description. See, e.g., Noel K. Weeks, "The Ambiguity of Biblical 'Background,'" *Westminster Theological Journal* 72 (2010): 219–36, and "The Bible and the 'Universal' Ancient World: A Critique of John Walton," *Westminster Theological Journal* 78 (2016): 1–28 (esp. 1–21). On the other hand, it is incontrovertible that contemporary pictures (plural) of the world are reflected in the biblical texts.

35. In *I Love Jesus*, 44, Lamoureux provides the example mentioned above of Jesus presupposing the "scientific" view of his day in the parable of the mustard seed (Matt. 13:31–32). He points out how concordistic considerations have seduced some Bible translators to add phrases like "as you think" or "as it appears" to the words "the smallest of all the seeds."

36. Lamoureux, *I Love Jesus*, 146. Of course, it can be discussed to what extent the ancient Near Eastern world picture also included certain *cultural* ideas, for example, as pertaining to polygamy, corporate thinking, the place of women in relation to men, the use of violence (cf. the "ban" in the Old Testament), the use of oracles or the casting of lots as revelatory means, etc. This is not the place to discuss these issues, but for a balanced evaluation, see Hendrikus Berkhof, *Christian Faith: An Introduction to the Study of the Faith*, rev. ed. (Grand Rapids: Eerdmans, 1986), 252–53.

37. Lamoureux, *I Love Jesus*, 148.

38. Lamoureux, *I Love Jesus*, 142–43.

emerged with the appearance of *Homo sapiens*. This happened tens of thousands of years ago, probably at various locations in the world, when groups of hominins more or less simultaneously made a radical jump in the evolutionary chain, developing into what we know as human beings.[39] It makes no sense to place Adam and Eve somewhere in this chain—that is like trying to fit the ancient Near Eastern three-layered universe into contemporary cosmology.[40] If we do this, we make a big *category mistake* by mixing up the perspectives of Bible and science, and we are guilty of a serious misuse of the Bible.[41]

In the meantime, Lamoureux continues to insist that his evolutionary creationism does not detract from the biblical message or from the authority of the Bible. We must carefully distinguish the Bible's embeddedness in an ancient worldview from its theological content. From this perspective we can discern the "eternal spiritual truths" that compose the proper message of the Bible.[42] This message does not deal with *how* questions, such as how we became a creature and, subsequently, a sinner. It just tells us that we *are* sinful creatures—and that, therefore, we are radically dependent on the grace of God, which is revealed to us in the person and saving work of Jesus Christ. In this way, Lamoureux's hermeneutics goes hand in hand with a doctrine of sin and grace that we can easily recognize as authentically Reformed. Thus, the question comes up how we should evaluate this perspectivist approach. It avoids the serious drawbacks of concordism, but does it solve all problems?

### 3.4 Where Science and Christianity Overlap: History

The perspectivism sketched above represents an attractive position. It goes beyond any artificial harmonizing attempt. Instead of trying to "salvage" as many individual biblical texts as possible, it makes a clear and unambiguous choice in favor of the dominant views in natural scientific discourse, while it recognizes in all honesty that often these are incompatible with (decontextualized) statements on the physical world that are made in the Bible. From

---

39. Lamoureux, *I Love Jesus*, 138.
40. Lamoureux, *I Love Jesus*, 140.
41. Cf. for the notion of a "category mistake" in this connection, Vincent Brümmer, "Introduction: A Dialogue of Language Games," in *Interpreting the Universe as Creation: A Dialogue of Science and Religion*, ed. Vincent Brümmer (Kampen: Kok Pharos, 1991), 4.
42. Lamoureux, *I Love Jesus*, 18; elsewhere he speaks of "inerrant spiritual truth" (45).

a hermeneutical and doctrinal point of view, that is no problem, since such factual statements simply do not belong to the message conveyed in the Bible. A big advantage of this approach is that apparently things cannot become "worse" time and again. As a result, there is no "ice-floe-hopping" going on here, since there is no risk that ten or fifteen years from now new scientific evidence will force Lamoureux to change his position. The same is true for authors who have opted for a similar position.[43]

But is not such a position at odds with the Reformed doctrine of Scripture, which strongly emphasizes the *literal* meaning of the Scriptures? We must realize that this emphasis was originally directed against all kinds of allegory that, at the time, were popular. For this reason, it was important to let the Bible speak for itself. This did not, however, imply a *literalistic* way of dealing with the Bible that ignored the genre and scope of specific texts. We already noted how Calvin made sure to take this scope into account in his exegesis of Genesis 1:16. In fact, an approach that operates in this way—that is, that takes into account factors like literary genre and scope—may well be called "literal."

Let us use the so-called "framework interpretation" of Genesis 1 as an example here. According to this interpretation, the "days" in Genesis 1 form a framework in which day one corresponds with day four, day two with day five, and day three with day six—in each case God first creates order out of chaos in three realms (the light on day one, the water and the heavens on day two, the land on day three) in order to then fill each of them with fitting inhabitants (sun, moon, and stars on day four; birds and fish on day five; animals and humans on day six). The author(s) did not want to suggest that God filled the earth with its inhabitants in six days but used these days as a framework to highlight the careful ways in which God made all things fit in with each other and to emphasize that God's creative work culminated in the introduction of his royal Sabbath rest on day seven. When this framework

---

43. The best known among them are geneticist Francis Collins, with his *Language of God: A Scientist Presents Evidence for Belief* (New York: Free Press, 2006), and from a Roman Catholic background evolutionary biologist Francisco J. Ayala, with, among other publications, *Darwin's Gift to Science and Religion* (Washington, DC: Joseph Henry, 2007)—a book that especially targets creationists but remains theologically too much on the surface to convince them. See also, e.g., Kenneth R. Miller, *Finding Darwin's God: A Scientist's Search for Common Ground between God and Evolution* (New York: HarperCollins, 1999); Keith B. Miller, ed., *Perspectives on an Evolving Creation* (Grand Rapids: Eerdmans, 2003); Darrel R. Falk, *Coming to Peace with Science: Bridging the Worlds between Faith and Biology* (Downers Grove, IL: InterVarsity Press, 2004).

interpretation indeed reflects the genre and scope of Genesis 1, then an exegesis based on it might be regarded as literal.[44] In such a scenario one should not say that "the days are not taken literally," for that is exactly what happens: the *litterae* (= letters) of the text display a particular genre, and this genre is taken seriously. It is the genre of an impressively crafted opening chorale, like the opening chorus of Bach's *Christmas Oratorio* where, in anticipation of what follows, God is already praised for the mighty works he has brought about.[45]

Moreover, the doctrine of accommodation, to which Lamoureux appeals, is not in opposition to the Reformed doctrine of Scripture. On the contrary, it was already applied by Calvin, who in turn stood in a long tradition here.[46] Opposing the so-called anthropomorphites, who concluded from the ascription of a mouth, ears, eyes, hands, and feet to God in the Bible the corporeality of God, Calvin argued: "For who even of slight intelligence does not understand that, as nurses commonly do with infants, God is wont in a measure to 'lisp' in speaking to us? Thus such forms of speaking do not so much express clearly what God is like as accommodate the knowledge of him to our slight capacity. To do this he must descend far beneath his loftiness."[47]

It has been argued that Calvin's use of accommodation was different from other early modern and modern applications of the same principle. For example, Faustus Socinus and representatives of the Enlightenment used it to explain away various Christian doctrines that they could not believe because of

---

44. The framework interpretation of Gen. 1 was introduced by the Dutch Old Testament scholar Arie Noordtzij in 1924 and has been adopted and elaborated by prominent evangelical biblical scholars such as Meredith Kline, Henri Blocher, Bruce Waltke, and Gordon Wenham. See, e.g., Lee Irons and Meredith G. Kline, "The Framework View," in *The Genesis Debate: Three Views on the Days of Creation*, ed. David G. Hagopian (Mission Viejo, CA: Crux, 2001), 217-304.

45. For a more recent interpretation of Gen. 1 that goes beyond the framework hypothesis (based on a closer comparison of Gen. 1 with other ancient Near Eastern cosmological texts), see Walton, *The Lost World of Genesis One*. Walton argues that Gen. 1 is not about the material origins of the cosmos but about the assignment of the *functions* God had in mind for each of its inhabitants, prior to his taking up residence in the cosmos, as in his temple.

46. Cf. Cornelis van der Kooi, *As in a Mirror: John Calvin and Karl Barth on Knowing God; A Diptych*, trans. Donald Mader (Leiden: Brill, 2005), 41-57; Van der Kooi points to Origen, Irenaeus, and Philo as earlier representatives of this hermeneutical tradition, and he clearly shows how crucial the idea of accommodation was to Calvin.

47. Calvin, *Institutes* 1.13.1.

their rationalist assumptions.[48] Indeed, accommodation is a risky hermeneutical category, since it can easily be used to turn the Bible into a ventriloquist. According to some, Calvin cannot be accused of opening the door to this path, since he used the principle in such a way that it did not detract from the *truth* of what the Bible writers claimed.[49] It is questionable, however, whether such a watertight distinction can be made here. Calvin's use of the principle of accommodation was inspired by his view of God's transcendence, which forbade him to ascribe, for example, passions to God. Strictly speaking, Calvin did not contend that the Bible makes any claims in this connection that as a matter of fact are false—he does not go any further than saying that they "do not so much express clearly what God is like." However, the result is the same, namely, that the biblical text is not reliable on such issues. For, according to Calvin, God does not *really* repent, become angry, or undergo other forms of change. As Huijgen writes: "Calvin rather pays the price of insufficient certainty that God's words are unequivocally true, than ascribing change of whatever kind to God."[50] Thus, Calvin used the accommodation principle to regulate nothing less than our thinking about *God*—that most crucial theological theme! It seems far more innocent to apply it to the way in which cosmographical world pictures appear in the Bible, as Lamoureux proposes.[51] For in such pictures we can quite easily distinguish between what is said and what is meant in the Bible. In more technical language: the authority of the Bible should not be found at the surface level of all its locutions but at the level of its illocutions—that is, it resides in what is *conveyed through* its locutions.[52]

---

48. Cf. Arnold Huijgen, *Divine Accommodation in John Calvin's Theology: Analysis and Assessment* (Göttingen: Vandenhoeck & Ruprecht, 2011), 28-33, on the erosive effect the doctrine of accommodation had on belief in the Bible's reliability during the seventeenth and eighteenth centuries (among Cartesians, German rationalists like J. S. Semler, etc.).

49. See, e.g., Hoon J. Lee, "Accommodation—Orthodox, Socinian, and Contemporary," *Westminster Theological Journal* 75 (2013): 335-48; Glenn S. Sunshine, "Accommodation Historically Considered," in *The Enduring Authority of the Christian Scriptures*, ed. D. A. Carson (Grand Rapids: Eerdmans, 2016), 238-65.

50. Huijgen, *Divine Accommodation*, 274.

51. For more analysis of the function and background of divine accommodation in Calvin's thinking, see Jon Balserak, *Divinity Compromised: A Study of Divine Accommodation in the Thought of John Calvin* (Dordrecht: Springer, 2006), and Huijgen, *Divine Accommodation*, esp. 106-54.

52. For this application of speech act theory (J. L. Austin and others) to the nature of biblical authority, see John Walton and D. Brent Sandy, *The Lost World of Scripture:*

Nonetheless, the approach of Lamoureux raises serious theological problems. The most important of these, it would seem to me, is that the biblical "message" cannot so easily be detached from the narrative form in which it comes to us and reduced to a couple of timeless messages, as he suggests.[53] Those who reduce this message to lessons or principles, however orthodox these may be, and leave the historical garment in which these have come to us behind them, as an empty cartridge, will be left with something else than the biblical view of how God relates to us human beings. For this biblical view is thoroughly historical in nature: God deals with us by going a particular *way* with us—a way that has a beginning, a certain course, and a goal. The temporal sequence of creation, sin, and salvation is crucial in this respect—at least (though not only) in Reformed theological appropriations of the biblical message. Moreover, the gospel hinges on God's involvement in history: the incarnation, cross, and resurrection of Jesus that constitute our salvation took place during specific moments in time and history. That renders the Christian faith vulnerable, because such events can by definition be contested and contradicted; from a Christian point of view, however, we cannot avoid such vulnerability by withdrawing into a sealed fortress of supertemporal principles and ahistorical truth claims.

When dealing with questions of human origin, reducing the biblical message to a set of lessons or principles or "eternal spiritual truths"[54] is inadequate for yet another reason. What we need here, it seems, is a comprehensive view or constitutive *story* of how life on earth and we human beings became what we are right now. Without such a story, our way of thinking will easily be stamped by atheistic stories according to which God does not play a role at all in the emergence and history of life on earth. From a methodical point of view, Lamoureux's attempt to remove all world-picture elements from the magisterium (domain of authority) of the Bible does not differ from Rudolf Bultmann's program of demythologizing the proclamation of the New Testament.[55] To be sure, Bultmann went much further than Lamoureux (who

---

*Ancient Literary Culture and Biblical Authority* (Downers Grove, IL: InterVarsity Press, 2013), 41–48, and passim.

53. In *Evolution: Scripture and Nature*, Lamoureux interchangeably speaks about "life-changing spiritual truths" (31), "inerrant spiritual truths" (110), and "life-changing messages of faith" (111).

54. Lamoureux, *I Love Jesus*, 18.

55. See Rudolf Bultmann, *The New Testament and Mythology and Other Basic Writings* (Minneapolis: Augsburg Fortress, 1984), 1–44 (the original German essay "Neues Testament und Mythologie" appeared in 1941).

does not deny God's involvement in the history of salvation), but Lamoureux's approach is structurally similar—and should perhaps, if consistently thought through, end where Bultmann's program ended. Bultmann as well had noble intentions with his proposals; in fact, his motives were strongly apologetic: as a Lutheran pietist, he wanted to defend the Christian faith in the context of his own time. In the process, however, a large part of the content of the Christian faith evaporated, because he considered talk of God's salvific acts as part and parcel of the ancient Near Eastern world picture. As a result, he detached the entire domain of *history* from God's active involvement.

Therefore, the question remains whether we should protect ourselves against any possible "collision" between the Bible and science by opting for a radically perspectivist approach. It is true that much is to be said for keeping faith and theology separate from science, as two distinct perspectives. The problems arise, however, where the two inevitably overlap: in the domain of *history*. There we see that the two perspectives are not completely incommensurable, like "organs and vacuum cleaners" (Barth), but cross each other at some point, illuminating one and the same reality. The text of Genesis 2–3 is a clear example here. As we will see more closely in chapter 6, these chapters do not allow for a completely ahistorical reading. Old Testament scholar C. John Collins has argued that the chronological sequence of creation and sin defines the biblical story of our humanness.[56] Indeed, according to the main thrust of the biblical narrative, sin is not inherent in creation. If it were, that would give us humans an excuse and would make God's anger about sin hard to understand. Sin goes back to a step taken by the first human being(s)—at whatever time or place that may have been. Sin is nonoriginal but corrupts the good life God had intended.[57] The view that sin is "not the way it's supposed to

---

56. C. John Collins, *Did Adam and Eve Really Exist? Who They Were and Why You Should Care* (Wheaton, IL: Crossway, 2011), 133–35. Collins deliberately uses the concept of "story" here to make clear that the Bible does not merely give us distinct messages but provides us with an overarching narrative, comparable to the way in which the self-understanding of other cultural and religious communities is shaped by their stories of origin. Cf. for a similar view, Henri Blocher, "The Theology of the Fall and the Origins of Evil," in *Darwin, Creation, and the Fall: Theological Challenges*, ed. Robert James Berry and Thomas A. Noble (Nottingham: Apollos, 2009), 149–72.

57. Cf. Cornelius Plantinga Jr., *Not the Way It's Supposed to Be: A Breviary of Sin* (Grand Rapids: Eerdmans, 1995), 16. Unlike John Schneider, "Recent Genetic Science and Christian Theology on Human Origins," *Perspectives on Science and Christian Faith* 62 (2010): 202, it seems to me that this is not just "Augustinian" but belongs to the constitutive pattern of the biblical narrative.

be" is of major importance for how we regard the relationship between God and humanity. Collins rightly begins by establishing this point—not on the basis of a biblicistic hermeneutics but from the desire to do full justice to the leading theological perspective of the Bible.

Only then does Collins take a look at what may, and must, be said scientifically about human origins. This does not lead him to a new harmonization in which the Bible and natural science confirm each other but leads him to develop guidelines for what must minimally be said about the issue of origin from a Christian point of view. These guidelines or criteria stipulate (1) that, because of our unique status as the bearers of God's image, we cannot just be the product of a natural process; (2) that Adam and Eve stand at the beginning of the human race—possibly together with others, among whom they may have occupied a special representative position; and (3) that, in some way, in the early period of the human race a decisive "fall" into sin occurred.[58] Collins agrees with C. S. Lewis, who graphically depicts the Christian story of creation and fall along similar lines.[59] In chapter 6 we will examine more closely whether such a picture is credible at all from a scientific point of view.

To conclude, we are well advised to reject concordism and take perspectivism as our default position. The particular genre of a biblical text is decisive, however, in determining its theological perspective. From the first chapters of Genesis onward, this theological perspective is indissolubly linked up with history. The Old Testament hinges on a series of salvific events that points forward beyond itself and culminates in the New Testament in the coming of Jesus Christ and the Spirit. This is how Christians, including Reformed ones, have read their Scriptures all along—at least ever since Irenaeus. As Reformed theologian Hendrikus Berkhof rightly claims: "Salvation depends on the historicity of . . . events."[60] Indeed, in this sense the Christian faith is not about "eternal spiritual truths" (Lamoureux) but about down-to-earth historical events. It is precisely on the historicity of events, however, that the

---

58. Collins, *Did Adam and Eve Really Exist?*, 120; as appears from (2), Collins allows for the possibility of polygenism.

59. C. S. Lewis, *The Problem of Pain* (New York: Macmillan, 1962 [1943]), 69-88 (= chap. 5, "The Fall of Man"). To cite only one sentence from this moving passage: "They [= the first humans] wanted to be nouns, but they were, and eternally must be, mere adjectives" (80).

60. Berkhof, *Christian Faith*, 274; Berkhof is referring here to "what has actually happened through and with Jesus," but this can be extended to other salvific events as, for example, those related in the Old Testament.

sciences—in our case especially the sciences of origin—have something to say as well. At this cross-section of science and faith, therefore, we cannot escape the search for harmony. Christian theology cannot take the easy way out by withdrawing to the spiritual realm in order to circumvent historical critique; it can only prove its value by showing that it can stand the test of historical research. There is no reason to fear such research, however, when we hold that the book of Scripture and the book of nature and history have, in the end, the same Author. If that is true, we may from time to time either have to reread the (alleged) data of science in the light of Scripture, or reread Scripture and reconsider established interpretations of it in the light of science. For even though the Bible does not contain scientific statements, as concordists think, neither can its theological meaning contradict what we know from the sciences.

### 3.5 In Search of Cocceians

The main upshot of this chapter is that when it comes to truthful biblical interpretation, we have to carefully differentiate between the scope or focus of what the biblical authors wanted to convey and the traces of an outdated model of the world in which they incidentally clothed their message. In doing so, we may learn that we will have to explain certain passages (for example, Gen. 1-3) in ways that differ from traditional exegesis. Surely this cannot mean that we should impose the results of contemporary science on Scripture, as if they had been stored there all the time. That would be a relapse into concordism. Given the nature and goal of Scripture, we should not expect it to anticipate the results of scientific research. Neither should we expect, however, that its theological content or focus or "message" is at odds with what we know from science.

Fortunately, there is a third way to go here. As G. C. Berkouwer argued, "certain results of science, be it natural science or historical research, can provide the *occasion* for understanding various aspects of Scripture in a different way than before."[61] That means that we may welcome scientific developments as far as they help us better understand what Scripture intends to teach. Berkouwer recalls that his predecessor Herman Bavinck rejoiced in the "excellent service" that geology may offer us "in the interpretation of the creation story."[62] Indeed, Bavinck held that

---

61. G. C. Berkouwer, *Holy Scripture*, trans. Jack B. Rogers (Grand Rapids: Eerdmans, 1975), 133.
62. Berkouwer, *Holy Scripture*, 133, referring to Bavinck, *Reformed Dogmatics*, vol. 2, *God and Creation* (Grand Rapids: Baker Academic, 2004), 496.

> Scripture and theology have nothing to fear from the *facts* brought to light by geology and paleontology. The world, too, is a book whose pages have been inscribed by God's almighty hand. Conflict arises only because both the text of the book of Scripture and the text of the book of nature are so often badly read and poorly understood. In this connection the theologians are not without blame, since they have frequently condemned science, not in the name of Scripture but of their own incorrect views.[63]

This reconsidering of traditional biblical interpretations in light of newly discovered scientific facts usually involves a slow and painful process. However, the rise and gradual acceptance of heliocentrism show us that it may well succeed over time. Even in orthodox Reformed theology it is possible to appropriate new understandings of the Bible, provided that we recognize that these are not violently imposed on the Bible because of science but do justice to its inner thrust. One might even argue that Reformed theology is *especially* open to such revisions, since, as we saw in chapter 1, its motto is to return to the Bible over and over again because of the expectation that in this way one may at all times learn surprisingly new and timely things.

Moreover, it was in the Reformed tradition that the "organic doctrine of inspiration" was developed by theologians like Kuyper and Bavinck—a doctrine that was to gain broad acceptance within the Reformed community and beyond.[64] According to this view, in guiding, enlightening, and inspiring the writers of the Bible, the Holy Spirit did not typically dictate what they should write down (as in the "mechanical" view of inspiration) but took them in his service in a much more organic way. That is, the Spirit put to use their personal skills and talents, characters, linguistic habits, biographical paths—see, for example, Luke's many references to medical issues—and cultural backgrounds.[65] If this view of divine inspiration holds water, it is not at all inappropriate to

---

63. Bavinck, *Reformed Dogmatics*, 2:396; interestingly, Bavinck himself goes on in his next section to condemn Darwinian evolution, partly on theological grounds (511–29).

64. Cf., e.g., Herman Bavinck, *Reformed Dogmatics*, vol. 1, *Prolegomena* (Grand Rapids: Baker Academic, 2003), 428–35.

65. "Their native disposition and bent, their character and inclination, their intellect and development, their emotions and willpower are not undone by the calling that later comes to them. . . . Their whole personality with all of their gifts and powers are made serviceable to the calling to which they are called." Bavinck, *Reformed Dogmatics*, 1:432.

assume that when writing the texts later included in the Bible, the biblical authors took for granted the picture of the world with which they were raised, and which most of their first readers would immediately and intuitively recognize. Thus, it seems to me that perspectivism is much more in line with the Reformed doctrine of Scripture in its organic articulation than concordism is; concordism presupposes a mechanical view of inspiration, God as it were dictating certain facts to the biblical writers out of the blue. For how else could the biblical writers have come to know present-day scientific facts than in a mechanical way, through special divine communication that would have overruled their own cultural embeddedness? Thus, if the doctrine of organic inspiration does not jeopardize the authority of the Bible, as most Christians hold, perspectivism with regard to world-picture issues does not do so either.

So why do Reformed Christians continue to reject evolutionary theory with an appeal to Scripture? No doubt, one part of the answer is that, for all of us, from a psychological point of view it is hard to reconsider certain Bible interpretations, especially when we have imbibed them from our youth onward.[66] The role of children's Bibles should not be underestimated here, especially when they have not been followed up at a later age by more open and mature interpretations of the Genesis texts. Another part of the answer, however, is that evolutionary theory has oftentimes been hijacked by naturalists in support of their atheist worldview, as a result of which in popular perceptions it became strongly associated with atheism.

This need not continue forever, though. It is illuminating to briefly return to the question how, in spite of the fierce and principled resistance of Voetius and his followers, the heliocentric model was eventually accepted by the Dutch Reformed orthodoxy. The studies of Rienk Vermij give us some insights. In the seventeenth century, the Copernican view became inherently linked to Cartesianism, which meant that in order to be a good Christian one had to reject it. When at a later stage, however, the theology of John Cocceius (1603-1669) came to prominence, things began to change.[67] Followers of

---

66. By the way, we also see such tenacity in science. The famous historian of science Thomas Kuhn has pointed out that paradigms usually don't disappear because their adherents become convinced by a new one but simply because they die out. Thomas S. Kuhn, *The Structure of Scientific Revolutions*, 2nd ed. (Chicago: University of Chicago Press, 1970), 150-51. Or as German physicist Max Planck (1858-1947) reportedly said: "Science advances one funeral at a time."

67. For an introduction to Cocceius, see Willem J. van Asselt, *The Federal Theology of Johannes Cocceius* (Leiden: Brill, 2001).

## Evolutionary Theory and the Interpretation of Scripture

Cocceius started to combine his piety and his deferential attitude toward the Bible with the recognition of the correctness of the heliocentric model. And they were able to hold on to this combination. At first, this attracted groups of Cartesians, but, in time, it also appears to have influenced the Voetians. In the end, heliocentrism became *communis opinio* among Christians, and it was generally perceived that the revised interpretations of some biblical texts did not detract from the authority of the Bible. If we ask how this process came about, Vermij's conclusion is telling: "Clearly, good and pious intentions counted more than rigorous ratiocination."[68]

Perhaps, then, this is what we also need today: new Cocceians—that is, Christians who show that it is possible to live faithful lives as Christians while accepting the results of evolutionary science. Already in 1964 Billy Graham gave a good example of this, when he argued:

> I don't think that there's any conflict at all between science today and the Scriptures. I think that we have misinterpreted the Scriptures many times and we've tried to make the Scriptures say things they weren't meant to say. I think that we have made a mistake by thinking the Bible is a scientific book. The Bible is not a book of science. The Bible is a book of Redemption, and of course I accept the Creation story. I believe that God did create the universe. I believe that God created man, and whether it came by an evolutionary process and at a certain point He took this person or being and made him a living soul or not, does not change the fact that God did create man. . . . Whichever way God did it makes no difference as to what man is and man's relationship to God.[69]

Some Christians, however, argue that the parallel drawn here between heliocentrism and evolutionary theory does not hold, because accepting evolutionary theory implies that so many *more* biblical texts than only one or two (as in the case of the heliocentric model of the universe) have to be interpreted differently. True as this may be,[70] we should ask here: Different from what? In most cases the answer will be "different from what we were used to because of

---

68. Vermij, *The Calvinist Copernicans*, 358; cf. also his "Debate on the Motion of the Earth," 621-23.

69. David Frost, *Billy Graham: Personal Thoughts of a Public Man* (Colorado Springs: Victor Books, 1997), 73; the quotation goes back to an interview of Graham by Frost on the BBC2-TV in 1964.

70. However, one should not underestimate the number of biblical texts involved at the time: opponents of heliocentrism sometimes mentioned no less than *ten* biblical

the tradition in which we stand." Here, however, Protestants have a clear advantage as compared to Roman Catholics. No tradition is sacrosanct to them, and especially traditions that try to bind the Word of God by prescribing how it should be interpreted are met with critical suspicion. For the Bible itself should have the final say, and no human traditions should stand in its way. That is why, as the Reformed motto has it, Reformed churches have to return time and again to the Bible, instead of satisfying themselves with fixed interpretations of what it is supposed to mean. In a particularly telling passage, the Belgic Confession puts this point as follows: "Therefore we must not consider human writings—no matter how holy their authors may have been—equal to the divine writings; nor may we put custom, nor the majority, nor age, nor the passage of times or persons, nor councils, decrees, or official decisions above the truth of God, for truth is above everything else."[71]

Therefore, the findings on which evolutionary theory is based may urge us to go back to the Bible, asking ourselves whether our traditional interpretations of various verses really captured their meaning or were partly inspired by a model of the world that has now become obsolete and should no longer be used as a hermeneutical lens.

Yet, while granting this point, one may still doubt whether the parallel between the case of evolutionary theory and that of heliocentrism fully applies. For it seems that evolutionary theory not only forces us to update our exegesis of a number of individual verses but also has much more serious *theological* consequences than adopting heliocentrism had at the time. As we have seen above (§3.4), this is a keen and fair observation indeed. I will therefore use the rest of this book to address it. Let us examine in the next couple of chapters which theological issues have to be revisited as a result of Darwinian evolution, focusing once again (though not exclusively) on the Reformed tradition. In deciding whether or not to take evolutionary theory on board, we need to know as exactly as possible how doing so will affect our theological outlook. Quite understandably, many Christians apply Luke 14:28 to the situation: before deciding to accept evolutionary theory (at least as a *possible* explanation for the natural world's biodiversity), they "sit down and estimate the cost." Let us see how we can do this. Could it be that we end up with a gain rather than a loss?

---

texts that in their view were in conflict with this new theory! (I am indebted to my PhD student Tera Voorwinden for pointing this out to me.)

71. Belgic Confession, art. 7, in *Our Faith: Ecumenical Creeds, Reformed Confessions, and Other Resources* (Grand Rapids: Faith Alive, 2013), 30.

CHAPTER 4

# Animal Suffering and the Goodness of God

> The universality of pain throughout the range of the animal world, reaching back into the distant ages of geology, and involved in the very structure of the animal organism, is without doubt among the most serious problems which the Theist has to face.
>
> —J. R. Illingworth[1]

## 4.1 Gradualism and the Fate of Animals

Imagine that you are skeptical about natural selection acting on random mutations as the principal mechanism that propels the evolutionary process. In your view, the evolution of life on earth might be explained along other lines. Also, imagine that you are not convinced of the theory of common descent; instead, you tend to think that God brought about human beings *de novo*, by a special act of his creative power and wisdom. In that case, you are not faced with the theological problems that will be discussed in the next couple of chapters (5–8). It is possible, however, that you are also impressed by the evidence for the geological timescale and for the gradual appearance of various forms of life over long periods of time. In other words, while you have difficulties with the second and third layers of evolutionary theory discussed in chapter 2, you tend to endorse its first layer on the basis of the fossil record and other (e.g., genomic) indications. That is, you accept gradualism: the view that life on earth gradually developed into ever more complex forms over millions of years.

---

1. J. R. Illingworth, "The Problem of Pain: Its Bearing on Faith in God," in *Lux Mundi: A Series of Studies in the Religion of the Incarnation*, ed. Charles Gore (London: John Murray, 1890), 113.

In that case, you have to face the problem that is the subject matter of this chapter: death and dying seem to be deeply ingrained in God's good creation, having been part and parcel of life on earth long before human beings appeared on the scene. For it is impossible that all organisms populating the earth before the arrival of humankind stayed alive over immense periods of time, all the time (or even only part of the time) giving birth to offspring that would not die either. Given common procreation rates, the earth would have soon been overcrowded with all sorts of herbivorous animals. Thus, unless one wants to invoke a huge change in the biological laws of nature after the arrival of humankind, it seems plausible that death accompanied life from the very days that "the earth [brought] forth living creatures of every kind" (Gen. 1:24). Similarly, unless one wants to assume that only pain-free ways of dying existed during these prehuman ages, it seems equally plausible that enormous amounts of animal suffering must have been going on since animals appeared on earth, and especially since in the course of time they started to develop nervous systems. Animals must have been suffering and dying from predators, illnesses, and natural catastrophes for ages, long before human beings surfaced on earth. As a result of such lethal events, the vast majority of animal species that ever existed have gone extinct already.

And here we meet one of the most egregious theological problems caused by evolutionary theory—a problem not only for so-called theistic evolutionists but also for most old-earth creationists: How could a loving and caring God, who created animals in such a way that "it was good" (Gen. 1:25), allow for so much pain and suffering, death and extinction, during countless ages in the animal world?[2] What is the point of the occurrence of so much "natural evil"?[3] Doesn't God care about animals after all? Or is it perhaps misguided to construe animal death and suffering as "evil" in any recognizable sense? In this chapter, having in mind Reformed theology's *tota Scriptura* principle, we first offer a cursory survey of the place and role of animals in the Bible (§4.2). In light of this, we then address the problem of animal suffering and death

---

2. In §6.6 we will pick up the question of what evolution means for the origin and significance of *human* death.

3. "The category of *natural evil* covers the physical pain and suffering that result from either impersonal forces or human actions"; Michael Peterson et al., *Reason and Religious Belief: An Introduction in the Philosophy of Religion*, 5th ed. (New York: Oxford University Press, 2012), 179–80. Examples mentioned include "the terrible pain and death caused by events like floods, fires, and famines as well as . . . diseases . . . defects, deformities and disabilities" (118); pain and death as a result of predation would be another example.

throughout evolutionary history from a theological point of view. First, we ask whether we can indeed ascribe pain and suffering to animals (§4.3). If so, is all animal suffering, death, and starvation perhaps to be explained as a result of human sin (§4.4)? Alternatively, should we ascribe it directly to God's will and purposes (§4.5)? Or should we attribute it to nonhuman evil forces that somehow infiltrated God's work (§4.6)? Finally, we evaluate these various options from a Reformed perspective (§4.7). We will venture the thesis that the problem of evolutionary suffering and evil is a specific instantiation of the more general problem of natural evil that we have to face anyhow; as a result, most of the defenses or theodicies that we use for natural evil will also apply to evolutionary evil. If this is true, evolutionary theory does not in and of itself cause a specific theodical problem.

### 4.2 The Appreciation of Animals in the Bible

In the history of Christian theology, the relationship between God and animals (or more broadly, the nonhuman part of creation) has often been neglected.[4] Influential interpretations of the creation story (especially of Gen. 1:28 and 2:19) and Psalm 8 defined animals exclusively in terms of their instrumental value for the well-being of human beings. As Reformed theologian Lukas Visscher has helpfully pointed out, such views can hardly be labeled anthropocentric but should more properly be called anthropomonistic.[5] That the Bible is anthropocentric in the sense that human beings fulfill a unique task in creation and are of central importance in the salvific purposes of God cannot and should not be denied. This is not to say, however, that *only* human beings play a role in God's plan, and animals—if at all—only for humans' sake. This kind of anthropomonism is clearly at odds with many strands of biblical literature. In more recent decades, theological discussions have highlighted the pervasive presence of animals throughout the Bible, also in contexts and narrative plots where they are not directly connected to humans and their well-being.

---

4. By "animals" I mean nonhuman animals. I won't add this modifier all the time (nor retreat to such equivalents as "beasts"), since even though, strictly speaking, human beings are *animales* as well, the connotation of animals as referring to nonhumans continues to be strongly embedded in everyday language and is clearly functional.

5. Lukas Visscher, "Listening to Creation Groaning: A Survey of Main Themes in Creation Theology," in *Listening to Creation Groaning*, ed. Lukas Visscher (Geneva: Centre International Réformé John Knox, 2004), 21–22.

A clear example is to be found in Psalm 104, a psalm that teems with animals of all sorts. Interestingly, animals that are tame from a human point of view are mentioned interchangeably with animals that can be threatening to us humans. Wild animals and wild asses (v. 11), the birds of the air (v. 12), the cattle (v. 14), the stork (v. 17), the wild goats and the coneys (or rock badgers, v. 18), all the animals of the forest (v. 20), the young lions that roar for their prey (v. 21), living things both small and great (v. 25), and, yes, even the dreaded Leviathan (v. 26)—all have been made by God in his wisdom (v. 24). Human beings are mentioned only in passing (v. 23). Whereas from an anthropocentric perspective it would make sense to give humans some special place and to divide animals into those that are harmful and those that are helpful or neutral to them, from the psalmist's theocentric perspective, this is less self-evident. Rather, from God's point of view all animals alike praise and glorify him by the sheer fact of their wonderful existence. It is God who created them all and keeps them alive, not because of their usefulness to human beings but apparently because God wanted to have a relationship with them as well. Psalm 104 gives us humans a modest place amid the enormous variety of living beings and displays the variegated ways in which these other beings can flourish to the honor of God, each in its own way testifying to his greatness and wisdom. More generally, many have argued that such a positive evaluation of animal life is typical of the entire Bible.[6] We will take a snapshot to illustrate this point.

6. Cf., e.g., Andrew Linzey, *Animal Theology* (London: SCM, 1994); Charles Birch and Lukas Visscher, *Living with Animals: The Community of God's Creatures* (Geneva: WCC Publications, 1997); Andrew Linzey and Dorothy Yamamoto, eds., *Animals on the Agenda: Questions about Animals for Theology and Ethics* (London: SCM, 1998); Stephen Webb, *On God and Dogs: A Christian Theology of Compassion for Animals* (New York: Oxford University Press, 1998); Michael J. Gilmour, *Eden's Other Residents: The Bible and Animals* (Eugene, OR: Cascade, 2014); Elizabeth A. Johnson, *Ask the Beasts: Darwin and the God of Love* (London: Bloomsbury, 2014). A nuanced account of the biblical data, also acknowledging texts that seem to disregard the value of animals, is provided by Robert N. Wennberg, *God, Humans, and Animals: An Invitation to Enlarge Our Moral Universe* (Grand Rapids: Eerdmans, 2003), 285–308. By now animals have also entered surveys of Christian doctrine; cf., e.g., David Fergusson, "Creation," in *The Oxford Handbook of Systematic Theology*, ed. John Webster, Kathryn Tanner, and Iain Torrance (Oxford: Oxford University Press, 2007), 84–86; Cornelis van der Kooi and Gijsbert van den Brink, *Christian Dogmatics: An Introduction* (Grand Rapids: Eerdmans, 2017), 228–33. Intensifying the tendency to rehabilitate animals theologically, David Clough, *On Animals*, vol. 1, *Systematic Theology* (London: Bloomsbury, 2012), argues that the

## Animal Suffering and the Goodness of God

The Torah speaks of the relative independence of land animals, in that, in the creation story of Genesis 1, they are created prior to us humans on the very same sixth day. The first human beings did not have animals as their food—the eating of animals began after the state of peace with them was broken by sin.[7] The story of Noah (Gen. 6-8) shows how God provides for the preservation of all animal species that were threatened with extinction as a result of the flood. In the wake of the flood, a new covenant was made not only with Noah and his offspring but also with "every living creature of all flesh that is on the earth" (9:16). So the very category of a covenant, which is prominent in the history of salvation, could be applied to animals as well. In the Mosaic law, special attention is given to livestock; lest they be abused, they should rest on the Sabbath day like their human possessors (Exod. 20:10; 23:12; Deut. 5:14). Also, "when you see the donkey of one who hates you lying under its burden and you would hold back from setting it free, you must help to set it free" (Exod. 23:5). The interests (or, if one wishes, "rights") of animals were further protected by rules forbidding the Israelites to plow with a donkey and an ox together (Deut. 22:10)—presumably because that would cause them a lot of physical pain—and to harvest their land every seventh year in order to leave its fruits for every living being who needed them, including "the wild animals in your land" (Lev. 25:7; cf. Exod. 23:11). According to Robert Wennberg, even the practice of animal sacrifice ordered in the Torah need not be seen as degrading animals; that animals could make reparation for human sin in a sense shows their high value. Accordingly, there is a clear sense of reverence for the blood of animals in the Old Testament (Lev. 17:10-16).[8]

In the other parts of the Old Testament, we observe an equally positive appraisal of the value of animals. God is praised because he takes care of the beasts, preserving "humans and animals alike" (Ps. 36:6), "satisfying the desire of every living thing" (Ps. 145:16), and thus also giving the animals their food (Ps. 147:9). Especially telling are, in addition to Psalm 104, the final chapters in the book of Job (39-41), where God's creative and sustaining relationship to even the most monstrous animals is professed. In the prophetic literature, we find the famous Isaianic visions of a peaceful world in which predators will live in harmony not only with human beings but also with their usual

---

purely instrumental view of animals stems mostly from nonbiblical Greek sources that were uncritically adopted into the Jewish-Christian tradition.

7. Ekkehard Starke, "Animals," in *The Encyclopedia of Christianity*, vol. 1, ed. E. Fahlbusch et al. (Grand Rapids: Eerdmans, 1999), 62.

8. Cf. Wennberg, *God, Humans, and Animals*, 297; for a balanced account of animals in religious rituals, see also Gilmour, *Eden's Other Residents*, 92-112.

prey (Isa. 11:6-9 and 65:25). We find the promise of a new covenant "with the wild animals, the birds of the air, and the creeping things of the ground" (Hos. 2:18). And we come across the climactic closing of the book of Jonah, where God's grace extends not only to the human inhabitants of Nineveh but also to its "many animals" (Jon. 4:11). God's benevolence toward animals is also reflected in what is seen as the proper human attitude toward domestic animals: "The righteous know the needs of their animals" (Prov. 12:10).

In the New Testament, in Jesus's teachings the animals are never far away. In fact, Jesus wasn't far away himself from the wild animals at the beginning of his earthly ministry, when he stayed with them in the desert (Mark 1:13).[9] In the Sermon on the Mount, Jesus uses birds as an example of proper human conduct (Matt. 6:26) because they do not plan and calculate their lives in advance (which ironically became a sign of human superiority in later philosophical and theological anthropology!),[10] but spontaneously seem to trust God without worrying. Later on in the Gospel of Matthew, some specific animals are praised for their prudence and innocence (10:16). Also, we hear that not even a bird as common as a sparrow will fall to the ground apart from the involvement of the heavenly Father (10:29).

That this involvement is not necessarily to be identified with God's will (an identification that was often made on the basis of a loose and incorrect translation of this text) is suggested by the story about the herd of swine drowning in the Sea of Galilee (Mark 5:11-13 par.). That is, according to a common explanation, it is the demons that make the herd rush into the sea, not God or Jesus.[11] Is there a mysterious connection between animal suffering and death on the one hand and satanic powers on the other? We will pick up this suggestion later in this chapter (§4.6). In any case, Jesus is only indirectly involved here, as the one who allows the demons to enter (not to drown) the swine. In a fascinating reading of the passage, Michael Gilmour has suggested that it is the pigs, cooperating with Jesus in an act of self-sacrifice, that plunge the demons into the sea—the place where in the biblical imagery the primordial chaotic forces are at home—rather than the other way around.[12]

---

9. Cf. Richard Bauckham, *Living with Other Creatures: Green Exegesis and Theology* (Waco, TX: Baylor University Press, 2011), 111-32.

10. Starke, "Animals," 62. In the Old Testament as well, animals had been set as an example to God's human covenant partners; see, e.g., Isa. 1:3; Jer. 8:7.

11. Richard Bauckham, "Jesus and Animals I. What Did He Teach?," in Linzey and Yamamoto, *Animals on the Agenda*, 47 (33-48).

12. Gilmour, *Eden's Other Residents*, 86-87.

*Animal Suffering and the Goodness of God*

As to the New Testament letters, in Romans 8:19-22, a passage that has received much attention in recent theological and eco-theological reflection,[13] animals are presumably included in the whole of creation that is now groaning but "will be set free from its bondage to decay" (8:21, 22)—as they are also included in Colossians 1:20. Examples of a more indifferent attitude toward animals are not entirely lacking, however. The most famous case is Paul's quite perplexing spiritualizing interpretation of Deuteronomy 25:4 in 1 Corinthians 9:9-10: "Is it for oxen that God is concerned? Or does he not speak entirely for our sake?" This interpretation is echoed in 1 Timothy 5:17-18, where Paul applies the command not to "muzzle an ox while it is treading out the grain" to Christian missionaries who should receive a living while proclaiming the gospel. Indeed, for those who want to build their case for animal compassion on the Bible, Paul's passing remark in 1 Corinthians "presents a considerable impediment."[14] Finally, the book of Revelation is replete with animals, both bad ones and good ones, which apparently play an important role in the apocalyptic end of time and beyond. Though such passages contain a lot of literary symbolism,[15] it is nevertheless significant that animals figure so prominently in them, not only in the present era but also in portrayals of the eschatological future.

This summary of some of the most conspicuous biblical texts on animals shows that the Bible mostly portrays animals in a positive way, as the objects of God's ongoing care and concern. Therefore, we cannot but be bewildered by the enormous amount of suffering that animals apparently have had—and still have—to undergo in the natural world. After all, as indicated, Genesis 1 ends its rendering of the creation of animals with the comment that "God saw that it was good" (Gen. 1:25). How can a natural world that contains such widespread "waste of life" as well as often horrible forms of suffering and predation, death and extinction, be called good? It is to this problem of animal suffering that

---

13. Cf., e.g., Christopher Southgate, *The Groaning of Creation: God, Evolution, and the Problem of Evil* (Louisville: Westminster John Knox, 2008), passim.

14. Gilmour, *Eden's Other Residents*, 28; given the strong language Paul uses, Gilmour's attempt to read the text in a more animal-friendly way ("Is God concerned *only* for oxen?") is hardly convincing (28-36, 35). It seems more reasonable to acknowledge that the valuation of animals in the Bible is not entirely uniform.

15. As is clear from the fact that the book of Isaiah can both envisage the messianic kingdom as a place where "the lion shall eat straw like the ox" (65:25) and as a place where "no lion shall be" (35:9); I owe this observation to Denis Alexander's excellent *Creation or Evolution: Do We Have to Choose?* 2nd rev. ed. (Oxford: Monarch Books, 2014), 381.

we turn now. Given that in the Jewish and Christian traditions the Creator of heaven and earth is maximally good, wise, and powerful, how is it possible that billions of animals have had to suffer from parasites and predators and die from starvation and diseases throughout the ages—often in apparently gruesome ways? How about insects that eat their mates, cats that literally play with mice before killing them, sharks that rip their prey to shreds, and polar bear mothers that devour their own young? How, to cite Darwin's famous example, about the Ichneumonidae, a wasp family whose larvae feed themselves from the inside out with the bodies of their host caterpillars, leaving the vital organs intact until they have consumed all the rest?[16]

### 4.3 Is There a Problem at All? Neo-Cartesianism

Before we discuss these questions, however, we should first consider whether the situation is really as drastic as suggested above. Perhaps we should be careful when considering phenomena like animal pain, death, and extinction as *evils*. For isn't this a grossly anthropomorphic way of thinking? When it comes to extinction, for example, it might be argued that there are no individual animals that suffer from it; it is just that we *humans* may regret the loss of so many species. Why shouldn't we be satisfied with the view that God has apparently allotted a limited span of time to each of the species he created? And when it comes to animal *death*, there is perhaps no need to see this as an evil either. According to Reformed theologian Henri Blocher, the ancient Epicurean arguments against the fear of death apply here:

> Atoms that were combined for a time now part company—so what? Epicurean reasoning has cogency for animals, although it must be rejected in the case

---

16. "I cannot persuade myself that a beneficent and omnipotent God would have designedly created the Ichneumonidae with the express intention of their feeding within the living bodies of caterpillars, or that a cat should play with mice." Darwin to Asa Gray, May 22, 1860, DCP no. 2814; see www.darwinproject.ac.uk. The sentence should be read in context, however—less well known are the subsequent sentences: "On the other hand I cannot anyhow be contented to view this wonderful universe and especially the nature of man, and to conclude that everything is the result of brute force. I am inclined to look at everything as resulting from designed laws." Cf. Denis O. Lamoureux, "Darwinian Theological Insights: Evolutionary Theodicy and Evolutionary Psychology," *Faith and Thought* 57 (October 2014): 6 (3–20).

of human death because of the human difference, the transcendence of the *imago Dei*. . . . The death of animals that are our close companions make us sad, but this is a subjective projection and must be kept within bounds.[17]

Blocher is a little more hesitant about animal *pain and suffering*, but he ponders that in this as well, our late modern Western sensitivity might not be the most reliable guide. Scripture doesn't always seem to care much about the suffering of animals (Blocher points to 1 Corinthians 9:9 in this connection—the Pauline text we highlighted above). Moreover, where should we draw the line? Should we also feel concern "for the 'suffering' of rats, flies or lice"? And finally, it is far from self-evident that we should attribute "selves" to animals: perhaps even the higher animals do not possess self-consciousness to such an extent that they are able to realize that what they experience is pain. "None can say 'I.'"[18]

Without mentioning it, Blocher is connecting here with what has been called the neo-Cartesian denial of animal pain. This denial is named after Descartes, who reputedly held that animals do not experience pain at all, since, lacking rational minds, they are unconscious automata rather than sentient beings. *Neo*-Cartesians usually do not go as far as that. Rather, while acknowledging that animals as a matter of fact are sentient beings, they deny that animals can suffer, because they lack something else, namely, self-consciousness or self-awareness.[19] As a result, animals can at best go through momentary sentient states of pain, but they cannot realize that they do so since there is no "I," no self-conscious reflective subject, to realize anything at all. Therefore, presumably, their momentary physical sensations of pain are not morally relevant. C. S. Lewis, one of the first to explore and (tentatively) defend this view with regard to the category of "merely sentient" animals, put the point this way: the nervous system of such animals "delivers all the *letters* A, P, N, I, but since they

---

17. Henri Blocher, "The Theology of the Fall and the Origins of Evil," in *Darwin, Creation, and the Fall: Theological Challenges*, ed. Robert James Berry and Thomas A. Noble (Nottingham: Apollos, 2009), 167.

18. Blocher, "Theology of the Fall," 168.

19. In fact, the neo-Cartesians have several slightly different options here, as shown by Trent Dougherty, *The Problem of Animal Pain: A Theodicy for All Creatures Great and Small* (New York: Palgrave Macmillan, 2014), 65. I leave out the subtle varieties here, since they do not basically affect the argument. The question whether or not animals (and especially primates) have self-awareness has been hotly debated ever since Gordon Gallup developed his so-called mirror test; cf. Gordon G. Gallup Jr., "Chimpanzees: Self-Recognition," *Science* 167, no. 3914 (1970): 86–87.

cannot read they never build it up into the word PAIN."²⁰ The fact that many animals seem to learn from their past experiences, can anticipate pain, avoid situations in which they might feel pain, etc., does not show that they have self-consciousness; it shows only that they behave *as if* they learn from what they recollect (i.e., as if they remember their past), whereas in fact they might just as well be responding instinctively to potentially damaging stimuli at a purely physiological level—as we do when we blink our eyes at the approach of an object.

From a philosophical point of view, it is not so easy as one might think to show that the neo-Cartesian approach to animal suffering is incorrect.²¹ The view that nonhuman animals do not feel pain cannot simply be dismissed out of hand as "absurd," as Peter van Inwagen does.²² Still, a couple of observations make it highly implausible.²³ First, animals of many species display nonreflexive behavior in avoiding the return of painful sensations to which they have previously been exposed. This so-called pain guarding—for example, limping, standing on one leg—continues for some time after the damaging stimulus has ceased. Admittedly, this does not *show* that such animals have

---

20. C. S. Lewis, *The Problem of Pain* (New York: Macmillan, 1962 [1943]), 138. In response to criticisms by C. E. M. Joad, Lewis maintained his view but emphasized that his chapter on animal suffering in *The Problem of Pain* was "confessedly speculative" and "guesswork about Beasts"; cf. C. E. M. Joad and C. S. Lewis, "The Pains of Animals: A Problem in Theology," *Month* 189 (1950): 95-104; reprinted in C. S. Lewis, *God in the Dock: Essays on Theology and Ethics* (Grand Rapids: Eerdmans, 1972), 161-71. The neo-Cartesian response has been refined and elaborated by Peter Harrison, "Theodicy and Animal Pain," *Philosophy* 64 (1989): 72-92. For careful critical discussions, see Michael Murray, *Nature Red in Tooth and Claw: Theism and the Problem of Animal Suffering* (Oxford: Oxford University Press, 2008), 43-49, and Dougherty, *Problem of Animal Pain*, 56-95.

21. Murray concludes his extensive discussion of neo-Cartesianism (*Nature*, 41-72) by stating that although "few will find the neo-Cartesian position to be compelling or even believable," "the evidence against the neo-Cartesian position is quite weak" (71). Likewise, Dougherty, *Problem of Animal Pain*, acknowledges that he can't "rule out" neo-Cartesianism, though he tries "to drive its probability down for the reader" as much as he can (69).

22. Peter van Inwagen, *The Problem of Evil* (Oxford: Oxford University Press, 2006), 131.

23. Cf. for this list Robert Francescotti, "The Problem of Animal Pain and Suffering," in *The Blackwell Companion to the Problem of Evil*, ed. Justin P. McBrayer and Daniel Howard-Snyder (Chichester: Wiley and Sons, 2013), 117-21. One might also add the ethical consideration that if neo-Cartesianism were true, it is difficult to see why we should be opposed to (and have laws against) animal abuse and maltreatment (cf. Wennberg, *God, Humans, and Animals*, 313-14).

self-consciousness, but it cannot be easily ruled out either. Second, the neural similarities between humans and other mammalian species (including the presence of a prefrontal cortex) offer further confirmation that our spontaneous inclination to conclude from certain types of animal behavior the existence of animal suffering may be correct. Third, the level of self-consciousness needed to experience not only damaging sensory stimuli but also real suffering need not be very high. Even invertebrates may experience conscious pain. And animals that lack self-consciousness over time may still experience momentary sensations of pain that are morally relevant. Fourth, recent research in ethology and primatology has brought to light that animals are much closer to us humans than we used to think, in that they exhibit a wide range of emotions, including happiness, joy, empathy, sympathy, fear, grief, depression, jealousy, etc.[24]

In light of this, it doesn't seem unduly anthropomorphic to attribute not only physical pain to them but also other instances of suffering.[25] To be sure, none of the considerations mentioned here (nor their cumulative force) is decisive. Despite all efforts to understand animals, they continue to be strange to us, at least to some extent, since we cannot feel what it is like to be one of them. We cannot say for sure in which ways their experiences of pain and suffering are comparable or incomparable to ours.[26] Yet, the most reasonable position to take here is to suppose that there is a continuum, with animals having increased capacities of experiencing conscious pain to the extent that their neuroanatomy and neurophysiology more closely resemble ours.[27] Al-

---

24. See, e.g., Marc Bekoff, *The Emotional Life of Animals: A Leading Scientist Explores Animal Joy, Sorrow, and Empathy—and Why They Matter* (Novato, CA: New World Library, 2007). Cf. Bekoff's plea for "critical anthropomorphism" in his *Why Dogs Hump and Bees Get Depressed: The Fascinating Science of Animal Intelligence, Emotions, Friendship, and Conservation* (Novato, CA: New World Library, 2013), 111-15.

25. For a summative study on all sorts of animal suffering, see Neville G. Gregory, *Physiology and Behaviour of Animal Suffering* (Oxford: Blackwell Science, 2004); as is clear from this book, "remarkable advances in comparative neuroanatomy, physiology and behavioral sciences have provided strong evidence that the capacity for subjective experience of unpleasant (and pleasant) feelings is not limited to humans only" (ix). I am grateful to Dr Jacques Schenderling for this reference.

26. Note, however, that it is often equally difficult to estimate the amount of pain a fellow human being suffers; those who cry and scream most are not by definition the ones that suffer most.

27. Cf. Lewis, "The Pains of Animals," in *God in the Dock*, 168: "It will hardly be denied that the more coherently conscious the subject is, the more pity and indignation its pains deserve."

though it is difficult to know where exactly to draw the line, it is irresponsible to take for granted that no other species than we humans can consciously and continuously experience physical pain and other forms of suffering. In 2012 an interdisciplinary group of scientists even officially declared that many animals, like birds and mammals, but also octopuses, are in possession of the neurological substrates that generate consciousness.[28] In brief, the empirical evidence for animal suffering seems much stronger than not only Descartes but even C. S. Lewis assumed.

Of course, if someone finds the neo-Cartesian position convincing, she could skip the rest of this chapter, since the problem of evolutionary evil does not arise at all. It seems to me, however, that denying the existence of animal suffering is too easy a way out—a way that the Christian tradition has rarely taken and that we won't take either. Therefore, we have to examine other ways of dealing with the problem of animal suffering throughout the course of evolutionary history. Could it be that somehow we humans are to blame for it?

### 4.4 The Cosmic-Fall Theory: Animal Suffering and Human Sin

If animal suffering is real, then how might it be understood from a theological point of view? This question has bothered Christians since long before Darwin. Evolutionary theory, or its basic idea of a progressive appearance of life-forms on earth, did not present Christian theology with a *new* question in this connection. For animal suffering and dying have been observed as being widespread ever since human civilization. And, of course, even those who reject evolutionary theory altogether can observe that animal suffering and death still occur on an immense scale. As Richard Dawkins writes: "The total amount of suffering per year in the natural world is beyond all decent contemplation. During the minute it takes me to compose this sentence, thousands of animals are being eaten alive; others are running for their lives, whimpering with fear; others are being slowly devoured from within by rasping parasites; thousands of all kinds are dying of starvation, thirst and disease."[29]

---

28. See "The Cambridge Declaration on Consciousness," accessed April 3, 2019, http://fcmconference.org/img/CambridgeDeclarationOnConsciousness.pdf. I owe this reference to Eva van Urk, MA.

29. Richard Dawkins, *River out of Eden: A Darwinian View of Life* (New York: HarperCollins, 1995), 131–32.

Although these phenomena were not discovered as a result of evolutionary theory, evolutionary theory did cast doubt on the most influential theological explanation of animal suffering that had been given for many centuries: animal pain and death are the consequences of *human sin*.[30] This explanation is sometimes labeled "the cosmic-fall theory," that is, after the first human beings lapsed into sin, and as a result of that fact, God's originally perfect creation was distorted to such an extent that the entire biosphere fell into disarray. Thus not only human beings but also, in their slipstream, all other forms of life on planet earth regressed into a miserable state.[31] As a result, animals gradually came to suffer from predation and other forms of natural evil as we know them today.

Let us discuss this cosmic-fall theory in more detail. Since it served as the most dominant explanation for animal suffering for a great many centuries, presumably it must have some very strong points. Indeed, it is instructive to sum up some of these strengths. First, the cosmic-fall theory forcefully underlines the strong interconnectedness of all forms of life on earth. Humans, animals, plants, the environment—it is an encompassing network of mutually dependent forms of life. Second, and more specifically, the theory highlights the disconcerting fact that many wrong and shortsighted choices we humans make affect not only our own species but also the natural world. Indeed, it is because of interventions by us humans throughout the so-called Anthropocene that many animals have been suffering and dying and entire species have become extinct. Third, the cosmic-fall theory keeps God "one step removed" from the production of animal pain.[32] To be sure, if we assume (as Christian theology has usually done) that God is omnipotent,[33] God must somehow have permitted animal suffering, at least to the extent that it actually takes place. It does not follow, however, that God *wanted* animal suffering and death to occur in the first place. The cosmic-fall theory highlights that God is

---

30. See, for recent endorsements of this traditional explanation, several contributions in Norman C. Nevin, ed., *Should Christians Embrace Evolution? Biblical and Scientific Responses* (Phillipsburg, NJ: P&R, 2011), e.g., 23, 66–67, 79–83.

31. One may take the word "cosmic" in "cosmic-fall theory" even more literally by suggesting that indeed the entire cosmos fell into a suboptimal state as a result of human sin, thus counting, for example, deadly meteorite impacts among the effects of sin. However, in what follows we will concentrate on the earthly biosphere.

32. Wennberg, *God, Humans, and Animals*, 332.

33. Cf. Gijsbert van den Brink, *Almighty God: A Study of the Doctrine of Divine Omnipotence* (Kampen: Kok Pharos, 1993).

not the direct cause of animal suffering and death, since it is we humans who are responsible for their entrance into the natural world. In this sense, the theory neatly coheres with the traditional caveat that God is not "the author of evil."[34] Fourth, given that the cosmic-fall theory is usually extended to the fate of *humans*, it keeps God one step away from human suffering as well. When a child dies from cancer, it is not God who is to blame in any direct sense; rather, that child has its own share in paying the price for our general human sinfulness. For, generally speaking, our death is "the wages of sin" (Rom. 6:23).

On the other hand, the cosmic-fall theory has a number of significant drawbacks. First, it is at odds with common notions of justice; for, clearly, it seems unfair that some humans have a much larger share in bearing the consequences of sin than others. And as to animals, it seems equally immoral that they have to suffer so badly from the consequences of *human* wrongdoing. Thus, even if God only permitted animal pain rather than intentionally willing it, the problem of natural evil still stands: How can a good God allow innocent creatures of all sorts to be so drastically affected by the wrong choices of one (and only one) other species? Would such a God "pass the moral test"?[35] Second, the cosmic-fall theory has to presuppose that all animal species were originally herbivores and that many of them developed into carnivores and became predators after the human fall into sin. It is hard, however, to see how such a transformation could have taken place—and even harder to imagine how the species involved could still be the same species as before. Ironically, young-earth creationists have to appeal to incredibly rapid forms of evolution to explain how efficient predatory structures (including intricate mechanisms of attack and defense) could emerge in herbivorous animals within several thousand years at a maximum.[36] Carnivorous digestive systems are so different from herbivore ones that the latter cannot just "degenerate" into

---

34. Because they held to a strong view of divine providence, Reformed theologians in particular were traditionally at pains to point out that evil is not in itself caused by God. See, e.g., Dolf te Velde, ed., *Synopsis Purioris Theologiae / Synopsis of a Purer Theology*, vol. 1 (Leiden: Brill, 2015), disputation XI, theses 20-23 (276-79).

35. Cf. Southgate, *The Groaning of Creation*, 146n48.

36. Alexander, *Creation or Evolution*, 174-75; cf. for some creationist ways to deal with the problem of animal death, Kenneth D. Keathley and Mark F. Rooker, *40 Questions about Creation and Evolution* (Grand Rapids: Kregel Academic, 2014), 255-62 (Question 26: "Was There Animal Death before the Fall?"). Some are committed to a young earth precisely because they want to hold on to God's introducing death on earth only after the human fall. The price they pay for this move is that "it is then impossible to accept the *prima facie* reading of the fossils" (258).

the former.³⁷ Therefore, such an overhaul would amount to a comprehensive re-creation after the Fall. Given that this must have been a major event, it is amazing that the Bible does not tell us anything about it.

Even if we were to invoke special divine intervention to solve these difficulties, we would face a third problem—and this brings us back to the heart of our topic. For it is hard to reconcile the cosmic-fall theory with what contemporary science tells us about the history of the natural world. Already in the course of the seventeenth century (i.e., almost two centuries prior to Darwin), the discovery of fossils in apparently ancient geological formations suggested that animals must have been suffering and dying long before human beings appeared on the scene. How could that be part of God's response to human sin, given that this sin still had to be committed? It is especially this discovery that makes it hard to hold on to the cosmic-fall theory in its traditional form. Therefore, one of the first responses to the discovery of fossils was to deny their authenticity: perhaps some unknown geological processes had produced animal-like structures in seemingly old strata—or perhaps even *God* had done so in order to mislead conceited scientists, or for some other reason. Alternatively, perhaps all fossils (or most of them) were the result of Noah's flood and therefore postdate the human fall into sin.³⁸ However, in time most of those who examined the ever-increasing fossil discoveries came to the conclusion that such theories were contrived, since the geological and paleontological evidence did make clear that animals had been suffering and dying—and even entire species had gone extinct—from ancient times onward.

The situation turned out to be even worse: the death of animals has always been the *condition* for the emergence of new life and the sustenance of existing life-forms, so that animal suffering and death are not only inextricably linked with, but even an integral part of, the whole fabric of life. As Christopher Southgate posits, "The very processes by which the created world gives rise to the values of greater complexity, beauty and diversity also give rise to the disvalues

---

37. Mark S. Whorton, *Peril in Paradise: Theology, Science, and the Age of the Earth* (Waynesboro, GA: Authentic Media, 2005), 124.

38. This theory, as popularized by John Whitcomb and Henry Morris in their book *The Genesis Flood: The Biblical Record and Its Scientific Implications* (Philadelphia: P&R, 1961), inaugurated the modern young-earth creationism movement. See more generally on the reception of the Noachian flood story throughout the ages, Norman Cohn, *Noah's Flood: The Genesis Story in Western Thought* (New Haven: Yale University Press, 1996).

of predation, suffering, and violent and selfish behaviour."[39] Even though we may question whether these processes can be fully explained by (or described in terms of) Darwinian natural selection,[40] they definitely predate the appearance of *Homo sapiens* on the scene by many millions of years.[41] Thus, the idea that human sin is responsible for such phenomena in the natural world is highly incongruous, to say the least, given the scientific picture of the development of the biosphere. Arthur Peacocke put this point more straightforwardly when he wrote: "Biological death can no longer be regarded as in any way the *consequence* of anything human beings might have been supposed to have done in the past."[42]

But doesn't this view go against the grain of the Jewish and Christian Scriptures? That remains to be seen. Indeed, some have argued that, upon closer scrutiny, the Bible does not suggest a cosmic fall at all. Old Testament scholar John Bimson, for example, contends that "most of the Bible is completely silent on the matter, and the doctrine [of the fall of the natural world] actually depends on the interpretation of a few key texts"—texts that he is keen to interpret differently.[43] Indeed, in the Old Testament, God himself is given credit for having created predatory animals all along (Ps. 104:19-21; Job 38-40; see also Gen. 1:21 on the "great sea monsters," which, elsewhere, e.g., in Job 7:12 and Ps. 74:13, turn out to be dangerous).[44] The idea of a cosmic fall gained much more traction and was elaborated in much more detail in various stages of *church history*. This process started already in pseudepigraphical texts of Second Temple Judaism.[45] From this later perspective a full-fledged cosmic-fall

---

39. Southgate, *The Groaning of Creation*, 29.

40. Cf. §2.4 above.

41. Already in 1839—so, twenty years before Darwin's *Origin of Species*—English theologian and geologist William Buckland (1774-1856) preached a famous sermon in which he argued that the sentence of death pronounced at the fall of humankind did not include the realm of animals (where death was already prevalent) but was restricted to the human race. Cf. William Buckland, *An Enquiry Whether the Sentence of Death Pronounced at the Fall of Man Included the Whole Animal Creation or Was Restricted to the Human Race* (London: J. Murray, 1839).

42. Arthur Peacocke, *Theology for a Scientific Age: Being and Becoming—Natural, Divine, and Human*, enlarged ed. (Minneapolis: Fortress, 1993), 222.

43. John J. Bimson, "Reconsidering a 'Cosmic Fall,'" *Science and Christian Belief* 18 (2006): 63-81 (quote on 67).

44. Cf. David Snoke, *A Biblical Case for an Old Earth* (Grand Rapids: Baker Books, 2006), 79-86; cf. David Snoke, "Why Were Dangerous Animals Created?" *Perspectives on Science and Christian Faith* 56 (2004): 117-25 (esp. 119-21).

45. Especially telling here is The Life of Adam and Eve; cf. *The Apocrypha and*

theory was often read back into the Scriptures (especially the Old Testament) in a way that goes beyond what is actually implied by them. Still, although F. F. Bruce—writing in 1963—surely overstated the issue when he claimed that "[the] doctrine of the cosmic fall is implicit in the biblical record from Genesis 3 . . . to Revelation 22,"[46] it is at least intimated in various layers of the biblical texts. As Michael Murray writes in his thoroughgoing philosophical study on animal suffering, "There is little doubt that the Hebrew and Christian Scriptures *provide encouragement* to accord . . . greater explanatory scope to the Fall" than seeing its consequences only in terms of human guilt and proclivity to sin.[47]

Especially noteworthy is the suggestion that the wrongdoing of the first human beings had consequences for their natural environment as well. The idea that the Fall ushered in a curse that affected the natural order is prevalent in the Fall narrative in Genesis 3. Even when we follow Bimson in translating the key text here (in Gen. 3:17; cf. 5:29) as "Cursed is the ground in regard to you" rather than "Cursed is the ground because of you," the reason for the curse clearly must be found in the disobedience of the first couple. A similar connection between human sin and a cursed earth (or land) is made in Isaiah 24:

> The earth dries up and withers,
>     the world languishes and withers;
>     the heavens languish together with the earth.
> The earth lies polluted
>     under its inhabitants;
> for they have transgressed laws,
>     violated the statutes,
>     broken the everlasting covenant.
> Therefore a curse devours the earth,
>     and its inhabitants suffer for their guilt. (Isa. 24:4-6)

---

*Pseudepigrapha of the Old Testament*, trans. R. H. Charles (Oxford: Clarendon, 1913). In addition to this, Bimson ("Reconsidering a 'Cosmic Fall,'" 63-64) also points to some of the church fathers (e.g., Theophilus of Antioch, *To Autolycus* 2.17) and further suggests that the idea of a cosmic fall "was particularly influential in the sixteenth and seventeenth centuries," referring here to Milton's *Paradise Lost* as an example. For the reception of Gen. 2-3, see also Stephen Greenblatt, *The Rise and Fall of Adam and Eve* (New York: Norton, 2017).

46. F. F. Bruce, *The Epistle of Paul to the Romans*, Tyndale New Testament Commentaries (London: Tyndale Press, 1963), 169.

47. Murray, *Nature*, 74 (emphasis added).

We need not deny that human sin leads to a cursed earth, however, in order to reject the cosmic-fall theory. For first, the texts cited above do not imply that human sin has, either spontaneously or through a miraculous intervention by God, led to distorted physical or biological laws and structures.[48] They only imply that at times (Isa. 24:4-6 is not a general statement but refers to the land of Israel at a certain stage) existing laws may start to work against human sinners in an unusual way as a result of divine punishment. Second, the reference in these texts is to the flora ("thorns and thistles," etc.) rather than to the animal world. Where the animal world is involved, such as in Genesis 3:14 and in eschatological passages like Isaiah 11:6-9 and 65:25, there is no indication that the suffering, death, and extinction of animals in the present dispensation are seen as the result of human sin.[49] And third, in places where death is seen as a result of human sin—as in Genesis 2-3, Romans 5:12, and 1 Corinthians 15:21—it is clear from the context that it was *human* death that the authors had in mind.[50] When we apply such texts to the animal world as well, we are extending their meaning beyond what they intend to say within their original context.

A similar tradition of overinterpreting can be observed for the famous passage on the groaning of creation in Romans 8:19-22. Michael Murray argues that Paul here "extends the scope of the effects of the Fall from the cursed ground . . . to the entirety of creation."[51] Indeed, the passage has often been interpreted in this way. The reference to the Fall is not in the text, however; it is rather read into it from our (Augustinian and Reformed?) theological perspective. As David Snoke argues, the phrase "until now" (8:22) suggests that Paul has in mind the time "from the beginning up to now," not from some intermediate point of time up to the present.[52] Also, the metaphor of birth pangs that Paul brings into play refers much more naturally to the beginning of life than to a sudden change in an existing way of life. What is conveyed in

48. Cf. C. John Collins, *Genesis 1-4: A Linguistic, Literary, and Theological Commentary* (Phillipsburg, NJ: P&R, 2006), 164: "The text [= Gen. 3:17-19], however, does not imply that the pain results from changes in the inner workings of the creation."

49. Murray, *Nature*, 75, overlooks this point.

50. Cf. Collins, *Genesis 1-4*, 165-66. In Rom. 5 this is even quite explicit; though the limitation to humans in Rom. 5:12 is not always visible in translations, it is clearly there in the Greek: "Therefore, just as sin came into the world through one human being, and death through sin, in this way death spread *to all human beings*, because all have sinned."

51. Murray, *Nature*, 76.

52. Snoke, "Why Were Dangerous Animals Created?," 122.

*Animal Suffering and the Goodness of God*

the text is that, for reasons that only God knows, God has initially subjected the earth to forms of decay and futility, such as suffering and death among animals, in order to redeem it and lead it to glory through the Spirit.[53] This ties in neatly with what Paul writes in 1 Corinthians 15:46, where he observes a logical order in the works of God with regard to the created world: the physical order (which is perishable, in dishonor, and weak; vv. 42–44) is first, the spiritual order (which is imperishable, glorious, and strong) comes next.[54]

In the past it may have been more reasonable to consider animal suffering and death as a consequence of human sin (as human death was) than as part of God's original plan.[55] However, as we have seen before, new developments in science may give us occasion to reread biblical texts and carefully check our intuitive interpretations as possibly influenced by taken-for-granted traditional ways of reading that may, after all, turn out to go beyond the sober text of Scripture itself.[56] If we do this here, taking gradualism as a "background belief," it may dawn on us that the relevant biblical texts were overinterpreted when they were read as connecting animal death to human sin.[57] Note that this is not a matter of concordistically squeezing the texts until they fit in with contemporary science. Rather, in the light of changed background beliefs, we realize that the texts, as a matter of fact, are silent about the reason(s) for animal death. Interestingly, some of the greatest biblical interpreters seem to have seen this all along. John Calvin, for example, already explained Genesis 1:29–30 (the most explicit text according to which in the beginning plants were given to eat to humans and animals) in a perspectivist way.[58]

---

53. Cf. Southgate, *The Groaning of Creation*, 92–96, who rightly points out that we should not read evolutionary theory into this passage either.

54. Interestingly, Paul literally refers back here (1 Cor. 15:45) to the creation of humanity in Gen. 2, not to its fall in Gen. 3.

55. Cf. on the theodicy function of the cosmic-fall theory in the early church, David Fergusson, *Creation* (Grand Rapids: Eerdmans, 2014), 43.

56. G. C. Berkouwer, *Holy Scripture*, trans. Jack B. Rogers (Grand Rapids: Eerdmans, 1975), 133.

57. For the notion of (data-)background beliefs, see, e.g., Nicholas Wolterstorff, *Reason within the Bounds of Religion* (Grand Rapids: Eerdmans, 1976), 63.

58. Collins, *Genesis 1–4*, 165. Collins refers to Calvin in this connection, according to whom the Genesis text was inconclusive with regard to prelapsarian vegetarianism and who "was not troubled by the possibility of meat-eating before the fall." Indeed, Calvin judged that the real meaning of Gen. 1:29–30 was that "by these words, he [= God] promises [humanity] a liberal abundance, which should leave nothing wanting to a sweet and pleasant life." Thus, Calvin read this text in a perspectivist, not in

Some who have argued that animal suffering (as part of natural evil) is the result of human sin have realized that this should not be spelled out in terms of a change in the biological constitution of animals that turned them into carnivorous—and sometimes even savage—beasts. Rather than suggesting that animal suffering and death entered creation only after human sin, they have hinted at the possibility that the Creator, being omniscient, might have structured creation in such a way that it included meat eating, death, and decay right from the beginning. As Reformed theologian Emil Brunner writes:

> If then God knew beforehand that the Fall of man would take place, should not His creation of the world have taken *this* sort of man into account? Is it unallowable to think that the Creator has created the world in such a way that it corresponds with sinful man? Is not a world in which, from the very beginning, from the first emergence of living creatures, there has been a struggle for existence, with all its suffering and its "cruelty," an arena suitable for sinful man? We cannot assert that this is so; still less have we any reason to say that it is not so.[59]

This suggestion can be unpacked in terms of a "retroactive Fall," according to which the punishment for sin already precedes the actual committing of it.[60] It can also be fleshed out, however, as Brunner suggests, in a less "harsh" way, in which animal suffering is not a direct consequence of human sin but part of a natural environment that is suitable for sinful human beings, who have to learn by living in a dangerous environment. They have to

---

a concordist, way. John Calvin, *Commentaries on the First Book of Moses called Genesis* (1554), trans. John King (Grand Rapids: Eerdmans, 1948), 99–100.

59. Emil Brunner, *The Christian Doctrine of Creation and Redemption* (Philadelphia: Westminster, 1952), 131. There is a weaker allusion to this view in Herman Bavinck, *Reformed Dogmatics*, vol. 3, *Sin and Salvation in Christ* (Grand Rapids: Baker Academic, 2006), 182: God took account of the Fall already in creating the world by including the possibility of "futility and decay"; it was only after the Fall, however, that "nature gradually became degraded and adulterated and brought forth thorns and thistles . . . and carnivorous animals" (181). So, according to Bavinck, these processes unfolded in a natural way—which is harder to conceive from a biological point of view than he suggests.

60. This has been done more recently by William A. Dembski, *The End of Christianity: Finding a Good God in an Evil World* (Nashville: B&H Academic, 2009). Note that this solution aggravates the moral problem, since it implies that animals had to suffer the consequences of human sins even ages before these sins actually took place.

*Animal Suffering and the Goodness of God*

struggle to subdue the earth and can fall prey to wild animals—just as tame animals can. The problem with this view, however, is that whereas it may make sense of a scenario in which the human species was created almost simultaneously with the animals, it can hardly account for the evolutionary scenario in which animals had to suffer for millions of years before humans appeared on the scene. For why would God choose such an extremely long route full of pain and suffering just to teach human beings their lessons? "Wouldn't God secure all the relevant goods and avoid a massive array of evil simply by creating the universe in much the way the young-universe creationist believes it was created?"[61]

Thus, given both its weak basis in Scripture and its manifold moral and scientific perplexities, the cosmic-fall theory should be abandoned. This leads us to wonder whether God may have had other reasons for taking such a long and troublesome route with animals as the notion of gradualism implies.

### 4.5 Animal Suffering as Part of God's Plan

Many of those who try to make sense of our evolutionary history from a theological perspective think that, indeed, God must have had other reasons for creating our evolutionary world than punishing us humans for our sins. But what reasons could there plausibly be? In his study on "God, evolution, and the problem of evil," Christopher Southgate argues that there can be only one such reason: there must have been no other way for God to create a world containing so many valuable things than by allowing animal suffering in all its gruesome dimensions as well: "I hold that the sort of universe we have, in which complexity emerges in a process governed by . . . Darwinian natural selection, and therefore by death, pain, predation, and self-assertion, is the only sort of universe that could give rise to the range, beauty, complexity, and diversity of creatures the Earth has produced."[62]

Note that there is no anthropomonism in this formulation, as there seems to be in Michael Murray's otherwise similar view. At the end of his insight-

---

61. Murray, *Nature*, 96.
62. Southgate, *The Groaning of Creation*, 29. In the original, the entire sentence is italicized to highlight that this is a core idea in the book. See also Southgate, *Theology in a Suffering World: Glory and Longing* (Cambridge: Cambridge University Press, 2018), 96–148, and Bethany N. Sollereder, *God, Evolution, and Animal Suffering: Theodicy without a Fall* (Abingdon: Routledge, 2019).

ful monograph, Murray claims that "animal suffering is necessary since . . . a spectrum of organisms with increasingly complex cognitive capacities is necessary in order to secure the emergence of beings capable of morally significant freedom."[63] Thus, Murray considers the emergence of "beings like us" the reason God may have had to create our cosmos, including the amount of evolutionary animal suffering it contains: "Adopting this position opens the door to a CD of animal suffering according to which that suffering is a necessary condition for securing outweighing goods, namely, the emergence of organisms capable of imaging God in the way Christians think human beings do."[64] It seems to me that Southgate has a stronger case here because he explicitly includes the values of nonhuman forms of life in his argument.[65]

It is not so clear, however, why God would need and allow so much pain and evil in order to achieve the goods he wanted to emerge—creation's beauty in all its diversity, including (but not restricted to) human freedom and flourishing. To be sure, we ourselves do similar things—things against which virtually nobody objects. For example, we all know that car driving inevitably leads to a multitude of casualties each year, many of whom die suddenly, at a young age, and also under heinous circumstances and in painful ways. Still, nobody proposes that we abolish car driving; at best, we take measures to reduce the number of casualties—but even limiting maximum speeds often raises protest among the public. Apparently, we almost unanimously consider the gains of car driving so much more important than its losses that we endorse the system as a whole almost without reflection. So perhaps we should pause before claiming that it would be morally wrong for a perfectly good God to create a world that includes so many pains alongside its gains. Nevertheless, there are two important disanalogies that have to be taken into account here.

First, in the case of automobile traffic, nobody is required to take part in it; it is up to the individual person to decide whether or not to risk the possibility of becoming involved in a serious accident. Therefore, except for

---

63. Murray, *Nature*, 191.

64. Murray, *Nature*, 184. A "CD" is a *causa Dei*, a phrase Murray borrows from Leibniz for a possible case to be made on behalf of God over against evil, as in a trial (40). To be sure, Murray also develops CDs that allow for "outweighing goods enjoyed by or affecting animals themselves" (130), but he argues that these cannot stand on their own feet (191).

65. A recent proposal that goes to great lengths to solve the problem of animal suffering entirely in terms of the ultimate well-being of these animals themselves is elaborated by Dougherty in his *Problem of Animal Pain*, 134–78.

situations in which the authorities have been negligent in road maintenance or proper sign posting, etc., we usually do not blame the government when accidents happen. Animals, on the other hand, never chose to become part of the biological cycle with all its hazards; many of them involuntarily became victims of the system. In ordinary life, only when an exceptionally clear and overwhelming greater good is at stake do we feel allowed to force people into a system they have not asked to be part of. For example, in many countries psychiatric patients can be confined to an institution against their will only when they form a life-threatening risk to themselves or others.

Second, in the mainstream of the Jewish and Christian traditions, God is confessed to be almighty. Whereas humans would presumably make sure that car traffic led to no casualties at all if they had the power to do so, it is difficult to see why an almighty God would have to accept (reluctantly?) so many casualties in creating the world we inhabit. One might object to this that, for all we know, it may be true in every possible world that human beings as we know them could come into being only through an evolutionary history that includes amounts of animal suffering and death as vast as they actually are. In that case, even an omnipotent God could not arrange things otherwise. Perhaps the laws of nature that apply in our world are not contingent but necessary—they could not be otherwise than they are. Even in that case, however, it is not clear why the history of the natural world as regulated by the natural laws could not have been different in relevant ways. As Robert Francescotti comments: "So even if the laws of nature are not contingent . . . it seems highly likely that some of the ways the universe might have progressed contain less animal distress. The fawn could have followed a slightly different path, thereby avoiding that horrible fire. The gazelle might have lost consciousness before being torn apart by the cheetahs. The bear cub might have been found by its mother before it starved to death. And with different initial conditions, fewer carnivorous species and more herbivores might have evolved."[66]

---

66. Francescotti, "Problem of Animal Pain," 124; the reference to the fawn hints at a famous example used by William Rowe ("The Problem of Evil and Some Varieties of Atheism," *American Philosophical Quarterly* 30 [1979]: 335–41) to show that at least some instances of severe animal suffering are absolutely pointless. The example concerns a fawn that is trapped in a forest fire, "horribly burned," and lying "in terrible agony for several days before death relieves its suffering" (337)—all this without any human being around who could possibly learn some lesson from the terrible event.

It may seem that "only way arguments" like those of Southgate, Murray, and also Van Inwagen[67] work only if such alternative scenarios can be ruled out. The rejoinder that the regularity of nature might become endangered if God from time to time prevented animals (or humans, for that matter) from pointless suffering or reduced the amounts of suffering they have to undergo is unconvincing, since enough regularity would remain for the natural world to be generally reliable and predictable. And especially for those who believe that God *does* intervene from time to time in the world—for example, by performing miracles—it is difficult to see why he could not do so more often.

Indeed, given the involuntary involvement of animals in the evolutionary "survival of the fittest" and so many seemingly needless instances of animal suffering—including highly troubling and disconcerting forms of predation, like those that involve the predator's agonizingly slow devouring of its prey—we may wonder what can compensate for all this. What kinds of goods intrinsically connected to them may be so great as to possibly outweigh these sufferings? It seems that by ascribing such horrendous events to God's permissive will we not only run the risk of declaring good what is utterly evil but also deeply implicate God in the occurrence of large amounts of suffering. Nevertheless, we should realize that we lack the cognitive resources to come to a final judgment here. We are simply not in a position to rule out the possibility that outweighing greater goods of the right sort are indeed involved in the process, even though we cannot begin to fathom what greater goods these might be and why they necessitate the occurrence of so much suffering. Arguably, it is the lesson of the book of Job that we humans just don't know and should not pretend to know. Instead we are called to trust God in faith, and though this first of all applies to God's dealings with us humans, it can easily be extended to his dealings with the animal world.

Thus, so-called skeptical theism—that is, the view that we should be skeptical of our human capability to identify God's reasons for allowing evil and suffering—may provide an adequate answer to the problem of animal

---

67. Van Inwagen, *The Problem of Evil*, 113–34: "We have no reason to accept the proposition that an omniscient and omnipotent being will be able so to arrange matters that the world contains sentient beings and does not contain patterns of suffering morally equivalent to those of the actual world" (123). Van Inwagen is keen to point out that two patterns of suffering can be morally equivalent even if one of them involves slightly less suffering than the other: the difference may just be morally irrelevant (124). Presumably, however, the leeway here is pretty small.

suffering.[68] Note that this answer applies to animal suffering both as we see it around us today and as it must have been occurring for many millions of years in evolutionary history. However, it is precisely the horrendous character of some instances of animal suffering that gives us reason not to stop our investigations here but to consider yet another answer—a fourth option next to the three discussed thus far: Could it be that at least some instances of animal suffering—for example, those that strike us as particularly gruesome—should not be attributed directly or indirectly to God but rather to demonic forces that attempt to thwart God's intentions?

### 4.6 Animal Suffering and the Demonic

Although, as we have seen above (§4.4), the traditional way of explaining animal suffering by an appeal to Adam's fall has become problematic, it does not follow that every interpretation of the natural world as *fallen* is out of the question. On the contrary, when there are reasons to see the suffering of animals neither as nonexistent (§4.3) nor as morally unproblematic (§4.5), the idea that the natural world in its present form has somehow fallen out of God's perfect intentions comes to mind again. As a matter of fact, many theologians and Christian philosophers continue to hold the traditional Christian view here, even when they realize that such fallenness can no longer be explained as the consequence of Adam's fall. But what could it then possibly mean to hold that our world is fallen, and that animal suffering is a sign of that?

Here, two answers have been given in theological reflection that may look different but are in fact identical from a conceptual point of view.[69] The first answer has the form of a vivid and highly pictorial, almost mythical, narrative: all that is awful in the natural world is brought about by a group of nonhuman spirits headed by *Satan*—Satan being the head of those angels who were created by God before all ages but who, before the dawn of human (and animal) history, rebelled against their creator. Sometimes this idea is related to the "gap theory," which we mentioned above (§2.3). According to this theory, a long period of time (a "gap") should be postulated between

---

68. For an overview of the ins and outs of this recent strand of thought, see Trent Dougherty and Justin P. McBrayer, eds., *Skeptical Theism: New Essays* (Oxford: Oxford University Press, 2014).

69. Another response would be to refer to the *tragic* dimension of life on earth; unless this dimension came in later, however, this is not a case of "being fallen."

Genesis 1:1, where God created an original world including animals but not humans, and the "formless void" of Genesis 1:2 that supposedly resulted from God's destruction of this original creation in response to its distortion by Satan and other fallen angels. Most geologic strata that modern geologists have found, and the many fossilized animals within them, stem from this original creation many millions of years ago. The undeniable traces of violence in the fossilized animals show that their original harmony and perfection must have been twisted and distorted by the workings of these fallen angels.[70] Others adhere to this story without dating it as precisely as the gap theorists do. Here is Alvin Plantinga: "Satan, so the traditional doctrine goes, is a mighty nonhuman spirit who, along with many other angels, was created long before God created man. Satan rebelled against God and has since been wreaking whatever havoc he can. The result is natural evil."[71]

So animal suffering, being part of natural evil, might be attributed to the free actions of Satan and other fallen angels. Indeed, C. S. Lewis, Michael Lloyd, and others have explicitly applied this view to animal suffering.[72] Lewis, for example, argues that it is "a reasonable supposition, that some mighty created power had already been at work for ill on the material universe, or the solar system, or, at least, the planet Earth, before ever man came on the scene.... If there is such a power, as I myself believe, it may well have corrupted the animal creation before man appeared."[73] To those who find such a view fanciful, Lewis replies that "the doctrine of Satan's existence and fall is not among the things we know to be untrue: it contradicts not the facts discovered by scientists but

---

70. The gap theory gained some traction in the twentieth century as a result of its inclusion in the Scofield Bible (1909). Cf. Keathley and Rooker, *40 Questions*, 111-18.

71. Alvin C. Plantinga, *God, Freedom, and Evil* (Grand Rapids: Eerdmans, 1974), 58. Plantinga not only described this view but also made it clear that he takes it seriously; he still did so thirty years later in his "Supralapsarianism, or 'O Felix Culpa,'" in *Christian Faith and the Problem of Evil*, ed. Peter van Inwagen (Grand Rapids: Eerdmans, 2004), 1-25. Cf. for a similar view Clark Pinnock, *Most Moved Mover: A Theology of God's Openness* (Carlisle, UK: Paternoster, 2001), 133-34.

72. Lewis, *The Problem of Pain*, 134-36; Michael Lloyd, "Are Animals Fallen?," in Linzey and Yamamoto, *Animals on the Agenda*, 147-60. Cf. also Gregory Boyd, *Satan and the Problem of Evil* (Downers Grove, IL: InterVarsity Press, 2001). In the Bible, the idea of an angelic fall is supported by two relatively late texts (both of which draw on earlier Jewish sources): 2 Pet. 2:4 and Jude 6. Murray, *Nature*, 97-98, ignores these texts but quotes at length two Old Testament passages (Isa. 14:12-15 and Ezek. 28:12-19) that, in my view, have been less influential in Christian theological thinking.

73. Lewis, *The Problem of Pain*, 134-35.

the mere, vague 'climate of opinion' that we happen to be living in."[74] In any case, from the perspective of the biblical worldview, it is not strange to connect animal suffering and death with the demonic. We already saw how, according to the synoptic story, demonic spirits were responsible for the drowning of a herd of swine in the Sea of Galilee (Mark 5:1-20 par.; cf. §4.2).

A more abstract and sophisticated version of what is basically the same approach has been put forward by theologians influenced by Karl Barth, such as Thomas Torrance and Neil Messer. Here, the intuition that evil forces have somehow intruded into God's creation is articulated in a more muted way. Torrance, for example, takes with great seriousness the scientific picture of the world of nature as being "red in tooth and claw," and he concludes from that picture that "far from evil having to do only with human hearts and minds, it has become entrenched in the ontological depths of created existence."[75] Torrance realizes that, given the second law of thermodynamics, important parts of the evolutionary process (e.g., death, decomposition, etc.) could hardly be other than they are, but he thinks things are different when it comes to such troubling aspects that are at the heart of this process, such as the predator-prey relationship and "indeed sheer animal pain."[76] Hence, "it is difficult not to think that somehow nature has been infiltrated by an extrinsic evil," which gave its functions and features "a malignant twist."[77] Rather than depicting these evil forces as fallen angels, however, Torrance, following Barth, speaks more abstractly about "anti-being," which because of God's rejection can have only an "improper existence." In this way, Torrance is more reticent about the origins of the dreadful distortions of nature, acknowledging that "evil remains an utterly inexplicable mystery."[78] Similarly, Neil Messer argues that "whatever in the evolutionary process is opposed to God's creative purpose is to be identified with 'nothingness'; it is an aspect of the chaos and disorder threatening the creation."[79]

---

74. Lewis, *The Problem of Pain*, 134.

75. Thomas F. Torrance, *Divine and Contingent Order* (Oxford: Oxford University Press, 1981), 116.

76. Torrance, *Divine and Contingent Order*, 122.

77. Torrance, *Divine and Contingent Order*, 123, 122; cf. Southgate, *The Groaning of Creation*, 32, and Paul D. Molnar, *Thomas F. Torrance: Theologian of the Trinity* (Farnham, UK: Ashgate, 2009), 88.

78. Torrance, *Divine and Contingent Order*, 118.

79. Neil Messer, "Natural Evil after Darwin," in *Theology after Darwin*, ed. Michael S. Northcott and R. J. Berry (Milton Keynes, UK: Paternoster, 2009), 139-54 (149).

In her rich monograph on animal suffering, Nicola Hoggard Creegan combines both lines of thinking, carefully arguing for what she calls a "modified dualism" that she thinks is in line with the New Testament. Finding Barth's talk about nothingness (*das Nichtige*) too abstract and impersonal, she points to the personal nature of the temptation that Jesus encountered in the wilderness and of the demons he exorcised.[80] The template she uses for interpreting the close coinherence of good and evil in the evolutionary process is Jesus's parable of the wheat and the tares (Matt. 13:24–30, 36–43). Surely the evolutionary process contains many good and valuable things that have largely been obscured by the picture of evolution as a "materialist, reductionist ... process"[81]—a picture that has held us captive for too long. Creegan points to recent developments in evolutionary theory that highlight the important role of cooperation, symbiosis, empathy, compassion, care, and even sacrifice among animals.[82] As a result, the picture of a cruel nature "red in tooth and claw" is changing "in ways that are more conducive to theological insight."[83] Despite all of nature's tragedies, we can sense the sheer beauty and goodness of the highly variegated forms of life that inhabit our planet. In this way, believers can still discern the hand of God in nature and see why created reality was called "good": it contains many intimations of the goodness and wisdom of its creator. Thus, to put this in the idiom of the Belgic Confession, which we explored in chapter 1, there is continuity between the book of nature and the book of Scripture. The goodness of God proclaimed in the latter is not at all absent from the former. It may even be observed in the ways in which predators catch and hold their prey; anyone who has ever seen a garter snake in the talons of a red-tailed hawk knows how majestic such a sight is.

At the same time, however, Creegan points out deeply troubling aspects to the mechanisms of the evolutionary process, such as the "harshness and callousness of natural selection" and the "relentless carnage" that has gone

---

80. Nicola Hoggard Creegan, *Animal Suffering and the Problem of Evil* (Oxford: Oxford University Press, 2013), 76.

81. Creegan, *Animal Suffering*, 97.

82. Creegan, *Animal Suffering*, 65. This is a significant new development indeed. Cf., e.g., Martin A. Nowak and Roger Highfield, *Super Cooperators: Beyond the Survival of the Fittest; Why Cooperation, Not Competition, Is the Key to Life* (New York: Free Press, 2011), as well as the subsequent volume that Nowak coedited with theologian Sarah Coakley, *Evolution, Games, and God: The Principle of Cooperation* (Cambridge, MA: Harvard University Press, 2013).

83. Creegan, *Animal Suffering*, 97. Hereafter, page references from this work will be given in parentheses in the text.

on since prehuman times (53). When we have become sensitive to these dark aspects of evolution, we "need some way of understanding animal suffering as more than just a means to a higher order of different creatures" (49). It will not do to suggest that God will ultimately redeem nature from these afflictions as long as God is regarded as their creator, since in that case "the values God reveals in creation are completely opposed to those shown in redemption" (53). Whoever baptizes all of life's modes of being as "God's will" runs the risk of minimizing evil and acquiescing to it (66). Instead, we should acknowledge that, distinct from the wheat, there are also tares in the natural world. Both resemble each other so much that we are ill advised to try to separate them. It takes so much discernment to tell the difference (apparently, the bad seed mimics the good seed and becomes intertwined with it), that this will be possible only in the eschatological future (73, 87). Still, the difference is real, and we should not be tempted to call good (because of its supposed contribution to some higher-order good) what is evidently horrendous.

Who or what is responsible for the tares? Somehow, the ontology of evil will always remain inscrutable to us (93). The Bible is full of images and metaphors, however, that help us take these evil forces seriously. Animal suffering, like much human suffering (cf. Luke 13:16), is part of the corrupted world, "caused in some sense by the evil known variously in Scripture as the Evil One, powers and principalities, or '*shadow sophia*'" (137).[84] Creegan is keen to grant that these demonic forces don't have any independent authority or life in themselves (76). Such a radical dualism would indeed be sub-Christian. Still, the New Testament suggests what might be called a "provisional" or "modified" dualism, in which evil, though elusive and hidden in its agency and ontology (52), operates as a very real destructive force that transcends the human dimension. In this way, Creegan argues for the rehabilitation of a theology of fallenness. "Why discard the element of the demonic when the Scriptures are so full of it? A theology of fallenness enables us to say 'no' to the idea that suffering is necessary or a part of God's kingdom in some way" (148).[85]

---

84. The reference to "*shadow sophia*" is based on Celia Deane-Drummond, *Christ and Evolution: Wonder and Wisdom* (London: SCM, 2009), 185-86, who in turn draws on the work of Sergei Bulgakov. It is not clear to me where this notion is mentioned in Scripture (as Creegan suggests). Bulgakov uses the term in a way that calls to mind Barth's notion of *das Nichtige*: the dark side of creation that is not willed by God but as such nevertheless exists.

85. The "ecological" passage in Rom. 8:19-20, which we discussed above, might also be interpreted along such lines, the futility to which nature is subjected in the

In sum, a revised version of the cosmic-fall theory, as developed by thinkers like Plantinga, Lewis, Torrance, and Creegan, may enable us to account for (the most troubling aspects of) animal suffering without somehow having to condone it or explain it away. We can simply acknowledge that there is something deeply amiss in creation, given the devastating violence and horrendous mercilessness of nature to which animals have been falling victim since long before humanity emerged. This theory also helps us to acknowledge that God is *opposed* to it, working all the time to redeem creation from its evil intruders. Is it this intuition that made the psalmist exclaim, "You save humans and animals alike, O LORD" (Ps. 36:6)? As Torrance reminds us, this divine intent has become most clear at the cross of Christ, where the demonic powers of this world were decisively conquered (Col. 2:15). It is from the vantage point of the incarnation, crucifixion, descent, and resurrection of Jesus Christ that we may come to see how God got at the heart of evil in order to destroy it from within its ontological depths and started to rebuild what he had once made to be good.[86] The gospel invites us to become God's fellow workers in this struggle until the final victory is reached.

Although this view of (animal) suffering and natural evil is indeed deeply informed by the Christian gospel, I cannot help but make a critical comment by way of closing this section. It seems that modified dualism thrives on an unresolved ambivalence that results from its refusal to answer the crucial question of the precise relationship between God and the evil forces that are responsible for the world's fallenness. For, either these forces have been created by God or they have not been created by God. If the former, God is at least indirectly responsible for all the havoc they wreak. Even if God intentionally granted them freedom, given his omniscience, God must have known what they would bring about. If the latter, they apparently fall outside the realm of his power and question (to say the least) his almightiness. In that case, however, the label *modified* dualism seems misleading, since it is unclear how this view differs from the *unmodified* metaphysical dualism that Christians have rejected ever since Marcion. Evidently, unmodified dualism is also rejected by Plantinga, Lewis, Torrance, Creegan, and the like, so that only the first horn of the dilemma is left to them. Therefore, we conclude that the difference between those who see suffering as a part of God's plan (§4.5) and those who ascribe it to demonic forces (§4.6) is a difference in degree, not a difference in kind.

---

present era being due to "the devil and his fallen angels." Thus Robert Jenson, *Systematic Theology*, vol. 2 (Oxford: Oxford University Press, 1999), 151.

86. Torrance, *Divine and Contingent Order*, 116.

## 4.7 Evaluation

How can this history of evolutionary evil be reconciled with the life-giving goodness and love of the God Christians have preeminently come to know through Jesus Christ? Whereas evolutionary theory blocks the cosmic-fall theory, it is largely indifferent to both of the other responses distinguished above. As a result, depending on the underlying assumptions that seem most plausible to them, Christians can either argue with Southgate and others that there was no other way for God to realize his ultimate goal and that, for all we know, this goal is worth the price of evolutionary evil; or they can argue with Creegan and others that evil forces—either demons, fallen angels, or forces that cannot so easily be labeled—have been opposing and thwarting God's life-giving intentions for creation ever since the origins of life on earth.

Taking a Reformed perspective, we may have a preference for the "only way" view. Reformed theology has always been very much impressed by God's all-encompassing sovereignty and providence. In the first chapter of this book we highlighted its strong theocentric focus. Although the existence of demons and other evil forces was not denied, these forces were usually not invoked for theodical purposes, that is to say, for explaining the occurrence of evil and suffering in a world created by a good God. A famous example here is the doctrine of providence as expounded in the Heidelberg Catechism, where any reference to the devil and his cohorts is conspicuous by its absence. According to this Reformed confession, "God upholds, as with his hand, heaven and earth and all creatures, and so rules them that leaf and blade, rain and drought, fruitful and lean years, food and drink, health and sickness, prosperity and poverty—all things, in fact, come to us not by chance but by his fatherly hand."[87]

Whereas the possibility of chance or randomness is in view here as a possible alternative, the devil is not mentioned. It is to God's rule and nothing else that everything that happens is ascribed—including events and situations that cause a lot of suffering and pain to human beings, such as drought, poverty, and sickness. We have argued elsewhere that this view should be qualified and criticized insofar as it should not be interpreted in a stoic way but from a biblical Trinitarian perspective, since it is always the Father of Jesus Christ who deals with us rather than a God who resembles blind fate.[88] Still, it may be seen as part of "the Reformed stance" not to be scrupulous when it comes to

---

87. *Our Faith: Ecumenical Creeds, Reformed Confessions, and Other Resources* (Grand Rapids: Faith Alive, 2013), 77–78.
88. Van der Kooi and Van den Brink, *Christian Dogmatics*, 233–44.

connecting God with whatever goes around, even with phenomena that may strike us as evil and abhorrent. This is combined with a strong awareness (e.g., in John Calvin) that we are not in a position to figure out the ways of God, since they are inscrutable to us. What is incumbent on us is to humbly trust God.

The Presbyterian natural scientist David Snoke applies this line of thought to animal suffering: here as well, no one can fathom the reasons God has. Rejecting the option that evil forces are responsible for the emergence of carnivorous animals ("Demons are never credited with creative power in the Bible, only destructive power"),[89] he further suggests that violent and dangerous animals may simply reflect the power and wrath of God—notions that are much more prominent in the biblical literature than many want to acknowledge. "If God's character is eternal and unchanging, as the Bible says, then if we see wrath in nature now, we should expect that God would reveal this aspect of his character from the very start."[90] In this way we learn what it is to fear God. Snoke points out that in the final part of the book of Job God takes credit for the birth pangs of wild goats (39:3); for creating the ostrich, which is "cruel to her young" (39:16-17); for the eagle, whose babies "drink blood" (39:30); and for the Leviathan, which has "rows of sharp teeth"(41:14). All such forms of life and ways of living belong to God's *good* creation. "We must marvel at the shark, even while fearing it. It is well designed, frighteningly so. So also are many parasites."[91] Perhaps the fact that other animals fall prey to them is not as bad as we, anthropomorphically, tend to think. "We may not like animal death and suffering, but the fact is that the Bible does not say anywhere that such things are bad, in and of themselves."[92]

Thus, from a Reformed perspective one might address animal suffering theologically in this way. Indeed, it has been argued that in the past liberal Christians had much more difficulty in coming to terms with Darwinism than orthodox Calvinists, since liberals had a more rosy view of the world,

---

89. Snoke, "Why Were Dangerous Animals Created?," 118; for a similar view, see Christopher Southgate, "Divine Glory in a Darwinian World," *Zygon* 49 (2014): 784-807. Ascribing predation to a rebellious power "would be to suggest that God set out to create straw-eating lions and was unable to do so" (Southgate, 785).

90. Snoke, "Why Were Dangerous Animals Created?," 124. This comment, however, seems to ignore the fact that in the Bible wrath is not an inherent but a *reactive* property of God, which is especially (or even exclusively) evoked by human sin.

91. Snoke, "Why Were Dangerous Animals Created?," 124 (defining sharks, along with viruses, as "killing machines," 119).

92. Snoke, "Why Were Dangerous Animals Created?," 123; note the typically Reformed way in which Snoke leans heavily on the Bible.

interpreting providence in terms of progress; the gloomy picture of the world sketched by Darwin resonated much better with orthodox Protestants.[93] Even from an orthodox Reformed perspective, however, there is no need to ascribe animal suffering as directly to God's purposes as Snoke does. Those who fear that Snoke's view borders on neo-Cartesianism might either resort to the third answer (ascribing all or the most worrying forms of animal suffering to evil forces) or flesh out the second answer in a different way. For example, following an old scholastic tradition, one might make a distinction between God as the *prima causa* and evolutionary mechanisms as the *secundae causae* of evolution. Reformed theologian Nico Vorster elaborates on this distinction as follows:

> As *prima causa* God's creative acts put in place processes that bring forth living creatures. In doing so he allows for the contingency and fortuitous nature of evolutionary processes in order to safeguard the integrity of creation. . . . God's good intent with evolutionary processes has a shadow side in the suffering it causes. Such suffering, though foreseen by God and part of God's providence, is not coerced by God but emanates from the complex operations of second causes. Yet, the suffering and intrinsic competition that accompanies evolutionary processes serve a good outcome in that they make possible complex forms of life able to transcend themselves and capable of relating to God.[94]

In fact, it seems that, apart from the traditional appeal to the Fall, all types of theodicy and defenses that have been put forward to deal with the problem of natural evil might also be used to address the problem of *evolutionary* evil, including that of animal suffering. That is to say, as far as one is convinced that some particular type of theodicy or defense works for natural evil in general, it is not clear in advance why it should not work for the problem of the evolutionary suffering of animals. In this way, one may resort to a "free will" or "free creatures defense" (e.g., by ascribing freedom to demons), to a soul-making theodicy (arguing that either we humans or the animals themselves have to learn important things from animal suffering that can only be learned

---

93. James R. Moore, *The Post-Darwinian Controversies: A Study of the Protestant Struggle to Come to Terms with Darwin in Great Britain and America, 1870–1900* (Cambridge: Cambridge University Press, 1979), passim.

94. Nico Vorster, *The Brightest Mirror of God's Works: John Calvin's Theological Anthropology* (Eugene, OR: Wipf & Stock, 2019), 57.

in this way), to other greater-good defenses (by specifying other supreme values that can be reached only through the long evolutionary history we have had), to skeptical theism (by arguing that we are not in a position to know which reasons God has for allowing evolutionary evil), to a christological response (according to which, the "cruciformity" of nature should not surprise those who have the cross of Christ as the center of their faith), to the idea of an eschatological compensation for animals that had to suffer deeply on earth—or to a possible combination of these.[95]

The point is not that all or even some of these forms of theodicy and defense actually work. The point is that *if* they work with regard to natural evil in general, why would they not work for evolutionary evil? And if they don't work, we should not blame evolution, since it must be for other reasons that they fail to do their job. After all, animal death and suffering throughout evolutionary history are not a different type of evil from animal death and suffering that we observe in the natural world today—and that we therefore have to face even when we don't accept evolution. It is just its *quantity* that dramatically increases when we accept a progressive (i.e., becoming more and more complex) creation over millions of years. As Southgate explains: "Though animal suffering was known before Darwin, the narrative of evolution that emerged in his work stretched the extent of that suffering over millions of millions of years and millions of species, most of them now extinct."[96] This may make us despair of the possibility of finding an adequate answer to the *why* question, but we must remember that the same feeling

---

95. A classic statement of the free will defense is Plantinga, *God, Freedom, and Evil*; the seminal work on soul-making theodicy is John Hick, *Evil and the God of Love* (Basingstoke, UK: Macmillan, 1966); and an application to animal suffering is to be found in Dougherty, *Problem of Animal Pain*. A more general greater-good defense can be found in Richard Swinburne, *Providence and the Problem of Evil* (Oxford: Clarendon, 1998), with an application to animals on 171-75, 189-92; for skeptical theism, see Dougherty and McBrayer, *Skeptical Theism*, and my comment at the end of §4.5 above; for a christological response, see Marilyn Adams, *Christ and Horrors: The Coherence of Christology* (Cambridge: Cambridge University Press, 2006); and for an application to evolutionary evil, see Deane-Drummond, *Christ and Evolution*; for an eschatological approach, see, e.g., Jay B. McDaniel, *Of God and Pelicans: A Theology of Reverence for Life* (Louisville: Westminster John Knox, 1989), 41-49, who addresses the question how even predators might have a place in heaven. For a combination of various lines of argument, see Southgate, *The Groaning of Creation*, 16.

96. Southgate, *The Groaning of Creation*, 1-2.

of despair often befalls us when we are confronted with *human* atrocities or miseries. Theodicies usually do not work on an existential level but only (if at all) on an intellectual one.

It may be countered that there is a structural difference between evolutionary evil and other forms of natural evil, in that the struggle for life that is ingrained in nature is especially poignant given the goodness and lovingkindness of the God of the gospel.[97] Doesn't this God show a special concern for the weak and vulnerable, whereas in the evolutionary process the weakest organisms are the easiest victims of predators and other competitors? First, however, it is not exactly true that evolution by definition sacrifices the weakest organisms of a herd or species or group; rather, it is most injurious to the *least well-adapted* organisms—which may as a matter of fact be very strong animals. Second and more importantly, that the least well-adapted individuals of a group are the most vulnerable ones has been observed since time immemorial and can still be observed today. We don't need evolutionary theory (or even its first layer: gradualism) to come to know this. Sometimes Christians reject evolutionary theory because in their view the God of the gospel would never allow for a world characterized by the struggle for life and the survival of the fittest. The right response to that argument is: just look around! Apparently God *does* allow for and even sustain such a world. The difference made by evolutionary theory is that whereas we used to blame this struggle for life on human sin, that has now become implausible. We have to ascribe it either to demonic forces or more directly to Godself. That is a relative difference, though, since in all scenarios we have to admit that God at least *allowed* this situation to arise, including the bleak prospects of the "weakest" specimens. Thus, the question how God's special concern for the weak and vulnerable as shown in the gospel relates to this struggle for life is still pertinent.

Do we have an answer to this question? Aren't the amount, diversity, and distribution of suffering in the animal world so baffling that we must doubt whether the God of the gospel is the creator of the universe? That type of question has accompanied—and from time to time afflicted—the church ever since it was confronted with Marcion's rejection of the creator God in the second century. The discovery of the progressive appearance of life-forms on the geological timescale and of the concomitant evolutionary history of animal suffering may be seen as lending more credence to Marcion's stance:

97. See on this, e.g., Rik Peels, "Does Evolution Conflict with God's Character?," *Modern Theology* 34 (2018): 544–64.

How can the God of the gospel be responsible for such a world? Over against this stance, however, the church has always understood itself as the community called into being by the triune God who created, redeems, and perfects our world as Father, Son, and Spirit. It seems to me that given the various responses reviewed in this chapter (esp. §§4.5 and 4.6), evolutionary suffering does not confront us with a compelling reason to give up the basic Christian belief that the world's redeemer and perfecter is none other than its creator. Readers can make up their minds as to which of the responses makes the most sense. Personally, I am convinced that skeptical theism entails a particularly credible view, especially to those of us with Reformed sensibilities: we small and sinful human beings are not in a position to evaluate what sorts of evils God may or may not permit, since we do not know which greater goods may be intrinsically connected to them.

But doesn't this long-standing history of evolutionary suffering compromise the goodness of God? As we saw in chapter 1, according to the Belgic Confession, we know God just a little bit from the book of nature and much "more clearly and fully" from the book of Scripture. What we have found in the present chapter confirms this picture. When it comes to God's goodness, we get a glimpse of it from its resonances and intimations in the natural world: the emergence of values like creativity and cooperation, empathy and sympathy, compassion and care, freedom and love in the course of evolutionary history. These values are mixed up with so many disvalues, however, that we more than ever need both the incarnate and the written Word of God to see what God's goodness really looks like and how radical it is: "God's love was revealed among us *in this way*: God sent his only Son into the world so that we might live through him" (1 John 4:9).

It is this same God who declared the created world to be good despite all appearances to the contrary. The conclusion of the writer of Genesis 1 that "God saw everything that he had made, and indeed, it was very good" (1:31), far from having to be denied or downplayed because of evolutionary suffering, becomes all the more vital for the Jewish and Christian faith. For this verse contains a profound affirmation of the positive value of created earthly life from a radically theocentric perspective. Even though we humans, impressed as we are by so many instances of evil, suffering, and tragedy, often cannot see this, we are told that *in God's eyes* our world is good, indeed "very good." That does not mean that it is perfect—the Hebrew word *tob* just does not imply that. But it means that from God's perspective—and clearly God's perspective is decisive, not ours—the created world is highly worthwhile, entirely fitting for the excellent purposes that God has in mind.[98] For even though at the end of Genesis 1 God's work of creation is finished, the world is not yet as God intended it to be but on its way to that destination.[99]

Therefore, when confronted with natural evils—either evolutionary or otherwise—we are called upon not to give in to despair or cynicism but to find guidance in this divine judgment. We may trust that one day the goodness and glory of creation will become manifest in an unparalleled and unambiguous way.

---

98. Indeed, biblical scholars have forcefully argued that "good" should be interpreted in functional terms here: creation is good because it is perfectly fit for God's purposes with it. See, e.g., Claus Westermann, *Genesis 1-11: A Commentary* (Minneapolis: Fortress, 1994), 645; John H. Walton, *The Lost World of Genesis One: Ancient Cosmology and the Origins Debate* (Downers Grove, IL: IVP Academic, 2009), 50, 148.

99. Cf. John Goldingay, *Genesis for Everyone: Part I* (Louisville: Westminster John Knox, 2010), 17.

CHAPTER 5

# Common Descent and Theological Anthropology

> The crux of the *imago Dei* is not "some thing"—but a unique calling.
>
> —Malcolm Jeeves[1]

## 5.1 Introduction

In the previous chapters we have discussed the theological challenges arising from the most basic layer of evolutionary theory discussed in chapter 2: the view that the various forms of life on earth appeared gradually over long periods of time on the geological timescale. We have examined how the notions of deep time and gradualism urge us to revisit both the doctrine of Scripture—or, more precisely, the practice of biblical interpretation—and the problem of natural evil. We have tried to establish that in both cases positions are possible that do full justice to this part of evolutionary theory and also are in line with the Reformed theological stance discussed in chapter 1.

It is now time to move on to the second layer of evolutionary theory, the notion of common ancestry or common descent, and discuss the issues this notion raises for a traditional Reformed view of life. In §2.3 we observed that at least two doctrinal themes have to be reexamined in this connection, namely, theological anthropology and the doctrine of the covenant, or (as it is sometimes called) the locus of the history of redemption. In theological anthropology we must answer the question to what extent the notion of common descent rules out or changes established views on the unity and uniqueness of the human race. Is it still possible to believe that humans and humans only

---

1. Malcolm Jeeves, afterword to *The Emergence of Personhood: A Quantum Leap?*, ed. Malcolm Jeeves (Grand Rapids: Eerdmans, 2015), 225.

have been created in the image of God? If so, what might this mean under evolutionary conditions? Or are we forced to give up this claim altogether? With regard to the covenant and the history of redemption, we will have to consider the difference evolutionary theory makes to the traditional view that human history started when God created the first couple, Adam and Eve, who then fell into sin and as a result brought about the reality of human death. Are the notions of a "historical Adam" and a primordial fall still feasible? If so, to what extent can this fall explain human death and human sinfulness? And what does this all mean for the nature of redemption as accomplished by Jesus Christ? We will address these two clusters of questions in this and the following chapter respectively.

**5.2 A Reformed Concern?**

I cannot introduce the problem of this chapter better than by relating a personal story. Almost a decade ago I was invited by a Dutch Reformed daily newspaper to discuss the topic of evolution with American young-earth creationist Terry Mortenson by writing some letters back and forth to one another that were then published over a couple of weeks.[2] In my letters I challenged Mortenson's views by arguing that creationism is neither tenable nor necessary for Christians to embrace, cautiously suggesting that instead Christians should come to terms with the facts of evolution. Apparently our exchange was closely followed by the readership, for it stirred quite a number of letters to the editor. Although I had tried to be as nuanced as I could, these letters were overwhelmingly critical of my position and supported Dr. Mortenson on all counts.

One of the letters was particularly harsh in tone, and actually a bit over the top. Because I was curious to know what was behind it, I decided to contact its writer. It turned out that he was a Reformed dairy farmer living in a rural area in the Dutch Bible belt. After taking some time to become familiar with him, I asked him what I honestly wanted to know: Why did he become so angry with me on the issue of evolution? His answer was most revealing. Clearly, he could have given a variety of answers. He might have answered

---

2. The exchange appeared in several issues of *Reformatorisch Dagblad* during April and May 2009. Mortenson is involved in Answers in Genesis (cf. https://answersingenesis.org/bios/terry-mortenson/, accessed April 4, 2019). For his more recent work, see Terry Mortenson, ed., *Searching for Adam: Genesis and the Truth about Man's Origin* (Green Forest, AR: Master Books, 2016).

that evolution challenges the authority of the Bible, or that it rules out the possibility of a historical Adam and thereby destroys the Christian view of salvation. In fact, he might have given any answer that is the topic of one of the chapters in this book. The farmer, however, raised his voice and exclaimed: "I don't want to stem from the apes!" From the discussion that followed it became clear that by far the most troublesome issue to him vis-à-vis evolution was the issue of *human dignity*. In his view, the notion of common ancestry jeopardized this human dignity.

Of course, this farmer was not the first one to voice this worry. Ever since the days of Darwin, it has been one of the most prominent concerns provoked by evolutionary theory.[3] And for many contemporary Christians, both Reformed and others, it is still as significant an issue as any others discussed in this book. The idea that evolutionary theory seems to threaten human dignity may even at least partially explain why creationist theories of various stripes remain so popular in contemporary Western societies. Presumably, it would not have helped much if I had explained to the farmer that, as a matter of fact, we do not stem from the apes but only share common ancestors with them.[4] For it is the more general idea of humanity as somehow emerging from the animal world that is a cause of concern about human distinctiveness and human dignity.

Does this "anthropological concern" vis-à-vis evolution reflect a typically *Reformed* worry? At first sight, it does not seem to. In fact, Reformed Christianity does not have a particularly high view of the human being. As we saw in chapter 1, the most fundamental distinction in Reformed theology is that between Creator and creation (here as well, the Reformed tradition intensifies an intuition that is deeply embedded in the Christian tradition as a whole). Re-

---

3. As early as 1860, the so-called Wilberforce-Huxley debate at Oxford became famous for its heated controversy on the issue of human dignity as affected (or not) by common ancestry. The debate became an almost iconic testimony of Christian narrow-mindedness, but only because a twisted account of the actual course of events became popular. See J. R. Lucas, "Wilberforce and Huxley: A Legendary Encounter," *Historical Journal* 22 (1979): 313-30; Alister McGrath, *The Foundations of Dialogue in Science and Religion* (Oxford: Blackwell, 1998), 15-16; and esp. David Livingstone, "Myth 17: That Huxley Defeated Wilberforce in Their Debate on Evolution and Religion," in *Galileo Goes to Jail and Other Myths about Science and Religion*, ed. Ronald L. Numbers (Cambridge, MA: Harvard University Press, 2009), 152-60.

4. Even when one doubts the credentials of the theory of common ancestry as an overall explanatory scheme for the development of all species, as we saw in chapter 2 the provenance of humans and other primates from a common ancestor is quite firmly established nowadays in the empirical data.

formed theology displays a strong awareness of human frailty, insignificance, and misery in the eyes of the Lord God, who surpasses all our understanding. Calvin, for one, points out that we humans tend to be "blinder than moles" when it comes to our comprehension of God.[5] Moreover, our dignity is seriously compromised by our human sinfulness. Though we continue to be distinct from the other species, sin makes us entirely unworthy before God. So why would we claim such a high status for ourselves within God's creation?

On closer scrutiny, however, it is precisely in this dialectic that we may find the key to the Reformed farmer's anger. For apparently it was their wish to emphasize the seriousness of human sin that brought Reformed theologians to ascribe a very high ontological, intellectual, and moral status to the created human being before the Fall. Obviously, the more perfectly the human being was equipped by God, the more wicked became his sin. Humans were created in the "state of integrity" (*status integritatis*), which means they were equipped with "true righteousness and holiness." This phrase from the Heidelberg Catechism (Lord's Day 3) is derived from Ephesians 4:24; sometimes "(true) knowledge" is added to it, on the basis of Colossians 3:10.[6] Interestingly, there can be no doubt that both texts actually refer to the eschatological renewal of the human being according to God's image, but in the Reformed tradition they were also applied, without any discussion, to the original prelapsarian situation of humanity.[7] In this way the heinous character of human depravity could be expressed in the clearest possible terms. For the more perfect the human being had been created, the more serious became his sin—and the more praiseworthy the saving grace of God (in the Augustinian tradition, safeguarding the grace of God is usually the deepest motif).

We can observe this pattern not only in founding sources of the Reformed tradition like the Heidelberg Catechism and the theology of John Calvin[8] but

---

5. Cf., e.g., John Calvin, *Institutes of the Christian Religion*, ed. John T. McNeill, trans. Ford L. Battles (Philadelphia: Westminster, 1960), 2.2.18: "When it comes to knowing God and especially God's fatherly favor in our behalf . . . the greatest geniuses are blinder than moles!"

6. Cf., e.g., Louis Berkhof, *Systematic Theology: New Combined Edition* (Grand Rapids: Eerdmans, 1996 [1932 and 1938]), 206-7.

7. For some criticism, cf. Cornelis van der Kooi and Gijsbert van den Brink, *Christian Dogmatics: An Introduction* (Grand Rapids: Eerdmans, 2017), 263-64.

8. See, e.g., Cornelis van der Kooi, *As in a Mirror: John Calvin and Karl Barth on Knowing God; A Diptych*, trans. Donald Mader (Leiden: Brill, 2005), 63-70, 87-88; Nico Vorster, *The Brightest Mirror of God's Works: John Calvin's Theological Anthropology* (Eugene, OR: Wipf & Stock, 2019), chap. 1. Mary Potter Engel, *John Calvin's*

perhaps even more clearly in Reformed theologians from the era of post-Reformation Protestant scholasticism (ca. 1565-1725). Let us by way of example briefly examine the disputation on the image of God in the *Synopsis Purioris Theologiae*—a seventeenth-century compendium of Reformed dogmatics written by four theology professors at the University of Leiden.[9] Here, the unique position of the human being is highlighted right from the beginning: "Man is clearly the high point and goal of nature's lower order, yet he also belongs to a higher order, he is the 'sum' of everything and the bond that links earthly and heavenly things" (13.2).[10] Further, "the sequence of creation is . . . an illustration of the dignity of humanity," since God first made all other things for the sake of humankind, "things that make his condition a good and happy one," in order to finally create man (13.6). In this way, "there would be a progression from less to more perfected things" (13.6). God enhanced the dignity of humankind even further by not creating him just like the other creatures through his word alone but only after having made a plan (cf. the phrase "let us make" in Gen. 1:26). "Better yet, He girded himself for the task, like someone about to fashion a unique and exceptional work with great effort" (13.7). This "superior dignity (*summa dignitas*) of mankind" is further shown by the fact that only humans were created "according to the image of God," which means that the human being is "a rather close copy" of God (13.8).

The moderator, Antonius Thysius, goes on to explain in detail the way God proceeded when creating man. What he claims to know about it goes far beyond the sober narrative of Genesis 2. He appears to be slightly embarrassed, however, that according to this narrative the human being was made "from the dust of the earth" (13.10). He concedes that the earthly material of which we were made "is a reminder of our weakness and . . . humbler nature" (13.12),[11] but he immediately adds that our "exceptional fashioning is a testimony to our

---

*Perspectival Anthropology* (Atlanta: Scholars Press, 1988), tried to explain the tensions between Calvin's negative and positive utterances on the human being by identifying some distinctive perspectives that govern his anthropology.

9. Dolf te Velde, ed., *Synopsis Purioris Theologiae / Synopsis of a Purer Theology*, vol. 1 (Leiden: Brill, 2015), disputation 13 (315-37). Numbers within brackets in this and the next paragraphs refer to sections of this disputation. The disputation was presided over by Antonius Thysius.

10. As Te Velde explains (*Synopsis*, 315n2), this refers to the classical trope of the human being as a "microcosm," mirroring the unity of the lower, material and the higher, spiritual part of reality in the unity of his body and soul.

11. The comparative "humbler" may refer to the angels, whose creation was discussed in the preceding disputation.

dignity" (13.12). Although "the whole man in both soul and body is created in the image of God" (13.36), man's most noble part is his immortal soul, which was created in "very close approximation . . . to the essence of God" (13.18). Most probably, the soul is situated in "the chest or heart," as the Peripatetics and Stoics already taught (13.21). As to his moral and intellectual status, being made in the image of God expresses "the goodness of man, his uprightness and perfection (or ideal state), his surpassing excellence over all living creatures and his close approximation to God" (13.37). He was "of outstanding knowledge" (13.38). Moreover, his "body and limbs formed a constitution that was holy; upright and wholesome actions were to arise from them all" (13.38).[12] In short, the ontological, intellectual, and moral perfections of Adam and Eve are hailed to such an extent that one wonders how they could have lapsed into sin at all.

In the next disputation, "on the Fall of Adam," then, precisely because of their perfection, the full responsibility for the primordial sin is laid on the first couple's shoulders. Although both God's permission (14.25) and the devil's instigation (14.27) are recognized, "the internal cause of the fall is the free will of both our parents" (14.30). It is wrong to think that the sin of Adam and Eve was minor and excusable. "For the less burdensome and easier it was to observe God's commandment, so much the more without excuse was each of our parents before God on account of that transgression, and guilty of temporal and eternal death" (14.37). Thus, their original perfections make their transgression and disobedience all the more serious. Given this stark but highly functional contrast, we can imagine why Reformed lay believers today are particularly shocked when reading that humans might have evolved from the "brute" animals rather than stemming directly from a sublime divine initiative.

Meanwhile, perhaps for different reasons but no less emphatically, the concern for human dignity and distinctiveness is also voiced in other Chris-

---

12. Other Reformed scholastics distinguished a natural or "intrinsic" part of the divine image that cannot be lost (viz., the "spiritual, immortal, rational substance of the soul") from an "extrinsic" part that could be lost (viz., heavenly wisdom, perfect righteousness, and holiness). See Heinrich Heppe, *Reformed Dogmatics: Set Out and Illustrated from the Sources*, rev. and ed. Ernst Bizer (Grand Rapids: Baker Books, 1978), 235, 236, and cf. Richard Muller, *Dictionary of Latin and Greek Theological Terms* (Grand Rapids: Baker Books, 1985), 145. The human being's (or, actually, "man's") dominion over the animals is sometimes mentioned as a third aspect of the *imago Dei* (Heppe, 236-37).

tian traditions. It is clearly manifest, for example, in the aforementioned speech on evolution that Pope John Paul II delivered to the Papal Academy of Sciences in 1996. In this speech the pope was very open toward evolutionary theory, arguing that "the convergence, neither sought nor fabricated, of the results of work that was conducted independently is in itself a significant argument in favor of this theory."[13] However, following up on earlier statements by his predecessor Pius XII, he made one proviso: we should hold that "if the body takes its origin from pre-existent living matter, the spiritual soul is immediately created by God."[14] Now why is this so important? Here is the answer: "Theories of evolution which . . . consider the spirit as emerging from the forces of living matter . . . are incompatible with the truth about man. Nor are they able *to ground the dignity of the person.*"[15] As Stephen Pope comments, "the pope's major reservation about the theory of evolution came from a moral concern—that nature not be interpreted in such a way as to deny or downgrade human dignity."[16]

Thus, the anthropological concern about evolutionary theory is not "typically Reformed" but is—like most of the themes discussed in this book—recognized by other Christian (and presumably even other monotheistic) believers as well. How exactly should this concern be understood? And how, if at all, could it be overcome? It is to these questions that we turn now. First, I will offer a more precise delineation of the problem (§5.3). How is the relation between human dignity, human uniqueness, and human origins construed by those who think that a full-blown theory of common ancestry jeopardizes human dignity? It will turn out that the doctrine of humanity as being created in God's image plays a pivotal role in this connection. Therefore, in the next sections I will focus on how this doctrine of the *imago Dei* should be interpreted. I first look at a recent proposal to reinterpret it in a way that downplays the differences between humans and other created beings and suggests that animals as well might bear the image of God (§5.4). I will conclude, however, that this proposal is problematic for several reasons. Next, I explore what seems to be a more fruitful avenue, namely, the attempt to reinterpret the

---

13. Pope John Paul II, "Message to the Pontifical Academy of Sciences on Evolution," *Origins* 26, no. 22 (1996): 351 (see, e.g., https://www.ewtn.com/library/papaldoc/jp961022.htm; accessed April 4, 2019).

14. John Paul II, "Message to the Pontifical Academy," 351.

15. John Paul II, "Message to the Pontifical Academy," 352 (emphasis added).

16. Stephen Pope, *Human Evolution and Christian Ethics* (Cambridge: Cambridge University Press, 2007), 196–97.

notions of human uniqueness and human dignity in theological rather than biological terms (§5.5). This attempt indeed enables us to adhere to notions of human uniqueness and dignity (as encapsulated in the doctrine of the *imago Dei*) that are not at odds with contemporary evolutionary theory. Finally, I will summarize my main conclusions (§5.6).

### 5.3 Human Dignity and the Challenge of Evolution

Why is it so important that human beings should be unique in a way that goes beyond the uniqueness of every organism and every species? After all, every organism—you and me and every other individual—is unique by definition because of being that specific organism; and every species is unique by the simple fact of being that particular species, characterized by the set of properties that define it as that species—a form of uniqueness sometimes called *species specificity*. So why should we humans want more than that? One might surmise that people like the Reformed farmer want us to be clearly distinguished from all animal species, so that we humans on the one hand are set apart from all other species on the other. But then again, it is self-evident that this is the case. We can draw an imaginary dividing line between every single species and all other ones, pointing to the specific characteristics that make this particular species different from the others. So clearly we can also draw such a dividing line between humans and all (other) animals—evolutionary theory does not change that fact.[17] I trust that the Reformed farmer and his sympathizers are smart enough to avoid the *genetic fallacy*: the false belief that our nature and identity are completely determined by how we came to be what we are. Even when we emerged from the animals, we still are a unique species of our own.

Therefore, I suggest that the real reason behind the anthropological concern is the issue of human *dignity* as it is based on what we might call "special uniqueness"—the specific form of uniqueness that makes us radically different from all other species and more special and precious than them. As we have seen, in traditional theology the notion of humanity's uniqueness is strongly connected with the affirmation of human dignity—that is, of the special worth

---

17. The only way to escape this conclusion is to call into question the very concept of a *species*—which is exactly what some ultra-Darwinians do. See, e.g., Steve Stewart-Williams, *Darwin, God, and the Meaning of Life: How Evolutionary Theory Undermines Everything You Thought You Knew* (Cambridge: Cambridge University Press, 2010), 156-58.

we have by the simple fact of being a member of the species *Homo sapiens*. Arguably, it was this connection between our special uniqueness and our human dignity that was captured and theologically grounded in the doctrine of the *imago Dei*: it is because we humans are special, endowed with unique characteristics that somehow resemble our creator (much more closely than the unique-making characteristics of other species), that we deserve a moral status that differs from all other species.[18] This status that conforms to our dignity consists of the right to be treated and not to be treated in particular ways. In other words, our dignity leads to the possession of a couple of inviolable natural human rights, such as, most of all, the right of life.[19] As (Reformed) philosopher Nicholas Wolterstorff argues: "The attempt to show that human rights are grounded in human dignity is . . . the attempt to pinpoint some property or relationship whose possession by all human beings gives them a certain worth—some property or relationship on which worth supervenes."[20]

From this perspective, we can see why evolutionary theory may be considered a disconcerting phenomenon. For it seems that evolutionary theory has largely discredited the idea that human beings have one or more of these unique-making properties or relationships. Claims that only humans walk upright, use tools and language, and have self-awareness, culture, or reason are no longer defensible in light of the empirical evidence.[21] Nor is it as self-evident as it used to be that humans alone have personhood and morality.[22]

---

18. Cf. Kevin Vanhoozer, "Human Being, Individual and Social," in *The Cambridge Companion to Christian Doctrine*, ed. Colin E. Gunton (Cambridge: Cambridge University Press, 1997), 163: "Christians ground their affirmation of human dignity and personhood in the special resemblance of the human creature to its Creator."

19. For the link between the image of God and the inviolability of human life, see esp. Gen. 9:6.

20. Nicholas Wolterstorff, *Justice: Rights and Wrongs* (Princeton: Princeton University Press, 2008), 320, and cf. 311-61 for his own proposal to ground human rights in human dignity.

21. Cf. Marc Bekoff, "The Evolution of Animal Play, Emotions and Social Morality: On Science, Theology, Spirituality, Personhood, and Love," *Zygon* 26 (2001): 616, and see more extensively on animal cognition Randolf Menzel and Julia Fischer, eds., *Animal Thinking: Contemporary Issues in Comparative Cognition* (Cambridge, MA: MIT Press, 2011).

22. See, e.g., Bekoff, "Evolution of Animal Play," 615-55, and Barbara J. King, "Are Apes and Elephants Persons?," in *In Search of Self: Interdisciplinary Perspectives on Personhood*, ed. J. Wentzel van Huyssteen and Erik P. Wiebe (Grand Rapids: Eerdmans, 2011), 70-82. As to animal (proto)morality, see, e.g., the work of primatologist Frans

## Common Descent and Theological Anthropology

In particular, the strong similarities between the human being and its closest relative, the chimp, have often been emphasized during the past decades, both because of ethological research and because of the stunning genetic correspondence between the two species that was discovered. As a result, it seems that any theory that sees a large gap between humans and the rest of creaturely reality has become problematic.

This raises the question whether we can still uphold an adequate notion of human dignity. In his book *Created from Animals*, naturalist philosopher James Rachels gives a succinct argument to the contrary:

> The doctrine of human dignity says that humans merit a level of moral concern wholly different from that to mere animals; for this to be true, there would have to be some big, morally significant difference between them. Therefore, any adequate defense of human dignity would require some conception of human beings as radically different from other animals. But that is precisely what evolutionary theory calls into question. . . . This being so, a Darwinian may conclude that a successful defense of human dignity is most unlikely.[23]

So here we have a third voice next to those of the farmer and the pope; it is that of an atheist neo-Darwinist who drives the problem home: if we have no special human uniqueness, which is what evolutionary theory shows (or so it seems), then in all probability we will have no special human dignity either. But is this argument correct? First, is it true that evolutionary theory has "debunked" the notion of human special uniqueness? And second, even if this were so, does it follow that belief in human dignity has to collapse as well? In what follows I will start with this second question and discuss a recent proposal to turn the tables: What if our close ties with the animal world as

---

de Waal, e.g., his *Good Natured: The Origins of Right and Wrong in Humans and Other Animals* (Cambridge, MA: Harvard University Press, 1996) and Frans de Waal et al., eds., *Evolved Morality: The Biology and Philosophy of Human Conscience* (Leiden: Brill, 2014); cf. §8.2 below.

23. James Rachels, *Created from the Animals: The Moral Implications of Darwinism* (New York: Oxford University Press, 1990), 171. Similar arguments can be found in, e.g., Williams, *Darwin, God, and the Meaning of Life*, 258-79; Peter Singer, *Animal Liberation*, 4th ed. (New York: HarperCollins, 2009), 1-24, 213-50 (on the alleged evil of "speciesism"). Cf. in this connection the quite disturbing comment of John Gray, *Straw Dogs: Thoughts on Humans and Other Animals* (London: Granta, 2002), 151: "*Homo sapiens* is only one of very many species, and not obviously worth preserving."

displayed by evolutionary theory in fact *upgrade the dignity of animals* rather than downgrading our human dignity? Translated into theological categories, could it be that we have to widen the concept of *imago Dei* so that it may include animals next to humans?

## 5.4 Animals in the Image of God?

Perhaps, as Ernst Conradie has put it, "an affirmation of human dignity . . . does not require a strong position on human uniqueness. . . . The inalienable dignity which we attach to human life may serve as a paradigm for the dignity (or integrity) of the whole earth community."[24] In the first place, this may pertain to animals. Thus, some blurring of the boundaries between humans and nonhuman animals, instead of degrading us humans, may help us to reevaluate animal life and appraise it for what it is worth. The recent rise of cultural and political movements in the West committed to the defense of "animal rights" and the liberation of animals shows that this suggestion is not as outrageous as it would have been deemed in the past. From a theological point of view, could it be that animals are to be included in the image of God rather than excluded from it? In this section we will briefly explore this line of thought.

An explicit plea for broadening the concept of *imago Dei* has been made by American Protestant theologian David Cunningham.[25] Cunningham starts his argument by observing that the most well-known reason for making a theological distinction between human beings and the rest of creation has been the

24. Ernst M. Conradie, *An Ecological Christian Anthropology: At Home on Earth?* (Aldershot, UK: Ashgate, 2005), 80.

25. David S. Cunningham, "The Way of All Flesh: Re-thinking the *Imago Dei*," in *Creaturely Theology: On God, Humans, and Other Animals*, ed. Celia Deane-Drummond and David Clough (London: SCM, 2009), 100-117. Cf. Markus Mühling, "Menschen und Tiere—geschaffen im Bild Gottes," in *Geschaffen nach ihrer Art. Was unterscheidet Tiere und Menschen?*, ed. Ulrich Beuttler et al. (Frankfurt: Lang, 2017), 129-43. A related proposal (taking its cue in a christological interpretation of the *imago Dei* and seeing the incarnation as God-taking-on-creatureliness) has been advanced by David Clough, *On Animals*, vol. 1, *Systematic Theology* (London: T&T Clark, 2012), 100-103. For a critical response, see Veli-Matti Kärkkäinen, *Creation and Humanity* (Grand Rapids: Eerdmans, 2015), 228-29. For the discussion as a whole, see also Joshua M. Moritz, "Does Jesus Save the Neanderthals? Theological Perspectives on the Evolutionary Origins and Boundaries of Human Nature," *Dialog* 54, no. 1 (2015): 51-60.

belief that human beings *alone* were created in the image of God. He argues that the traditional interpretation of the *imago Dei* has been influenced by widespread cultural assumptions on the superiority of the human race, based on the Aristotelian view that rationality and the use of language were exclusively human faculties (indeed, Aristotle famously defined the human being as the "rational animal"). Since Darwin, however, these assumptions have been seriously called into question. Thus, "Darwin reminded us that we had placed a great deal of faith in Aristotle."[26] If this turns out to have been misguided, we have reason to reexamine the *imago Dei* tradition by testing its biblical roots.

In doing so, Cunningham first points out that the very few biblical texts that use the notion of the image of God and ascribe it to humankind do not explicitly deny it to other entities. Therefore, we need an "argument from silence" to make a case for its exclusive ascription to human beings—and arguments from silence are always "somewhat hazardous."[27] Even if we grant that the notion of the image of God is applied only to human beings, its precise meaning "has been one of the more contested sites of theological reflection throughout the centuries," as a result of which "we are unable to determine its meaning."[28] Therefore, the notion is not of much help in gaining a clear theological understanding of what is so special about humanity. Further, Cunningham reminds us of the many ways in which the Genesis stories reflect our intimate material connectedness with (other) animals. Darwin did not come up with something totally new when he reinforced this point. Indeed, according to the biblical texts, not only humankind but also the rest of creation somehow reflects the "mark of the Maker."[29] In the famous opening verse of Psalm 19, for example, it is said that the heavens tell the glory of God and the firmament proclaims his handiwork. And the more sober prose of Romans 1:20 asserts that God's power and divinity are seen in the things God has made.[30]

In response to Cunningham, it has to be acknowledged that the Bible contains only a few references to the image of God. In fact, there are only

---

26. Cunningham, "Way of All Flesh," 103. Cunningham points out that Aristotle himself was more nuanced on the humans-nonhumans distinction, but that this did not apply to the later tradition.

27. Cunningham, "Way of All Flesh," 106.

28. Cunningham, "Way of All Flesh," 107, 110.

29. Cunningham, "Way of All Flesh," 100.

30. In chapter 1 we saw how these verses (and one could also mention the final chapters of the book of Job here) inspired the Christian tradition to regard nature as the first book through which God reveals himself.

three of them in the Old Testament (Gen. 1:26-28; 5:1; 9:6), whereas the New Testament has seven—most of which assume Jesus Christ rather than humanity to be the true image of God. This is not to say that it cannot be a central notion in theological anthropology. René Descartes uses the famous formula *cogito ergo sum* only three times in his oeuvre, but because of its many conceptual relations with other parts of his thinking, it still is one of his most pivotal ideas. The same may apply to the notion of the image of God.[31] Next, following Daniel Miller, it seems to me that "we do not belittle other animals by excluding them from the designation *imago Dei*,"[32] since they may have their own specific relationships to God of which we know little.[33] However, the statement that the Genesis story is silent about whether or not the image of God also belongs to nonhumans is slightly misleading. For whereas the creation of both plants and animals is said to be "after their pattern," for the creation of humankind we hear the words "in God's image" at the very same place. The linguistic similarity is even stronger, since the phrases are invariably repeated at least two times. So some kind of distinction clearly seems to be implied, and this distinction is made more explicit when in Genesis 9 it is said to Noah and his descendants that "every moving thing that lives shall be food for you" (v. 3) but that the blood of a human being shall not be shed since "in his own image God made humankind" (v. 6). Clearly, we cannot escape the conclusion here that nonhuman beings do not share in the image of God, and neither do they enjoy the same dignity—that is, the same rights to be treated in certain ways (which is not to say that they have no rights at all). As Veli-Matti Kärkkäinen writes, "the point of the biblical narrative [in Genesis] is to stress the uniqueness of humanity among the creatures."[34]

Of course, we might decide to disconnect our own theological views from the Bible here, arguing that we should extend the notion of the image of God to the nonhuman realm even though the Bible does not do so. However, given the

---

31. I am grateful to René van Woudenberg for this example. For relevant considerations on the *imago Dei* in this connection, cf. Kärkkäinen, *Creation and Humanity*, 269-71, over against David Kelsey, *Eccentric Existence: A Theological Anthropology*, vol. 1 (Louisville: Westminster John Knox, 2009), chap. 4.

32. Daniel K. Miller, "Responsible Relationship: Imago Dei and the Moral Distinction between Humans and Other Animals," *International Journal of Systematic Theology* 13 (2011): 334 (323-39).

33. Cf. Karl Barth, *Church Dogmatics* III/2 (Edinburgh: T&T Clark, 1960), 138: "The glory of other creatures lies in the concealment of their being with God, no less than ours in its disclosure. For all we know, their glory may well be greater."

34. Kärkkäinen, *Creation and Humanity*, 283.

paradigmatic status of Scripture in Reformed theology, it becomes difficult to align such a view with the tradition or to recommend it to the contemporary Christian community as a viable option.

Finally, it is true that there is a wide variety of interpretations of the notion of the *imago Dei* both in the Christian tradition and in contemporary theology. But that tells us only that we should carefully study these interpretations in order to determine which ones are the most plausible. Some of them may, as Cunningham notes, be largely inspired by reigning cultural assumptions and should therefore be distrusted. Others are perhaps not as mutually exclusive as one might think at first sight. However this may be, it seems just lazy hermeneutics to conclude from a plurality of interpretations that it is impossible to determine the meaning of a certain phrase or doctrine. In Christian theology, such a conclusion would be particularly disastrous, since hardly any doctrinal (or ethical) tenet has only one clear and generally accepted interpretation. We would be unable to establish the meaning of any doctrine if we had to wait until consensus on its interpretation had been reached. Nor is this typical for Christian theology, since debate on the right interpretation of beliefs and practices is characteristic of all living religious and philosophical traditions (even the *absence* of such debates might raise concerns about a tradition's viability; for example, no one debates how Donar produces thunder).

Thus, we still find ourselves at the very place where we started: Christians typically ground their affirmation of human dignity in the doctrine of the *imago Dei*, which requires a conception of the human species as somehow having "special uniqueness." The question continues to be, therefore, how we can make sense of this claim in light of the challenges of evolutionary theory, which casts doubt on the assumption that we humans have some faculty or faculties that all other species lack. This question becomes even more urgent when we acknowledge that the image of God, even though this phrase only rarely occurs in the Bible, is nevertheless a "central, canonical idea."[35] Drawing on work of Old Testament scholar James Luther Mays, Wentzel van Huyssteen argues that "what it stands for theologically does in fact become the structural theme of the biblical account of God and humankind."[36] Indeed, the dramatic

---

35. J. Wentzel van Huyssteen, *Alone in the World? Human Uniqueness in Science and Theology* (Grand Rapids: Eerdmans, 2006), 149.

36. Van Huyssteen, *Alone in the World?*, 122; cf. James Luther Mays, "The Self in the Psalms and the Image of God," *God and Human Dignity*, ed. R. Kendall Soulen and Linda Woodhead (Grand Rapids: Eerdmans, 2006), 38-39: "While 'image of God' is

story in which the Bible seems most interested is the narrative of God's dealings with both the people of Israel and the entire human race. Thus, the Bible implicitly underlines the special uniqueness of the human species from beginning to end, even when it does not explicitly mention it or link it to the notion of the *imago Dei*. But what does this special uniqueness consist in? A way to answer this question that is not discredited by evolutionary theory and is therefore still viable, I suggest, is by giving a theological account of human uniqueness rather than a biological one—a theological account, though, that is compatible with the empirical data. In the next section, I will spell out this suggestion in some detail.

## 5.5 Human Uniqueness as a Theological Category

If we want to retain the notion that we are special, that humankind somehow constitutes a category of its own, evolutionary theory helpfully reminds us that we should avoid superficial appeals to empirical characteristics that seem to set us apart from all other beings. Although more sophisticated proposals to this effect should not be ruled out in advance,[37] it is true that the "most frequently offered markers of difference—rationality, intelligence, and language—are unable to identify a qualitative difference between humans and other creatures."[38] There might be, however, more distinctively *theological* reasons for considering humanity as somehow radically different from the rest of creation.

A promising way forward here is to reexamine the doctrine of *imago Dei* along lines suggested in both recent biblical scholarship and systematic theol-

---

no longer used for the human being in the biblical story, its actuality is a structural theme of the biblical account of God and humankind."

37. For example, it can be argued that quantitative differences may become so large that a qualitatively new phenomenon arises; or that quantitative differences suffice to undergird human uniqueness. For a sophisticated contemporary defense, see Aku Visala, "*Imago Dei*, Dualism, and Evolution: A Philosophical Defense of the Structural Image of God," *Zygon* 49 (2014): 101–20. Visala argues that "although there might not be a clear-cut set of capacities that all humans share, we could still have a notion of human distinctiveness that is sufficient for the structural image of God" (101); he does not discuss the exegetical pros and cons, though.

38. David Clough, "All God's Creatures: Reading Genesis on Human and Non-Human Animals," in *Reading Genesis after Darwin*, ed. Stephen C. Barton and David Wilkinson (Oxford: Oxford University Press, 2009), 145–61 (152).

ogy. As we saw in §5.2, classical accounts of this doctrine tried to define some unique natural characteristics that allegedly distinguished human beings from other creatures. For example, as opposed to (other) animals, only human beings were supposed to possess a soul, or rationality—to mention only the two options proposed most frequently in the theological tradition. Independently of their vulnerability to evolutionary claims, however, it has become clear that such "substantive" or "structural" views of the image of God (finding the image in some unique substance or structure) cannot stand the test of exegetical scrutiny: quite simply, the identification of the *imago Dei* with either the soul or reason (or whatever other capacity) does not occur in the Bible. The background of these views is found in various forms of Greek philosophical thought. Both biblical scholars and systematic theologians have therefore come up with proposals to reinterpret the nature of God's image in humanity by studying its biblical ramifications—and they did so independently of any scientific considerations. Indeed, the biblical concept of *imago Dei* has been the object of intense scholarly attention in recent years.[39] Despite Cunningham's complaint about the large variety of interpretations, these recent examinations disclose a lot of overlap in that most of them point in one of two directions; the biblical notion of the image of God should be understood either in functional or in relational terms. Usually, biblical scholars, who primarily or even exclusively focus on the first chapters of Genesis, endorse a functional interpretation, whereas systematic theologians, who take into account the entire canonical framework, end up with a relational view.[40]

According to the functional view, it is the divinely conferred responsibility or function to take care of creation that constitutes the image of God. While in surrounding ancient Near Eastern traditions it was common for kings to represent the deity by ruling over their territory, the Genesis narrative "democratizes" this function by extending it to all humanity: according to Genesis 1:28, humankind as a whole is called to have dominion over the

---

39. For masterful surveys and evaluations of the history of its interpretation, see Van Huyssteen, *Alone in the World?*, 111-62, and Richard J. Middleton, *The Liberating Image: The Imago Dei in Genesis 1* (Grand Rapids: Brazos, 2005). See also F. LeRon Shults, *Reforming Theological Anthropology: After the Philosophical Turn to Relationality* (Grand Rapids: Eerdmans, 2003), 217-42, and Stanley J. Grenz, *The Social God and the Relational Self: A Trinitarian Theology of the Imago Dei* (Louisville: Westminster John Knox, 2001).

40. This is rightly observed by Marc Cortez, *Theological Anthropology: A Guide for the Perplexed* (London: T&T Clark, 2010), 30. See also Van der Kooi and Van den Brink, *Christian Dogmatics*, 264.

created world, thus acting as "God's vice-regent on earth."[41] According to the relational view, which came to prominence in the twentieth century, the image of God is located in the relationship that God establishes with human beings—a relationship that is then reflected in the manifold and profound ways in which humans interrelate with each other.[42] Perhaps these two views are not as mutually exclusive as is often suggested—and the same definitely applies to the dynamic or eschatological view of the *imago Dei* that is sometimes distinguished as a fourth "model."[43] In any case, the seminal Genesis text links the notion of the image of God both to a relationship ("male and female God created them") and to a particular function or task, namely, to have dominion over creation. Since the functional reference, which is spelled out twice (Gen. 1:26 and 28), is arguably the more prominent and certainly the most explicit, let us start there.[44] We might then suggest that what is theologically distinctive about God's human creatures according to this line of interpretation is that they are being called to rule and take care of the earth.

Understandably, from an ecological point of view there continue to be worries about the harsh language used in these verses ("subdue," etc.). Following up on previous research, however, Richard Middleton has argued that the *dominium terrae* should not be understood as including the use of violence. Rather, "the human task of exercising power over the earth is . . . modeled on God's creative activity, which, in Genesis 1, is clearly developmental and formative, involving the process of transforming the *tohû wabohû* into an ordered, harmonious cosmos."[45]

---

41. Theodore C. Vriezen, *An Outline of Old Testament Theology* (Oxford: Blackwell, 1962), 208; see also Ps. 8:6-8.

42. Particularly influential accounts of the relational view were developed by Reformed theologians Karl Barth, Emil Brunner, and Gerrit C. Berkouwer; as to Berkouwer, see his *Man: The Image of God* (Grand Rapids: Eerdmans, 1962).

43. This view locates the image of God ahead of us rather than behind us; insofar as human beings become united with Jesus Christ, the full image of God (2 Cor. 3:18; 4:4; Col. 1:15), by faith, they are being transformed into God's image—a process that will reach its culmination point in the eschaton. As Van Huyssteen, *Alone in the World?*, 141, rightly points out, this view does not exclude other interpretations of the *imago* but can incorporate them.

44. Here I part company on exegetical grounds with Miller, "Responsible Relationship," who gives the relational interpretation pride of place while delegating the functional one to a footnote. Still, the title of his essay captures the coherence of both interpretations very well.

45. Middleton, *The Liberating Image*, 89.

So it is precisely in being called to be God's image that the human being is summoned to mirror the wisdom, generosity, and creative love by which God made the earth into a habitable place. There is even a striking chiastic parallelism in God's first forming and then filling the earth and humanity's vocation to first fill and then organize the earth. However this may be, the Genesis account stands in strong contrast with the creation-by-combat myths from the *Umwelt* that, to be sure, we also find in other parts of the Old Testament.[46] So here, then, we have a theological account of what makes human beings unique among all other creatures, in terms of their royal and priestly God-given task to represent God on earth by being a blessing to its many inhabitants.[47] Understood in this way, the image of God is not a substantial quality (though, as we will see, it may presuppose such qualities) but an ethical challenge. Indeed, it doesn't lie somewhere behind us but is rather situated ahead of us.

While acknowledging that this might be the case, David Clough has nevertheless countered that this interpretation of the image of God should not be seen as a ground for "human separatism," since "the Bible repeatedly affirms that . . . each living thing has a part in God's purposes."[48] Thus, he suggests that being in the image of God does not make humanity more unique among God's creatures than other species. It is rather a kind of species specificity—the kind of distinctiveness that every species has by the very fact of being a species of its own—rather than a unique characteristic that sets humankind apart from the rest of creation. Whereas there certainly is a humbling element of truth in this argument—as already indicated, we should not underestimate the special relationships God has with the nonhuman parts of creation—it seems to me

---

46. Yet, even the well-known thesis that other Old Testament creation texts display traces of such a *Chaoskampf* has come under pressure in biblical scholarship; see Koert van Bekkum et al., eds., *Playing with Leviathan: Interpretation and Reception of Monsters from the Biblical World* (Leiden: Brill, 2017), xix, 48–51.

47. We should keep in mind that this cannot always be done without using violence, especially in relation to "wild nature"; this is what resounds in the strong vocabulary of Gen. 1:26 ("have dominion" and "subdue"). I am indebted to Jan J. Boersema for this insight.

48. Clough, "All God's Creatures," 153. Indeed, other creatures as well develop and take care of their natural environments by building ecological niches, as, for example, earthworms do by making the soil more habitable. Michael Burdett, "The Image of God and Human Uniqueness: Challenges from the Biological and Information Sciences," *Expository Times* 127 (2015): 3–10, rightly argues, however, that such examples are "entirely local and limited," whereas the cultivation task bestowed on humans "is unique in that it is universal . . . and divinely imparted" (9).

that as a whole it is unconvincing. Of course, one can always ask how specific a species must be to be "really" unique; but clearly, none of the other species has been given such a huge and wide-ranging calling for the whole of creation, including the other animals, as humankind has.

It is from this calling, or so I suggest, that our human dignity and our special moral status are to be derived. When it is declared in Genesis 9 that human life, in contrast to the lives of animals, is inviolable, this is justified by an appeal to the fact that the human being is made in the image of God. Therefore, from a Christian point of view, a complete moral identification of humans and animals should be rejected. It is important, however, not to exchange the order here: our calling precedes and grounds our dignity, rather than the other way around. In this way, our dignity can never legitimately become a cause for arrogance or boasting, nor for maintaining a purely anthropocentric view of the universe. Still, both our calling by God to represent God on earth and the corresponding dignity that God has granted us make us into unique beings on earth.

This point can be further corroborated by the *relational* interpretation of the image of God. In fact, this interpretation can be seen as closely linked with the functional one. For not only are we destined to be God's representatives on earth, continuing and maintaining God's work of creation, but we are also the only beings who are addressed by God so that we might know this, and so that we might live in loving relationship with one another as well as with God (Gen. 1:28). As Robert Jenson has argued, from a theological point of view humans are unique in that God speaks not only about them but also *to* them, thus establishing a relationship with them.[49] This relationship turns out to be reciprocal, in that human beings have been enabled to respond in person to their being addressed by God by means of ritual and prayer. Of all species, we humans have been enabled to fulfill our task by imaging God in reaching out to God and other creatures in profound relationality and loving commitment.[50]

Our capacity to engage in personal reciprocal relationships also extends to our fellow human beings, and this as well should be seen as an essential element of what it means to be made in God's image ("male and female God created them"). As Reformed theologian Daniel Migliore writes, "Being cre-

---

49. Robert Jenson, "The Praying Animal," *Zygon* 18 (1983): 311-25 (esp. 321-22).

50. The concept of love does not have a romantic or sentimental connotation in this connection but refers to one's commitment to pursue what is in the other's interest. For analysis of its various layers and meanings, see Vincent Brümmer, *The Model of Love: A Study in Philosophical Theology* (Cambridge: Cambridge University Press, 1993).

ated in the image of God means that humans find their true identity in coexistence with each other and with all other creatures. . . . Being truly human and living in community are inseparable."[51] Still, whereas we obviously share relationality and communality with the social animals (that is, those animals who live together in groups), if anywhere, it is in our *religious* propensity that a distinctive characteristic of the human species can be found.[52] Paraphrasing Reinhold Niebuhr, Van Huyssteen writes that "humans are in fact driven to something beyond reason, a self-transcendence, and it is this existential longing for a God who transcends the world that really sets human beings apart from other creatures."[53] Indeed, paleontologists and archaeologists continue to look for signs of religiosity when they want to ascertain whether a particular site has been occupied by human beings or by other primates.[54] In a sense, then, it seems that John Calvin had it right when he exclaimed that "it is worship of God alone that renders men higher than the brutes."[55]

So do we after all encounter here a natural, empirical characteristic that defines our human uniqueness and marks us off from the rest of creation? This question should be answered in a very careful way. For it is only from the theological perspective we adopted that we came to discern this distinctively human empirical characteristic. It may even be that, starting from this theological perspective, we *define* the human being in terms of its remarkable capacity for answering God's call in spirituality and religion, thereby simply denying hominins in whom this is missing the status of being "really" human. In fact, it is almost inescapable to work with such a stipulative definition, since from an empirical point of view the boundary lines of the human species are fuzzy: we cannot tell, for example, where exactly in the phylogenetic line we might first speak of human beings, or exactly which hominins were

---

51. Daniel L. Migliore, *Faith Seeking Understanding: An Introduction to Christian Theology*, 2nd ed. (Grand Rapids: Eerdmans, 2004), 144-45.

52. See chap. 8 for theories on how this religious propensity might have emerged.

53. Van Huyssteen, *Alone in the World?*, 133; cf. Reinhold Niebuhr, *The Nature and Destiny of Man: A Christian Interpretation* (New York: Scribner's Sons, 1941), 161-62.

54. For some discussion of possible traces of spirituality among nonhuman primates, see Jane Goodall, "Primate Spirituality," in *Encyclopedia of Religion and Nature*, ed. Bron Taylor (London: Continuum, 2005), 1303-6; Marc Bekoff, "Reflections on Animal Emotions and Beastly Virtues: Appreciating, Honoring and Respecting the Public Passions of Animals," *Journal for the Study of Religion, Nature and Culture* 1, no. 1 (2007): 72-73. I am grateful to my PhD student Eva van Urk for pointing my attention to these pieces.

55. Calvin, *Institutes* 1.3.3.

in and out. Neanderthals, for example, occasionally could reproduce with humans but mostly not. Therefore, questions like "Did Christ also die for the Neanderthals?" are unanswerable. We can say, though, that Christ died for all human beings[56]—that is, for all those who were enabled to respond to the divine calling by being created in God's image and who were thus destined for fellowship with God.

It is this destiny that comes with the divine image that also grounds our human dignity.[57] We therefore do not need some other, empirically based unique-making faculty or attribute to warrant this dignity—not even, I would suggest, a soul that was supernaturally implanted in us by God, since, as indicated above, who we are as human beings is not determined by how we came to be.[58] And we are who we are because of being called to share in God's image by living in fellowship with God and caring for other creatures. It is these God-given tasks and relationships that make us special enough to have inviolable rights. As Wolfhart Pannenberg says in a moving and powerful passage: "A feature of the dignity that accrues to us by virtue of our being destined to fellowship with God is that no actual humiliation that might befall us can extinguish it. In a special way, because they have nothing else that commands respect, the faces of the suffering and humbled and deprived are ennobled by the reflection of this dignity that none of us has by merit, that none of us can receive from others, and that no one can take from us."[59] This, then, is the answer to James Rachels, to our Reformed farmer, and to

---

56. See the exploration by Moritz, "Does Jesus Save the Neanderthals?" I leave aside here typically Reformed discussions on the scope of the atonement vis-à-vis predestination, as well as contemporary discussions on whether also nonhuman animals might need salvation.

57. Cf. Wolfhart Pannenberg, *Systematic Theology*, vol. 2 (Grand Rapids: Eerdmans, 1994), 176, and Kärkkäinen, *Creation and Humanity*, 287. Note that this definition of the *imago* in terms of our destination for fellowship with God, unlike certain substantive definitions, also includes the life of those with mental limitations.

58. It was suggested by Brian E. Daley, SJ, in personal communication that Pope John Paul II's insistence on "the spiritual soul as being created immediately by God" (cf. above, n. 14) might be interpreted as referring to the relationship God creatively establishes with human beings by enabling them to consciously relate to God and one another. If this charming interpretation was to be approved by the Vatican, this not only would benefit the science-religion dialogue but would also be of ecumenical significance.

59. Pannenberg, *Systematic Theology*, 2:177.

others who think we need some unique way of coming into being or some unique property to buttress our human dignity.

Meanwhile, it is clear that our God-given tasks and the profound ways in which we are relational beings cannot be conceived of without specific capacities. Thus, to be able to live in accordance to our calling, we have been endowed with many special substantive and structural characteristics that, though not unique in kind, are still unique in degree. As, for example, John Polkinghorne argues, "the consciousness of our animal cousins seems to be different from human self-consciousness."[60] Likewise, when it comes to symbolic and linguistic behavior, some have compared the difference between animals and humans to the difference between a two-dimensional area and a three-dimensional space. For example, only humans can produce symbols about what does not exist or is not there.[61] Similar arguments can be produced with respect to rationality (which only in humans leads to science, technology, sociopolitical organization, etc.), art, literature, music, economic behavior, and other elements of so-called cultural evolution. Even if we are willing to ascribe features such as personhood and morality to animal primates—as some ethologists do but most don't—it is clear that these function at a much more complex, much more developed, and much richer level in human beings. We can easily observe these enormous differences.[62] It is God who endowed us with a stunningly unique assemblage of personal, aesthetic, moral, and religious characteristics. God, as it were, evoked them in us, in order to call us into his image.[63]

---

60. John Polkinghorne, "Anthropology in an Evolutionary Context," in Soulen and Woodhead, *God and Human Dignity*, 95.

61. Celia Deane-Drummond, *Christ and Evolution: Wonder and Wisdom* (London: SCM, 2009), 165; see also J. Wentzel van Huyssteen, "When Were We Persons? Why Hominid Evolution Holds the Key to Embodied Personhood," *Neue Zeitschrift für Systematische Theologie und Religionsphilosophie* 52 (2010): 341-42 (on the "human capacity for the symbolic coding of the 'non-visible,'" 342).

62. Here, the old creationist bumper sticker saying "Did you ever see an ape drive a car?" has a point.

63. Michael Burdett rightly points out that capacities such as language, culture, self-awareness, and rationality also "exist in other creatures (even to a considerable degree of sophistication), but the relative human capacities are still superior and special," so that "a quantitative difference invites a qualitative distinction." He even argues that substantive views of the *imago* might still "hold up against modern challenges to human uniqueness." Burdett, "Image of God," 9-10, and cf. Visala, "*Imago Dei*, Dualism, and Evolution."

We might even argue that in all these cases the levels are so different that "the differences in degree . . . amount to differences in kind."[64] But also when we refrain from once again specifying biological or other empirical features that make us unique in kind, we have ample reason to continue to distinguish between human beings and animals. For even without being unique in kind, we are still special. As one evolutionary scientist has phrased it: "*Homo sapiens* is not simply an improved version of its ancestors—it's a new concept."[65] Thysius was not entirely wrong after all.

### 5.6 Why We Are Still Special

So where does all this leave us? What sort of answer should I have given the farmer who was criticizing my endorsement of evolutionary theory because he saw human dignity threatened by it? First of all, neither evolutionary theory in general nor the theory of common ancestry in particular has to be interpreted in such a way that it deprives us of our sense of human dignity. For we humans are still special, even after everything we now know (or think we know) about our close biological ties with all other created beings.

But, secondly, we are special for theological reasons rather than for biological ones. Theologians should learn from scientists to no longer ground their theories of the *imago Dei* and of human dignity on some supposedly unique-making biological or otherwise empirical characteristic that the human species has and others lack. And scientists should learn from theologians (insofar as they do not recognize this by themselves) that this state of affairs does not exclude all notions of human uniqueness and dignity, nor the particular meaning and value of human life. For we are special for theological reasons. That is, we are special because God has called us to be stewards of God's earthly creation, taking care of it in a myriad of ways. God has uniquely endowed us with the capacity to reach out to Godself and each other in profoundly relational ways, characterized by loving commitment. Moreover, in order for us to live in this profound relationality and to fulfill these tasks,

---

64. Polkinghorne, "Anthropology," 96; cf. Cortez, *Theological Anthropology*, 143n10. Cf. Alfred North Whitehead, *Modes of Thought* (New York: Macmillan, 1938), 37-38: "The distinction between men and animals is in one sense only a difference in degree. But the extent of the degree makes all the difference."

65. Ian Tattersall, *Becoming Human: Evolution and Human Uniqueness* (New York: Harcourt Brace, 1998), 188.

God has given us a mixture of capacities that are, if not unique in kind, at least unique in degree as compared to all other species. It is by this mixture of capacities that we continue to be objectively recognizable as members of the human species.

In brief, Christians can continue to uphold the notion of human special uniqueness as well as the belief that only humans were created in God's image in ways that are not arrogant to other creatures. The notion of God's calling us into his fellowship is strong enough to ground the classical doctrine of human dignity (as well as the derivative notion of human rights) as an articulation of the special status human beings have been endowed with by their creator. Some scholars have claimed that this theistic explanation may even be the only grounding of human dignity that holds water.[66] In any case, to the extent that it is no longer possible to convincingly develop the notion of human dignity along phenomenological lines (by grounding it in supposedly unique-making empirical traits), we will all the more need theology if we do not want to lose it. Thus, rather than making Christian faith and theology largely redundant, it can be argued that the theory of common descent makes them all the more important. Why we humans matter cannot simply be read off from the natural world, so we need the Word of God to know this. Let us hope that this idea at last may offer some consolation to our Reformed farmer and to all other believers who are troubled by the implications of evolutionary theory for the topic of human dignity.

---

66. Cf. Pannenberg, *Systematic Theology*, 2:175-79; Wolterstorff, *Justice*, 342-61; Nicholas Wolterstorff, *Understanding Liberal Democracy* (Oxford: Oxford University Press, 2012), 177-226; Kärkkäinen, *Creation and Humanity*, 285-89.

CHAPTER 6

# Evolution and Covenantal Theology

> The Fall is the place above all where theology and biology conflict.
>
> —Robert James Berry[1]

## 6.1 The Covenant as a Reformed Key Concept

Though not every systematic exposition of Reformed theology has the covenant as its key theme, the Reformed stance does take the covenantal nature of God's dealings with humanity seriously. Even though many people would associate Reformed theology, first of all, with a focus on (or even obsession with) God's eternal decrees, it has always highlighted the historical pattern of God's ongoing relationship with human creatures. In describing this pattern, many Reformed theologians single out the notion of the covenant that permeates the Bible—as *berith* in Hebrew, *diatheke* in Greek, and *pactum* in its Latin translations—as an organizing principle for interpreting the overall narrative flow of the Bible. Though not all Reformed theologians have been adherents of covenantal or federal theology in a strict sense (that label is usually reserved for a particular stream of theology originating in the second half of the sixteenth century and coming to maturity in the seventeenth century), it is no overstatement to claim that "all Reformed theology involves attending to the nature of God's covenantal life with humanity."[2]

---

1. Robert James Berry, "Did Darwin Dethrone Humankind?," in *Darwin, Creation, and the Fall: Theological Challenges*, ed. Robert James Berry and Thomas A. Noble (Nottingham: Apollos, 2009), 72.
2. R. Michael Allen, *Reformed Theology* (London: T&T Clark, 2010), 34; it is significant that the covenant is the theme of Allen's second chapter, right after his chapter

*Evolution and Covenantal Theology*

Although much internal discussion has ensued on the exact number of covenants as well as on the nature and conditions of each, it is widely agreed in the Reformed tradition that God made a covenant with humanity as soon as the first human being appeared on the scene. Reformed theology typically posits that this covenant (either as a "covenant of works" or as a "covenant of grace") was already made with Adam. It was then broken by Adam and Eve in that momentous event that came to be known as "the Fall" and that led to "original sin" in Adam's posterity. The good news of the gospel, however, is that God did not leave it at that but sent his only begotten Son Jesus Christ to suffer and die for the sins of humanity and to grant them eternal life through grace alone. This is the historical sequence or pattern of God's actions in his relationship with humanity that Reformed Christians—along with many others—have derived from the biblical narrative. Now, what happens to this grand soteriological story line when we accept the theory of common descent? In that case, has there ever been a first couple of human beings called Adam and Eve (§6.3)? Could it be held that the first human beings all of a sudden fell into sin (§§6.4, 6.5), as a result of which they as well as their descendants had to die at some stage (§6.6)? And how about God's redemptive agency in and through the work of Jesus Christ (§6.7)?

It is to this interrelated nexus of questions, which is key in many current evolution debates both in the United States and elsewhere, that we turn in this chapter. First, however, to see how evolutionary theory—and common descent in particular—affects these issues, we will briefly rehearse the account of human origins as it emerges from contemporary science (§6.2).

## 6.2 The Scientific Story of Human Origins

What scenario emerges from the contemporary scientific study of human origins? Roughly speaking, its main lines can be sketched as follows. The human species came into being through long-standing processes of evolution. Sharing a common ancestry with the apes, our ancestors looked less like modern humans and more like the apes the further we go back in time. This is clear from the fossil record as it has been established during the last century or so. Starting from the oldest hominins called *Ardipithecus* (dated 4.4

---

"The Word of God." Classic statements of covenant theology can be found in the Westminster Confession of 1646 (cf., e.g., its chap. 7) and in the work of theologians such as Heinrich Bullinger, John Calvin, Caspar Olevian, John Owen, Johannes Cocceius (cf. §3.5 above), and Francis Turretin.

million years ago; according to current estimations, at around 6 or 7 million years ago the human and the chimpanzee lines separated), the brain size of our ancestors gradually increased when it went from *Australopithecus afarensis* to *Homo habilis*, *Homo erectus*, *Homo heidelbergensis* (the so-called archaic human), and finally *Homo sapiens*, that is, the anatomically modern human.[3] The Neanderthals (or *Homo neanderthalensis*), like the so-called Denisovans, are generally believed to be *Homo sapiens'* "cousins," having evolved from *Homo heidelbergensis* through separate branches that did not survive the evolutionary process. The oldest fossils of *Homo heidelbergensis* (like so many others) have been found in Africa, and they are currently believed to date back to around 700,000 years ago. The first members of our species, *Homo sapiens*, can be recognized from the fact that they had skeletal features fully indistinguishable from ours. They also appeared in Africa, about 200,000 years ago.[4] Having spread rapidly throughout Africa, the first humans must have started to foray out of Africa around 100,000 years ago.[5]

Recent studies of genetic diversity in the present-day human population confirm these timescales. They further suggest that our African ancestor population was relatively small, consisting of some 10,000 to 12,500 reproductively active individuals.[6] Obviously, however, this is still many more than the two in-

---

3. Though the nomenclature is still under discussion, "hominins" has recently become the main term to denote the genus *Homo*; "hominids" is usually considered to include humans and the great apes. How exactly all hominin (and hominid) species and subspecies relate to each other is not known; most writers therefore refrain from constructing an evolutionary tree of them.

4. Cf., e.g., Robin Dunbar, *Human Evolution: A Pelican Introduction* (London: Penguin Books, 2014), 3–23, 217–22; Darrel R. Falk, "Human Origins: The Scientific Story," in *Evolution and the Fall*, ed. William T. Cavanaugh and James K. A. Smith (Grand Rapids: Eerdmans, 2017), 3–22. Falk points to the fact that although "hints of the story began to emerge late in the nineteenth and early in the twentieth century, most aspects of it have only become fully apparent in the past forty or fifty years" (6). Obviously, this "knowledge-explosion" (7) makes the theological task that we face in this chapter more urgent than ever.

5. Thus Falk, "Human Origins," 7; other sources mention seventy thousand or sixty thousand years ago (may I, as a theologian, complain about such considerable dating divergences in current scientific literature on human origins—and especially about the fact that most authors seem to take their own numbers for granted without properly discussing those given by their colleagues?).

6. Dunbar, *Human Evolution*, 14; Denis Alexander, *Creation or Evolution: Do We Have to Choose?*, 2nd rev. ed. (Oxford: Monarch Books, 2014), 265; and Dennis R.

*Evolution and Covenantal Theology*

dividuals with whom it all started according to the traditional view based on the Genesis narrative! Another so-called population bottleneck, once more comprising several thousand individuals, occurred around 50,000 to 70,000 years ago, and according to geneticists, "all of today's non-Africans are descended from the relatively small number of individuals that left Africa at about that time."[7] Unlike what the "multiregional hypothesis of human origins" assumed, the low level of genetic variety in contemporary humankind strongly suggests that we all descend from one and the same African population rather than from several geographic regions. In that sense, it seems that we humans can still be said to stem "from one," as Acts 17:26 has it.[8] From Africa the offspring of these first groups of *Homo sapiens* spread into Asia and Europe, Australia and the Pacific islands, and finally to the Americas.[9] Thus, the monophyletic account of human origins, according to which the various human races ultimately stem from one and the same population, has become more plausible than the polyphyletic model, that is, the theory that humans emerged from various hominin populations at different geographical locations.[10] So the evidence suggests

---

Venema and Scot McKnight, *Adam and Genome: Reading Scripture after Genetic Science* (Grand Rapids: Brazos, 2017), 43–55.

7. Falk, "Human Origins," 11. With reference to Eugene E. Harris, *Ancestors in Our Genome: The New Science of Human Evolution* (New York: Oxford University Press, 2015), Falk adds that there was never a time when the number of reproducing human individuals was fewer than ten thousand.

8. Many translations add a noun after "one" (e.g., "blood," or even "ancestor," NRSV), but the oldest Greek manuscripts don't have one. Therefore, the text cannot serve as a proof text for the belief that all humans stem from a single couple (or "monogenism"). John H. Walton, *The Lost World of Adam and Eve: Genesis 2–3 and the Human Origins Debate* (Downers Grove, IL: IVP Academic, 2015), 186–87, argues that Acts 17:26 may well refer to Noah instead of Adam, but it can be doubted whether Paul's audience in Athens would have understood such a reference.

9. Again, there are dating differences here. According to Falk, "Human Origins," 7–8, the first human fossils found in Australia are from some sixty thousand years ago, those in Europe from about forty-two thousand years ago, and in America from around eighteen thousand years ago (Chile). See also, e.g., Deborah B. Haarsma and Loren D. Haarsma, *Origins: Christian Perspectives on Creation, Evolution, and Intelligent Design* (Grand Rapids: Faith Alive, 2011), 232–37; Berry, "Did Darwin Dethrone Humankind?," 56–57. For the genetic details, see Venema and McKnight, *Adam and the Genome*, 60–65; Graeme Finlay, *Human Evolution: Genes, Genealogies, and Phylogenies* (Cambridge: Cambridge University Press, 2013); Alexander, *Creation or Evolution*, 252–305.

10. Note that what we may call monophyletism (i.e., the theory that we stem from "one stock") is to be distinguished from monogenism, which is usually taken to

a strong biological bond among modern humans: it seems that indeed we are all one big family.

In the evolutionary processes that led to the human species as we now know it, natural selection played an important role. Our ancestors had to compete with others for food, shelter, and other scarce resources. They had to protect themselves from predators and other threats, and so they developed sophisticated mechanisms of defense and aggression in order to survive as a species in the struggle for life. These mechanisms have been studied in animals that are relatively close to human beings from a genetic point of view, such as chimpanzees and bonobos, and it turns out that almost all types of acts "regarded as 'sinful' in humans are part of the normal, natural repertoire of behavior in other species" such as these.[11] Bullying, deception, infanticide, cannibalism, lethal conflicts with neighboring groups—all these types of aggressive, self-serving behavior not only have been observed but also seem to be pervasive among many species.[12] As we saw in chapter 4, sometimes the disposition to cause extreme pain and slow deaths to other animals is even built into the very body plan of an animal. To be sure, we also saw that evolution fosters empathy, social cooperation, and other forms of "virtuous" conduct among animals; but even these patterns of behavior are often believed to serve the survival of their own species (or genes, or whatever the "unit of selection" is supposed to be), showing what we might call a form of collective self-interest.[13]

Let us assume that we human beings emerged from this realm of violent primates, inheriting many if not all the character traits and behavioral patterns just mentioned. Then it seems that there has never been a paradisiacal time when everything we did was still all right. By consequence, it seems that this scenario does not allow for something like a historical or primordial *Fall*. Not only were death and decay, suffering and starvation part of the natural world

---

mean that we stem from a *single pair*; unlike monophyletism, monogenism is ruled out by contemporary scientific findings. Cf. Karl Rahner, "Theological Reflexions on Monogenism," in *Theological Investigations* 1 (London: Darton, Longman & Todd, 1961), 229-96, who argues that monophyletism is sufficient to uphold classical theological claims on original sin so that "monogenism in the strict sense" (286) may be dropped.

11. As says Catholic paleontologist Daryl P. Domning in a book coauthored with Monika K. Hellwig, *Original Selfishness: Original Sin and Evil in the Light of Evolution* (Aldershot, UK: Ashgate, 2006), 102.

12. See for examples Domning and Hellwig, *Original Selfishness*, 102-4.

13. Domning and Hellwig, *Original Selfishness*, 106-7; cf. John R. Schneider, "Recent Genetic Science and Christian Theology on Human Origins," *Perspectives on Science and Christian Faith* 62 (2010): 202.

*Evolution and Covenantal Theology*

long before humanity appeared on the scene, so that nature did not "fall" into such a state as a result of human behavior. But even—and this may be the hardest part of the scientific story—we human beings ourselves seem to have been aggressive savages from the very beginning of our existence. Rather than *having* fallen, it seems that we were "created fallen."[14] No wonder Arthur Peacocke bluntly says: "The traditional interpretation of the third chapter of Genesis that there was a historical 'Fall,' an action by our human progenitors that is the explanation of biological death, has to be rejected. . . . There was no golden age, no perfect past, no individuals, 'Adam' or 'Eve' from whom all human beings have descended and declined and who were perfect in their relationships and behaviour."[15] In short, things have never been substantially better than they are today, so there is no reason to speak of a "fall" into the present state. Or is there?

Though the picture of human descent sketched above is more or less generally agreed upon by scientists, perhaps in hindsight some time spans or numbers will turn out to be exaggerated and further research will lead to more nuanced conclusions. Perhaps some of the estimations were put forward by scientists with a strong view in the creation-evolution debate, who were subconsciously led by the wish to come up with "hard facts" that are devastating to their antievolution opponents. Perhaps new paradigms will emerge that shed a surprisingly new light on human descent. At any rate, there are still uncertainties when it comes to human origins and a great many things we just don't know.

Questioning or qualifying widely held scientific assumptions, however, is too easy a way out here—especially when we only retreat to such a strategy when the reported data do not please us. So let us bite the bullet by asking our lead question: What if it is true? First, we examine what the scientific picture summarized above would imply for the so-called historical Adam and the interpretation of the story about Adam and Eve in Genesis 2-3.

### 6.3 Genesis 2-3 and the Historical Adam

The discussion on how to read Genesis 2 and 3 has recently received fresh impetus by Old Testament specialists and other scholars who have launched

---

14. Jeff Astley, "Evolution and Evil: The Difference Darwinism Makes in Theology and Spirituality," in *Reading Genesis after Darwin*, ed. Stephen C. Barton and David Wilkinson (Oxford: Oxford University Press, 2009), 173.

15. Arthur Peacocke, *Theology for a Scientific Age: Being and Becoming—Natural, Divine, and Human*, enlarged ed. (Minneapolis: Fortress, 1993), 222-23.

novel theories, drawing on newly gained insights from the ancient Eastern world. For example, during the past couple of years biblical scholars Marjo Korpel and Johannes de Moor published their theory that two old clay tablets from Ugarit contain the Adamic myth, which in their view has been creatively and critically adopted by the writers of Genesis 2 and 3.[16] Their colleague John Walton applied his theory that Genesis 1 is not about the material origin of the world but rather about its order and function to Genesis 2 and 3 as well. In both Scripture passages, particular roles and functions are assigned to entities and living beings in order to secure an orderly world. Walton argues that many of the claims made in these chapters are archetypical rather than historical: Adam and Eve "embody all people and the affirmations of the forming accounts are affirmations made of everyone, not uniquely of them."[17] Evolutionary anthropologist Carel van Schaik and historian Kai Michel argue that the first chapters of Genesis (like many more in the Bible) reflect the transition in the cultural evolution of *Homo sapiens* from gatherer-hunter societies to a sedentary and agricultural way of life, some twelve thousand to ten thousand years ago.[18]

This is only a small sample of the available exegetical theories—and only from the past couple of years. It would be a daunting task to describe and assess all the options for interpreting Genesis 2-3 that have been put forward in the recent and not-so-recent past. Fortunately, we do not need to engage in this endeavor here. Nor do we need to enter protracted debates on the sources and dates of these chapters. Remember that our leading question is whether evolutionary theory is compatible with Reformed theology, a theology that interprets the Bible in terms of a historical pattern of divine actions ranging from creation through fall to redemption and consummation. The question I will focus on in this section is this: Which group of available exegetical theories—if any—allows us to remain faithful to this deeply covenantal character of Reformed theology while doing justice to the scientific data on human origins reviewed above? If we know the answer, we can then make up our mind about the plausibility of the basic hermeneutical assumptions of this category.

Thus, following up on Denis Alexander's insightful typology, I will give an

---

16. Marjo C. A. Korpel and Johannes C. de Moor, *Adam, Eve, and the Devil: A New Beginning* (Sheffield: Sheffield Phoenix Press, 2014).

17. Walton, *Lost World of Adam and Eve*, 199.

18. Carel van Schaik and Kai Michel, *The Good Book of Human Nature: An Evolutionary Reading of the Bible* (New York: Basic Books, 2016). Their theory suggests a time gap of no less than nine thousand years between this "Neolithic revolution" and its reflection in the biblical texts!

overview of five different *types* of interpretation of Genesis 2-3, each of which may encompass various exegetical theories. In doing so, I will focus on the most relevant issue for our purposes, namely, how each of them considers the role of history in these chapters.[19] I will discuss these types (or models, or approaches) in such a way that each subsequent approach assumes more historical elements in the narrative than the preceding one. Though it would be artificial to draw a sharp dividing line here, in this section I will focus mostly on the historicity of Adam in relation to Genesis 2-3 and in the next one on the notion of the Fall in relation to both the Genesis story and Paul's argument in Romans 5.

### The Ahistorical Approach

This approach—also sometimes called the *paradigmatic* view—tells us to read Genesis 2 and 3 as a myth (in its technical sense)[20] with a strong theological thrust: the chapters answer some of humankind's (or Israel's) perennial existential questions from the perspective of our relationship and that of our world with God. They picture humans as created by God but also as disobedient to God. At a narrative level we note a sequence—the human being is a creature before he becomes a sinner—but this does not determine the exegesis. In certain literary genres, indications of sequence must not be taken literally. If one applies this method, for example, to the book of Revelation, one soon discovers that it does not work; this is also true for other prophetic texts. In many ways Genesis 1-3 resembles a prophetic text, but it is projected in the reverse direction: what we are as human beings before God is not projected on the screen of the *future* but on that of the *past*. Paradise symbolizes how life could be when lived with God, whereas the "Fall" symbolizes our coming of age, including the unavoidable loss of innocence that takes place as soon as we realize that we can perform both morally good and morally evil actions (cf. Gen. 3:5, 22).[21]

---

19. See Alexander, *Creation or Evolution*, 288-94; my list basically corresponds to Alexander's, but I sometimes use different labels and elaborations.

20. Cf., e.g., Robert Segal, *Myth: A Very Short Introduction* (Oxford: Oxford University Press, 2004). The main reason why Gen. 2-3 is qualified as myth is the occurrence of many parallels with ancient Near Eastern texts whose mythological character is undisputed.

21. See, e.g., Lyn M. Bechtel, "Genesis 2.4b-3.24: A Myth about Human Maturation," *Journal for the Study of the Old Testament* 67 (1995): 3-26; and more moderately Dan Harlow, "After Adam: Reading Genesis in an Age of Evolutionary Science," *Per-*

In this way, Genesis 2 and 3 are not about the first human beings but about all men and women, about being human in general. It is common in this group of interpretations that the word "Adam" does not just refer to a particular human being. Almost everywhere in these chapters the name Adam could also be translated as "man" or "humankind." Adam is *Everyman*, a main character in late medieval morality plays: he represents every human being. Genesis 2 and 3 want to tell us that the story of Adam and Eve is repeated in the reality of the lives of all of us. Genesis 3 describes in narrative form the estrangement from God, which is—again and again—the result of the choices we make because of our lustful desires, hubris, and unbelief. This estrangement is highlighted by the entrance of death as a punishment for the sin that was committed. This narrative trait is not to be taken literally, however; it expresses how we will remain unable to flourish and will miss our destination when we live our lives closed off from God.

In brief: these chapters must be understood as *theological* rather than historical texts. From a historical point of view, physical death has always been part of our humanness, and suffering, death, and extinction are inherent to the history of our evolutionary ancestors. However, in mythical form Genesis 2 and 3 do tell us something (and perhaps what is most essential) about who we humans are before God. *We* are Adam and Eve. We have been made by God in a wonderful way, but at the same time we are sinners. However, we also receive the remarkable promise that God is going to do something about this situation (Gen. 3:15).[22]

### The Prehistorical Approach

Those who support this second type of exegesis concur with the theological reading of type one but cautiously situate the main events related in Genesis 2–3 "beyond the text" in prehistory—that is, in the time that precedes the appearance of written sources in a specific culture. In particular, they underline

---

*spectives on Science and Christian Faith* 62 (2010): 179–95 ("Genesis 2–3 can be read on a certain level as a coming-of-age story" [189]).

22. An example (out of many) of the ahistorical approach is Peter Enns, *The Evolution of Adam: What the Bible Does and Doesn't Say about Human Origins* (Grand Rapids: Brazos, 2012). Enns interprets the "Adam story" as a symbolic "preview of Israel's history, from exodus to exile" (141), which is intended to answer the question of Israel's "national identity" (142).

*Evolution and Covenantal Theology*

that the *couleur locale* (culture, geography, etc.) of Genesis 2 and 3 points to the Neolithic period (literally the "new stone age"), somewhere in the Middle East. These chapters describe the period in which people began to cultivate the earth. According to archaeologists, this period, in which humans left their nomadic hunter-gatherer way of life behind, began at around 10,000 BC. Around this time also the first cities were established in what is now Turkey (cf. Gen. 4:17).[23] This time frame is long after the emergence of the first anatomically modern humans, which is usually dated between 200,000 and 120,000 years ago.

It must, therefore, be recognized that Adam and Eve did not form the first human couple. At their time a considerable number of humans were already living and dying—their total is estimated at between 1 million and 10 million people. But Adam and Eve did exist, presumably as the chiefs of a Neolithic community, and at some moment they were called by God to live in a conscious relationship with him. The calling was in many ways similar to the later election of Abram and Sarai, the only difference being that the latter couple was already part of a tradition in which God had a relationship with humankind, while Adam and Eve stand at the beginning of this chain. "It is not that there were no settled farmers beforehand, but from now on there would be a community who would know that they were called to a holy enterprise, called to be stewards of God's creation, called to know God personally."[24] John Stott aptly labeled the group of humans that came on the scene in this way *Homo divinus* (the godly humans).[25] Genesis 3 then informs us about the intentional rejection of God's calling by these humans. The story of the Fall symbolizes this rupture in the relationship with God. This did not only have its "vertical" consequences through later generations but also "horizontal" effects among the many contemporaries of Adam and Eve. Thus the Fall was a kind of spiritual nuclear bomb: from the moment Adam and Eve sinned, the inclination to follow their sinful course spread to all people on earth. Or at least: to all those people who were included in the covenant with God (presumably Adam's "clan"), since these were the only ones who could break it.

Adherents of this view usually see many figurative, metaphorical, or symbolic elements in Genesis 2-3. For example, the creation of Eve out of

---

23. For an anthropological (nontheological) reading that belongs in this category, see the study of Van Schaik and Michel referred to above, *Good Book*, 29-57.

24. Alexander, *Creation or Evolution*, 290.

25. Cf. John Stott, *Understanding the Bible* (London: Scripture Union, 1972), 63; for a later elaboration, cf. R. J. Berry, "This Cursed Earth: Is 'the Fall' Credible?," *Science and Christian Belief* 11 (1999): 29-49.

one of Adam's ribs is vivid pictorial language that points to the intimacy of the marriage relationship, in which the partners physically belong together. The paradisiacal situation (just like the word "fall," the word "paradise" does not occur in Gen. 2-3) symbolizes how good it was to know God and to live with him. One might compare this to the honeymoon at the beginning of a marriage that is later broken. As in the first model, however, since people had died all along, the reference to the entrance of death in the human world (Gen. 2:17; 3:19) should not be taken literally. It reflects the fact that from that time forward death was experienced as a sign of divine judgment. In short: the similarity of this view with the ahistorical approach is that many elements in Genesis 2 and 3 have a symbolic meaning. The difference is that Adam and Eve were historical persons who lived in a community at the end of what we refer to as prehistory—and who fell into sin.[26]

*The Primordial or Protohistorical Approach*

The third exegetical approach argues, in contrast to the first two options, that in Genesis 2 and 3 we do in fact meet with the first anatomically modern human beings—that is, the first members (along with their clan) of the species *Homo sapiens*. This may, for instance, be deduced from the fact that in Genesis 3:20 Adam calls his wife "Eve," since she is the "mother of all living." Thus, these chapters deal with the very beginning of human history. Yet it is not just a historical record in the ordinary sense of the term, because our conventional historiographic tools do not enable us to reach as far back as this period. Therefore, the knowledge mediated by Genesis 2 and 3 comes to us through revelation—or through "divination," that is, through a prophetic vision of the constitutive events that must have shaped the historical reality in which we live.[27] This vision cannot be verified or falsified by any historiographical means; it can just be believed—or not believed. To underline the special, unique character of the history depicted here, it is often, in particular

---

26. For a further elaboration and defense of this view, see Alexander, *Creation or Evolution*, 289-94, 300-304, 317-18.

27. Cf. Karl Barth, *Church Dogmatics* III/1(Edinburgh: T&T Clark, 1958), 83; whereas Barth connects this divination with poetry, I would rather link it to the phenomenon of prophecy; indeed, the genre of the first chapters of Genesis is not poetical but could be referred to as "backward prophecy": reality—in this case primeval reality—is critically depicted from the perspective of what really matters.

by Old Testament scholars, referred to as primordial or primeval history (the German equivalent is *Urgeschichte*).²⁸

The language that tells us about this "edge" of history must necessarily use images (as does the language in which the Bible describes the other "edge" of history, the end times). It deals with facts, but these facts are communicated in a figurative manner. Karl Barth spoke in this connection of a "saga," which he clearly distinguished from a "myth."²⁹ A saga is (in this view) a story in which human imagination plays a major role but which is still based on a real event (or series of real events). Sometimes the supporters of this primordial approach link these events to the data provided by modern paleoanthropology. This would mean that—when we disregard all figurative "embellishments"—the events described in Genesis 2 and 3 occurred between 200,000 and 120,000 years ago, somewhere in Africa. As we saw, that is when and where the first anatomically modern human beings came into existence. However, to help the first readers of the Genesis story understand it, the story was placed in the cultural setting of agricultural people in the Middle East during the Neolithic period. This differs from the primordial view in that the events lying "beyond the text" are dated as far back as the rise of the human species, rather than only at the end of prehistory when *Homo sapiens* had been around for a long time already.

Some adherents of this approach insist that Adam and Eve, or their clan, somehow became the sole progenitors of contemporary humankind. Others allow for more flexibility here, suggesting that other *Homo sapiens* communities may well have emerged and survived next to them (thus coming close to the prehistoric view in this respect). Still others opt for a gradualist version of the primordial view, in which both the emergence of the *imago Dei* and the Fall took place incrementally (here, Adam and Eve as individuals disappear from view).³⁰

---

28. Cf. Claus Westermann, *Genesis 1-11: A Commentary* (Minneapolis: Fortress, 1994), who refers to Gen. 1-11 as "the story of the primeval events" (1). In the German original (Neukirchen-Vluyn, 1974), he calls Gen. 2-3 "ein unsere Geschichte jenseitiges Geschehen" (376).

29. Barth, *Church Dogmatics* III/1, 81. Barth defines a saga as "an intuitive and poetic picture of pre-historical reality"; as to myth, its content and object are "general realities of the natural and spiritual cosmos which, in distinction from concrete history [*Geschichte*], are not confined to definite times and places" (84). Since Gen. 1 and 2 are "portrayals of concrete events," they are "pure saga" (84).

30. See for these varieties Alexander, *Creation or Evolution*, 288-89 (the subdivisions have been inserted in this second edition). Adherents of this view are, e.g., Allan John Day, "Adam, Anthropology and the Genesis Record: Taking Genesis Seriously in the Light of Contemporary Science," *Science and Christian Belief* 10 (1998): 115-43;

### The Old-Historical Approach

The fourth type of exegesis proceeds even further on the path that the supporters of the third type have chosen. Just like the primordial view, this approach links Genesis 2-3 to the first human beings, but it considers the text to a greater extent as a journalistic, more or less "literal" account of what happened around them. Those who hold this view believe that humanity started with the couple mentioned in this text, Adam and Eve, who spent a more or less harmonious period of time before they fell into sin. At that time, they were the only humans around. Even though as far as I know nobody supports a literal interpretation of Genesis 2:7 (God using his hands to form the human body from the dust), adherents of the old-historical approach argue that this verse points to God's direct intervention in the creation of man. Thus, the first physical human beings did not evolve in an evolutionary process from animals but came about through a special creative act of God. Similarly, adherents of this view typically hold that God specially intervened at different points on the time line in order to create the first instantiations of each of the main biological groups.

Despite what the book of Genesis seems to suggest, this all happened a very long time ago—about as long ago as contemporary scientists date the appearance of the various biological groups on the earth. Thus, the nature and meaning of Genesis 2 and 3 are not primordial but old-historical: the text describes what went on in history in a way that is not structurally different from usual historical reports.[31] Perhaps the details have been transmitted orally over time, or alternatively, revealed to the writer(s) in some supernatural way. Anyway, the text refers to events that must have happened tens or perhaps hundreds of thousands of years ago. The garden of Eden refers to a place of relative rest and harmony within a world that otherwise closely resembles our natural world with all its threats and dangers. The genealogies that follow Genesis 3 do not provide information that allows us to date the events of Genesis 2 and 3. We don't know how many generations may have been skipped.

---

Francis S. Collins, *The Language of God: A Scientist Presents Evidence for Belief* (New York: Free Press, 2006), 208-9 (drawing on C. S. Lewis); John J. Bimson, "Doctrines of the Fall and Sin after Darwin," in *Theology after Darwin*, ed. Michael S. Northcott and R. J. Berry (Milton Keynes, UK: Paternoster, 2009), 116-19.

31. It is often nuanced in this connection that the style in Gen. 2-3 is not "*exact-historical*"; yet, most of the details in the text are usually considered to be reality-depicting.

And at any rate, using these registers to count years would fail to do justice to their *genre* and *scopus* (intent).

Representatives of this approach vary on how literally to take the different elements of the text. But in general we may conclude that, at the very least, God created humankind and the main biological groups *de novo*, while Genesis 3 shows how these human beings not only brought a spiritual but also a physical death upon themselves by disobeying God. Some hold that, prior to this moment, animals already died, for example, because of predation; other proponents of the old-historical approach deny this.[32]

### The Young-Historical Approach

The fifth approach has become the option that is most familiar to many people, Christians and non-Christians alike.[33] It takes as its point of departure—mainly on the basis of the genealogies in Genesis—that the events related in Genesis 2 and 3 must have occurred between six thousand and, at most, ten thousand years ago. Everything described in Genesis 2 and 3 has happened more or less literally as described. This, however, does not detract from the fact that these chapters also tell the story of all humankind and communicate important theological truths. Unless the text gives clear indications to resort to a figurative, symbolic, or only exemplary interpretation of certain elements, however, we must stick to a literal reading.

The data in Genesis 2 must also be related to those in Genesis 1. This means that Adam and Eve were created on the sixth day, counted from the moment heaven and earth came into existence. These days were ordinary days of twenty-four hours, as we know them today (even though some minor fluc-

---

32. C. John Collins, who is a representative of the old-historical approach, espouses the first view. See his *Genesis 1-4: A Linguistic, Literary, and Theological Commentary* (Phillipsburg, NJ: P&R, 2006), 165: "it is a mistake to read Genesis 2:17 as implying that physical death did not affect the creation before the Fall." See also his *Did Adam and Eve Really Exist? Who They Were and Why You Should Care* (Wheaton, IL: Crossway, 2011) and his "A Historical Adam: Old Creation View," in *Four Views on the Historical Adam*, ed. Matthew Barrett and Ardel B. Caneday (Grand Rapids: Zondervan, 2013), 143-75 (with responses by representatives of most of the other views discussed here, 176-95).

33. One may wonder whether that is because it is the most straightforward reading, because it is pushed by young-earth creationism, or (from the atheist side) because it can most easily be rejected as fairytale-like. Probably all three of these play a role.

tuations toward, say, twenty-five hours are not excluded). As the consequence of this linking of Genesis 1 and 2 (where we see a different sequence in the creative acts of God), we must in some cases in Genesis 2 use the pluperfect in our translation. Thus, not: "So out of the ground the LORD God formed every beast of the field" (RSV) but rather: "When God *had* made all the beasts of the field" (Gen. 2:19). This is necessary since, according to Genesis 1, the creation of the animal world preceded the creation of human beings. Moreover, the word "day" in Genesis 2:17 cannot be taken in a literal sense, since Adam and Eve did not die on the very day they ate from the tree of knowledge but much later.[34] In this young-historical approach, it is also assumed that death and suffering did not exist in paradise. Instead, after the first sin—either suddenly or in a process of degeneration—new natural laws and processes were established. These would, for instance, ensure that animals were no longer vegetarian but started to devour each other. Usually an exception is made for the death of plants, but animals began to die only after (and as a consequence of) the human fall into sin.[35]

This completes our short survey of the five basic exegetical approaches to Genesis 2 and 3. I have tried to describe their basic features as neutrally as possible. Let us now return to the question that formed our point of departure: Which approach, if any, is compatible with the basic tenets of Reformed (covenantal) theology while at the same time taking the scientific data about human origins seriously? Here is my assessment. Option 5 obviously does not do justice to scientific estimations of the age of the earth. Whereas option 4 does attempt to do this, it denies the theory of common descent. So if we take common descent for granted, as we do for the sake of argument in this book, then the old-historical view is no option. It is important to see, however, that both ap-

---

34. As Alexander, *Creation or Evolution*, 197–98, shows, all Christians, including those who prefer a young-historical interpretation, interpret parts of Gen. 2–3 figuratively. He uses *The Genesis Record* (Grand Rapids: Baker Books, 1976) of Henry M. Morris, founder of modern young-earth creationism, to illustrate this point.

35. Though he does not want to endorse the adjective "exact," James K. Hoffmeier comes close to the young-historical view in his essay "Genesis 1–11 as History and Theology," in *Genesis: History, Fiction, or Neither? Three Views on the Bible's Earliest Chapters*, ed. Charles Halton (Grand Rapids: Zondervan, 2015), 23–58. In the Netherlands, an important exponent of this view is the Old Testament scholar Mart-Jan Paul; see esp. his *Oorspronkelijk. Overwegingen bij schepping en evolutie* [Originally: Considerations on creation and evolution] (Apeldoorn: Labarum Academic, 2017).

proaches also have important internal problems. For example, neither of them explains how it is possible that the order of God's creative acts in Genesis 2 is clearly different from their order in Genesis 1. Coming from the other side, it is hard to see how option 1 can be reconciled with the doctrine of the Fall, which (as I will set out to demonstrate in the next section) forms an integral part of the overall scheme of Christian theology, both Reformed and Catholic. According to the paradigmatic view, the biblical text that serves as *locus classicus* for the Fall, Genesis 3, cannot be adduced in support of it—and indeed adherents of option 1 usually deny a "historical fall."

This leaves us with options 2 and 3—the prehistoric and the primordial approach. How should we decide between them? Both seem compatible with the doctrine of the Fall, although we must still sort out how such an event (or series of events) might actually be envisaged. Both can also be taken in such a way as to uphold a "historical Adam," although this individual must have had quite a number of conspecifics. It seems, however, that neither option 2 nor option 3 does full justice to the scientific data on human origins. For the emergence of personhood occurred much later on the time line than, say, 200,000 to 120,000 years ago with the first appearance of *Homo sapiens*, and much earlier than in the Neolithic period (ca. 12,000 years ago). Personhood in the sense of having the ability to engage in reciprocal personal relationships with other people and to communicate with them in the intricate ways that we humans do is usually thought to derive from the distinctive capacity to think and act in a symbolic way. The capacity for symbolic thought enables us to use an astonishing number of symbols that we can combine into ever new constellations that help us interpret the world; communicate (using highly sophisticated forms of language); reflect upon ourselves, the past, and the future; discern others as having minds just like we do; engage in abstract thinking; etc. Whereas we do not see the first inklings of this capacity until about 100,000 years ago, its full flourishing is documented only in the Upper Paleolithic cave art in southwestern France and northern Spain—that is, not later than some 45,000 years ago. It is quite understandable that we became anatomically modern humans *before* the capacity for symbolic thought and action—either gradually or in a more discontinuous way through "emergence"[36]—could start to evolve.

36. For this debate, see Malcolm Jeeves, ed., *The Emergence of Personhood: A Quantum Leap?* (Grand Rapids: Eerdmans, 2015), as well as the creative philosophical study of Jacob Klapwijk, *Purpose in the Living World? Creation and Emergent Evolution* (Cambridge: Cambridge University Press, 2008), esp. 160-71.

Given all this, it makes little sense to situate Adam and Eve at the beginning of the speciation event that led to anatomically modern human beings, as option 3 does. Instead, they might better be associated with "the dramatic so-called cultural big bang, the Upper Paleolithic revolution and the explosive growth of human creativity around 45,000 years ago."[37] It is after this transition that an artistic and religious awareness can be demonstrated that corresponds to people experiencing themselves as being addressed by God, or the transcendent. This is not to say that Adam and Eve and their group "really" lived in Europe (that would be a suspiciously Eurocentric view!). It is just to suggest that Genesis 2-3, using the later cultural setting of agricultural people in the Middle East during the Neolithicum, by way of prophetic divination and critical adoption of older ancient Near Eastern materials, tells the story of humans—wherever and whenever they may have lived—who for the first time came to hear and understand the voice of God.

Is such a view at all compatible with a sound exegesis of the biblical text? Aren't we stretching the text on the Procrustean bed of science in this way? Not necessarily. It is often thought that the young-historical reading of the text was universally accepted as self-evident before the nineteenth century. Many people think that only when the geological timescale and the Darwinian theory of evolution took hold, some Christians desperately started to twist the data in order to reconcile the science of the day with their faith. This historical construction is demonstrably false, however. From the earliest centuries onward, Christians have been struck by the interpretative challenges elicited by the first chapters of Genesis and have debated their correct interpretation. In doing so, no one less than Augustine held that these chapters must offer a nonliteral representation of God's creative work, since, for example, obviously day and night cannot have existed before the sun was created (as Gen. 1 suggests). This led Augustine to look for the literal

---

37. J. Wentzel van Huyssteen, *Alone in the World? Human Uniqueness in Science and Theology* (Grand Rapids: Eerdmans, 2006), 66; cf. Ian Tattersall, "Human Evolution: Personhood and Emergence," in Jeeves, *The Emergence of Personhood*, 44-45: "Significantly, even the earliest anatomically recognizable *Homo sapiens*, which show up in the African fossil record at under 200 thousands years ago . . . left nothing that would lead us to suppose that they possessed the symbolic faculty." For criticism of the alleged sudden character of the transition, see, e.g., Joshua M. Moritz, "Does Jesus Save the Neanderthals? Theological Perspectives on the Evolutionary Origins and Boundaries of Human Nature," *Dialog* 54, no. 1 (2015): 54-55, and Colin Renfrew, "Personhood: Towards a Gradualist Approach," in Jeeves, *The Emergence of Personhood*, 51-67.

reality behind the text.[38] In doing so, he "argued for a tentative approach allowing a latitude of conceptions but not the literal sense."[39] Thus, when we proceed in a similar way today, that is not at all new or artificial, even when we might not necessarily follow Augustine in assuming a special creation of Adam and Eve.

Moreover, by ascribing a paradigmatic or archetypal function to the text (according to which much of it applies to all people, not just the first ones), we do justice to its theological meaning, thus acknowledging the crucial insight of those who endorse option 1. On the other hand, by discerning its etiological function—that is, its intention to elucidate the cause (Greek: *aitia*) of the current human condition—we take seriously, just as adherents of options 3-5 rightly do, that it is a story of human origins. In this sense, the text indeed presupposes a "historical Adam."[40] This is clear, for example, from the fact that ancient Near Eastern genealogies such as we find in Genesis 5 did not contain the names of fictitious persons but "the names of real people who inhabited a real past."[41] It is also clear from the fact that Eve is called "the mother of all living" in Genesis 3:20. There is no reason to read this as "the mother of all human beings." It is not necessary to assume a relationship of biological inheritance between Eve and all other humans, since in Hebrew the word "mother" does not necessarily imply that (cf. Gen. 4:21, where Jubal is called "the father of all those who play the lyre and the pipe").[42] It does indicate, however, that

---

38. As is correctly observed by Peter van Inwagen, "Genesis and Evolution," in his *God, Knowledge, and Mystery: Essays in Philosophical Theology* (Ithaca, NY: Cornell University Press, 1995), 133, Augustine, in his *De Genesi ad litteram* ("On Genesis according to the letter"), was not investigating what we call "the literal meaning of the text" but "the literal reality behind the nonliteral mode of presentation." To Augustine this reality was not characterized by evolution but by one instantaneous act of creation, which was then followed by a gradual unfolding of the created "seed principles."

39. William VanDoodewaard, *The Quest for the Historical Adam: Genesis, Hermeneutics, and Human Origins* (Grand Rapids: Reformed Heritage Books, 2015), 32; VanDoodewaard (who cannot be suspected of being biased here since his goal is to show how deeply the "historical Adam" is entrenched in church history) also points out that Clement and Origen similarly rejected a literal interpretation (23-24).

40. We do not base this conclusion on Rom. 5, as is often done. The bare fact that Paul compares Christ and Adam in this chapter is less decisive here, since in argumentative texts it is common to make comparisons with literary figures. Cf. James D. G. Dunn, *Romans 1-8* (Dallas: Word, 1988), 289.

41. Walton, *Lost World of Adam and Eve*, 102.

42. Walton, *Lost World of Adam and Eve*, 187-88.

Eve is a beginning point, an ancestor.[43] For example, she might be seen as the beginning point of all who are born to be really alive, in the sense of being aware of the transcendent and being able to live in a personal relationship with God. Somehow what is most significant about us contemporary humans must have started at some concrete moment in time with some concrete people, bearing concrete names—such as Adam and Eve.

Meanwhile, as we said, Adam and Eve cannot have been the sole humans around at that time. Why then are they so important? It is precisely here that Reformed theology comes to our aid. Taking their cue from Romans 5 and 1 Corinthians 15, Reformed theologians in particular considered Adam and Eve to be the *federal heads* of humankind—that is, of all those to whom human personhood could be ascribed. As such, they acted as representatives of their group or groups, being addressed by God as stand-ins for all others. Reformed theology has seen that the relevant relationship here is of a covenantal rather than a biological nature: Adam and Eve were offered the terms of a covenant, with both duties and prerogatives (Gen. 2:16-17), to which all those represented by them should live up.[44] They acted as our "heads," as Christ acted as our federal head in the covenant of grace.[45] Clearly, this does not require that Adam and Eve were a solitary couple. Perhaps, the typical shibboleth question during interviews for positions in churches and seminaries, "Do you believe in a historical Adam?" might best be answered by pointing out that believing in just one historical Adam is a bit meager, since there must have been a whole bunch of other first humans on earth.

Thus, the message of Genesis 2-3 is both paradigmatic and historical—or, if the latter term bears too many modern overtones (the writers of Genesis were not historians in any contemporary sense), *event depicting*. The text relates the most decisive events that happened during the time that *Homo divinus* came into existence: its creation from already-existing material elements and its subsequent fall as a result of an act of disobedience to God's calling.[46]

---

43. Cf. Bimson, "Doctrines of the Fall," 116, over against Berry, "This Cursed Earth," 38-39, and Alexander, *Creation or Evolution*, 300-304.

44. C. John Collins, *Reading Genesis Well: Navigating History, Poetry, Science, and Truth in Genesis 1-11* (Grand Rapids: Zondervan, 2018), 177-78, argues that calling the arrangement with Adam a "covenant" fits well with the literary features of the Genesis text.

45. Cf. on the place of this "Adamic administration," "covenant of life," "covenant of creation," or "covenant of works" in Reformed federal theology and confessional literature, e.g., Allen, *Reformed Theology*, 41-42.

46. Cf. Tryggve N. D. Mettinger, *The Eden Narrative: A Literary and Religio-*

From the perspective of Reformed theology, one can stay noncommittal as to whether there are more elements in the text that should be interpreted literally rather than symbolically or archetypally. To the Christian faith, this does not make a huge difference, since the texts are highly meaningful anyway. For example, even though God did not literally form humanity from the dust of the earth, we can learn from Genesis 2:7 that God took special care to make sure that we modern humans came into existence. We were, as it were, formed by his very hands. Clearly, the symbolic nature of this part of the narrative does not in the least detract from its theological significance.

Note that in coming to these conclusions we did not in a concordistic way impose any (purported) scientific discoveries on the text. It is true that the *occasion* for going back to the texts was the scientific account as spelled out in §6.2. But we have subsequently listened to the various voices about how to interpret the text, and in light of them we have tried to determine how the texts present themselves.[47] That we juxtapose scientific facts with our reading of Genesis 2-3 and connect these two is neither concordistic nor otherwise inappropriate. It can be compared to what is done when exegetes use extrabiblical historical records made available through scholarly research to shed light on the whereabouts of, for example, Emperor Augustus as mentioned in Luke 2.

In the next sections, I will continue to follow this method by examining more closely questions surrounding the Fall. First, I will argue that the ways in which the biblical texts speak about the human "fall" into sin and about the entrance of human death in the world should not be interpreted as figurative embellishments or merely symbolic representations of eternal spiritual truths, but as references to concrete events—whether these took place in a sudden, punctiliar way or more gradually (§6.4). In fact, it seems that their theological significance largely depends on their "evental" character.[48] We cannot just "drop" the Fall and fancy that the overarching scheme of Christian soteriol-

---

*Historical Study of Genesis 2-3* (Winona Lake, IN: Eisenbrauns, 2007), 123: "The theme [of the Eden narrative] is disobedience and its consequences." I disagree with Mettinger, though, that the writer "must have known that he was not referring to a previous, one-time event in time and space" so that "his ambition was not referential" (126).

47. Thus, we have followed G. C. Berkouwer's approach as mentioned in §3.5.

48. For an elaboration and defense of this notion with regard to the Fall, see James K. A. Smith, "What Stands on the Fall? A Philosophical Exploration," in Cavanaugh and Smith, *Evolution and the Fall*, 48-64 (esp. 63). Though Smith uses both, we may prefer the term "evental" to "historical" so as to avoid the modern overtones that easily accompany the latter concept.

ogy remains virtually unaffected—and this is all the more impossible from a Reformed theological perspective, given its intensification of the doctrine of sin. Next, I will attempt to sketch how such events can be envisaged within a full-blown evolutionary framework (§6.5). It will become clear that, unlike what is often suggested, evolutionary theory does not force us to jettison the classical Christian notion of the Fall.

## 6.4 The Fall: Biblical and Theological Backgrounds

By the Fall—usually written with a capital letter to highlight its special and momentous character—we mean humanity's transition from a state of innocence to one in which its first sin was followed by many others, since this "original sin" caused a fundamental rupture in humanity's relationship with God, other humans, and the world.[49] Is this notion rooted in the Bible? Opinions vary, but many say no. The figure usually seen as its fountainhead is Augustine. Accordingly, those who want to dispense with the notion of the Fall, either in the name of human autonomy or in the name of evolutionary science,[50] usually consider Augustine the main culprit. However, many pre-Augustinian readings of Paul on the issue resemble Augustine's views.[51] Augustine himself had "a deep-seated sense of his role being one of guarding the faith against innovations" and was adamant that the doctrines of the Fall and original sin were part and parcel of catholic orthodoxy throughout the preceding ages.[52] Whether or not Augustine was mistaken here, his basic

---

49. Leaving aside its further ramifications, we may provisionally define a "sin" as a choice of the human will that goes against God's will.

50. An example of the first is Elaine Pagels, *Adam, Eve, and the Serpent* (New York: Random House, 1988), 98-126; an example of the second is Schneider, "Recent Genetic Science," 202-3.

51. Alan Jacobs, *Original Sin: A Cultural History* (New York: HarperCollins, 2008), 32-33 (who especially takes to task Elaine Pagels here). On the historiographical pitfalls in this connection, see Peter Sanlon, "Original Sin in Patristic Thought," in *Adam, the Fall, and Original Sin: Theological, Biblical, and Scientific Perspectives*, ed. Hans Madueme and Michael Reeves (Grand Rapids: Baker Academic, 2014), 85-107.

52. Sanlon, "Original Sin," 91; over against Sanlon, however, nobody before Augustine had developed the doctrines of the Fall and sin in such stark terms. For an excellent account of Augustine's reasons for doing this, see Ian McFarland, *In Adam's Sin: A Meditation on the Christian Doctrine of Original Sin* (Oxford: Wiley-Blackwell, 2010), 32-35.

## Evolution and Covenantal Theology

assumption that sin came into the world rather than being inherent in it all along was definitely Pauline, as is clear from Paul's argument in Romans 5 (see esp. Rom. 5:12). This is so even though James Dunn is right that the real point of Paul's argument here "is not so much to historicize the individual Adam as to bring out the more than individual significance of the historic Christ."[53] Indeed, like Paul, Augustine clearly developed his notions of the Fall and sin from the perspective of his theology of salvation in Christ as a free gift of divine grace.[54]

So, alternatively, we might try to blame *Paul* for the doctrine of the Fall. Perhaps Paul's line of reasoning in Romans 5 stems from the infelicitous fact that he was "not a careful reader" of the Genesis narrative on the garden of Eden.[55] Again, however, this remains to be seen. In any case, the two texts bear important continuities, such as that both Genesis 2–3 and Romans 5 picture the first sin as a strange and inexplicable intrusion in God's creation rather than as its inevitable shadow side. As Cornelius Plantinga writes, "Sin is an anomaly, an intruder, a notorious gate-crasher. Sin does not belong to God's world, but somehow it has gotten in."[56] It is true that the notion of *original* sin—that is, the idea that the sin of Adam somehow drew his descendants into bondage to sin—is absent from Genesis 2–3. The notion of a fundamental rupture in the relationship between God and humanity as a result of the first human sin, however, is strongly suggested by the Genesis text itself. In the course of the subsequent chapters (4–11), the human choice to turn against the Creator is not only repeated time and again but also aggravated and radicalized: starting from something as "innocent" as eating a forbidden fruit, it goes on with a murder (Gen. 4) and ends up with a situation in which "the LORD saw that the wickedness of humankind was great in the earth, and that

---

53. Dunn, *Romans 1–8*, 290. Contrary to Dunn, however, Paul's argument does presuppose that somehow sin was brought into the world by conscious human beings, rather than being ingrained in creation; otherwise, Paul's talk of justification through Christ would not make sense.

54. Cf. Ian McFarland, "The Fall and Sin," in *The Oxford Handbook of Systematic Theology*, ed. John Webster, Kathryn Tanner, and Iain Torrance (Oxford: Oxford University Press, 2007), 142, 155, and his *In Adam's Sin*, 32–33.

55. Patricia Williams, *Doing without Adam and Eve: Sociobiology and Original Sin* (Minneapolis: Fortress, 2001), 42, who hardly substantiates this claim; in fact, as in Pagels, *Adam, Eve, and the Serpent*, her extensive criticism of Augustine (40–47) stands in contrast with the near absence of any serious discussion of Paul.

56. Cornelius Plantinga Jr., *Not the Way It's Supposed to Be: A Breviary of Sin* (Grand Rapids: Eerdmans, 1995), 88.

every inclination of the thoughts of their hearts was only evil continually" (Gen. 6:5).

It is because of these disastrous consequences that the very first human choice for sin has traditionally been described as a *fall*. To be sure, the use of this metaphor can be criticized for various reasons. There is nothing in the text of Genesis 3 that entails it; it is open to misinterpretation in Greek and gnostic ways, as the metaphysical fall from a higher spiritual level of being into the lower realm of the material world; and last but not least, the image of a downward movement easily conjures up the romantic idea of a pre-Fall state that was perfect. Still, already in the Genesis account the first sin implies a taking of the wrong track that cannot so easily (if at all) be undone. It involves a transition to another state of being-in-the-world, which henceforward became characterized by the pervasiveness and the serious consequences of sin. The alternative reading of Genesis 3 as symbolizing the coming-of-age of humanity and thus portraying an upward rather than a downward movement[57] fits ill with the severe God-given punishments in response to the first couple's acts (vv. 14-19), as well as with the gradual increase in evildoing and suffering in the subsequent chapters of Genesis (from which Genesis 2-3 is all too often isolated). As John Bimson rightly notes, "the story is one of catastrophic losses rather than gains."[58]

If indeed sin is an intrusion in God's creation in this way, the Fall should be seen as *in some way* historical. By this I do not mean that its occurrence can be demonstrated using the tools and standards of contemporary historiography— it cannot, and that is why we might prefer terms like "primordial," "primeval," "protohistorical," or, following Jamie Smith, "evental."[59] Nor do I mean to suggest that the Fall narrative should be read "literally" (i.e., taking the text at face value); as indicated already, it is widely acknowledged by biblical scholars that the text displays symbolic imagery, figurative elements, and

---

57. As, e.g., in Ellen van Wolde, *A Semiotic Analysis of Genesis 2-3* (Assen: Van Gorcum, 1989), 216-29, and (independently of Wolde) in Bechtel, "Genesis 2.4b-3.24," 3-26. Over against such interpretations, Mettinger, *The Eden Narrative*, convincingly argues that "everything [in Gen. 2-3] revolves around the divine commandment and the issue of obedience versus disobedience" (97).

58. Bimson, "Doctrines of the Fall," 112.

59. Cf. Smith, "What Stands on the Fall?," 63: "the Fall is still historical, temporal, and even 'evental' though it is something like an episode-in-process.... Such a historical picture seems to be required in order to retain a sense of sin as 'not the way it is supposed to be.' And *that* seems essential to the tradition to me."

other literary conventions of the time. What I do mean is that the story "deals with an originative event that gave a catastrophic twist to our relationship with God."[60] The Genesis narrative does not portray human sin as coincidental and equitemporal with God's good creation—an observation that debilitates the paradigmatic reading of the passage mentioned before. Thus, it does not just symbolize "the tendency in human nature to grasp at more than is freely given, to seek to elevate our status beyond what is appropriate . . . , to seek to be 'as Gods.'"[61] No doubt the passage has this paradigmatic function, but apart from that, it also has an etiological function: it attempts to answer the question of how we humans came to be this way. It does so, *not* by giving an explanation for human sin—which, after all, remains inexplicable—but by tracing back human sin to a first act of willful disobedience to God's voice.[62]

Why is this so important? It is because the notion of a primordial fall into sin signifies a watershed between different views of life. By endorsing it, we acknowledge that, despite the undeniable role played by forces outside ourselves, at the end of the day we can blame neither God nor the devil for our moral evil, but only ourselves. In this way the doctrine of the Fall helps us to interpret moral evil as human sin rather than as part and parcel of the fabric of life. If we humans fell into a state of sinfulness by a wrong choice of our own free will, we cannot refer to "the way we were built" as an excuse for our conduct. Nor can we point to some cosmic power to condone ourselves. In this way, the doctrine of the Fall enabled Christianity to navigate the narrow path between the Scylla of metaphysical monism (according to which good and evil equally stem from God) and the Charybdis of metaphysical dualism (according to which the devil, or some other external force, is to blame for all

---

60. Bimson, "Doctrines of the Fall," 112; cf. Henri Blocher, "The Theology of the Fall and the Origins of Evil," in Berry and Noble, *Darwin, Creation, and the Fall*, 159: "The issue is not whether we have a historical account of the Fall, but whether we have the account of a historical Fall."

61. Christopher Southgate, *The Groaning of Creation: God, Evolution, and the Problem of Evil* (Louisville: Westminster John Knox, 2008), 101-2; cf. 102: "the Fall account in Genesis reflects a general condition rather than a historical chronology." So here is where I part company with this thought-provoking study.

62. Cf. Bimson, "Doctrines of the Fall," 109, who refers to various recent contributions to biblical scholarship in support of this etiological interpretation. Wolfhart Pannenberg, *Systematic Theology*, vol. 2 (Grand Rapids: Eerdmans, 1994), 263, unconvincingly limits the etiological function of the passage to its offering of an explanation for human death and for the painfulness of human childbearing and laboring, excluding human sin from the *explanandum* here.

evil). Creation and sin do not coincide, but sin spoils creation and as such, to quote Cornelius Plantinga's well-known formula, it is "not the way it's supposed to be."[63]

As soon as we conflate creation and Fall, we move to a substantially different view of the nature of evil, giving it a metaphysical rather than a contingent status.[64] To be sure, in one way this metaphysical view takes evil more seriously. For if sin is a historically contingent rather than a metaphysically necessary phenomenon, it is not bound up with human nature and therefore we can in principle be liberated from it without losing our human nature. In fact, the gospel tells us that there has already been a person with a fully human nature who did not get contaminated by evil, and by whose saving work we can indeed be liberated from sin and its consequences. Fortunately, the gospel nowhere requires us to take our moral evil seriously to the point of considering it irredeemable. It belongs to "the sunny sides of sin"[65] that it is a historically contingent and therefore redeemable reality. Precisely because it is a historically contingent phenomenon, sin is not the consequence of the way we were built, but we are the ones to blame for it.[66] Otherwise, God's indignation about human sin, palpable throughout the Bible, would be entirely unintelligible. Even though we are often misled by seductive powers (cf. the

---

63. Cf. Plantinga's title, *Not the Way It's Supposed to Be*, and its background in the film *Grand Canyon* (7-8).

64. Strictly speaking, from a philosophical point of view, sin may be equitemporal with creation but still contingent. I am grateful to Jeroen de Ridder for reminding me of this possibility. Still, I consider it a largely theoretical possibility that is difficult to uphold in practice. The Genesis narrative highlights the contingency of sin by first portraying creation and only then introducing sin. Hendrikus Berkhof, *Christian Faith: An Introduction to the Study of the Faith*, rev. ed. (Grand Rapids: Eerdmans, 1986), skillfully walks the tightrope here by on the one hand claiming that "sin . . . is deeply rooted in the creaturely structure of the risky being called man" while on the other hand acknowledging that "sin does not belong to created reality and does not issue from it. On the contrary, it is unnatural. It is not a tragic fate put on our shoulders against our will" (193).

65. Dutch theologian Arnold van Ruler (1908-1970), who was famous for the playfulness of his theological style, once wrote a short paper under this title; cf. "Zonnigheden in de zonde [1965]," in *Van schepping tot Koninkrijk. Teksten (1947-1970) uit het theologisch oeuvre van A. A. van Ruler* [From creation to kingdom], ed. Gijsbert van den Brink and Dirk van Keulen (Barneveld: Nederlands Dagblad, 2008), 121-24.

66. Cf. Henri Blocher, *In the Beginning: The Opening Chapters of Genesis* (Leicester: IVP, 1984), 160-70.

role of the snake in Gen. 3), in the end it is we humans who are responsible for our wrongdoings.[67] Therefore, we should not "drop the Fall," since we need this notion if we want to continue to tell the Christian story.[68]

In Reformed theology in particular, but also in (parts of) other Christian traditions, this insight has been kept alive throughout the centuries. At this very moment it seems that both in Reformed circles and beyond a new articulation of Christian orthodoxy is emerging around the novel doctrinal rule: "Accept evolutionary biology but hold on to the Fall."[69] In a recent monograph, Reformed theologian Ernst Conradie points out that only a limited number of ways are available to conceptualize what went wrong in the world. There is the pessimistic narrative, exemplified by, for example, Manichaeism on the religious side and social Darwinism on the nonreligious side, according to which evil and conflict are deeply ingrained in our biological makeup and will never come to an end. And there is the much more optimistic narrative—labeled the "Pelagian-liberal plot" by Conradie—according to which we humans, helped by some grace (in the religious version of the story) or luck (in its secular version) but most of all using our own free will, will be able to gradually improve ourselves and overcome all evils in the world. For anyone who does not want to believe the first narrative and who can't believe the second, there is basically only one alternative: the Augustinian view that created reality is fundamentally good but has become infected by evil to such an extent that only God can save us.[70]

Here we may find the most profound theological motive for holding on to the notion of a historical Fall: it is not just "because the Bible says so," but because without such a notion we end up with a non-Christian worldview that is either deeply pessimistic or utopian. If the intrusion of sin and evil in

---

67. Thus, correctly, C. John Collins, *Did Adam and Eve Really Exist?*, 48.

68. Cf. already the classic study of N. P. Williams, *The Ideas of the Fall and of Original Sin* (London: Longmans, Green, 1927), xxxiii: "We conclude, then, that . . . it is impossible to lift the Fall out of the time-series without falling either into Manicheism or unmoral monism."

69. See in this vein, e.g., William T. Cavanaugh and James K. A. Smith, eds., *Evolution and the Fall* (Grand Rapids: Eerdmans, 2017).

70. Ernst M. Conradie, *Redeeming Sin? Social Diagnostics and Ecological Destruction* (Lanham, MD: Lexington Books, 2017), 61–105. Conradie discusses Marxism as the secular counterpart of Augustinianism, because according to Marxism class conflict was not original but due to a "fall into capitalism"; just as in the liberal plot, however, redemption has to come from us humans (or from "history"), and it can be questioned, to say the least, whether that is not too optimistic.

the world is a serious but contingent phenomenon, however, then we can be redeemed from it without losing our human identity. Thus, we may be happy that the Fall is "only" historical, because that means that our fallenness is contingent and accidental rather than essential and eternal. The Fall is not just our gradually emerging awareness of the fact that we are sinners and have been sinners all along.[71] Rather, it is our willful and contingent—but, in the end, redeemable—choice for evil over against the good to which God enabled and summoned us. That, it seems to me, is the bottom line, or "theological message," of the story of Genesis 3. As Paul Ricoeur perceptively observed about what he called the "Adam myth": "The myth proclaims the purely 'historical' character of . . . radical evil; it prevents it from being regarded as primordial evil."[72]

In what follows I will therefore rely on this Augustinian view. It is not unreasonable to think that Augustine went beyond Paul when elaborating his influential views on the Fall and original sin. Similarly, we may think that Paul, drawing on forms of exegesis current in Second Temple Judaism, developed and intensified the notion of the Fall in his own way, going beyond what is implied by the Genesis story.[73] However, even if that is true, it is still, as Christopher Southgate argues, "perfectly appropriate for a community of interpreters prayerfully to decide that a certain text is a 'hermeneutical lens' that allows a particular theme in scripture to be understood in a particular way."[74] Thus, for all Christians the way in which Paul elaborates on Genesis 2–3 in his letter to the Romans is such a lens for interpreting the nature of the Fall and sin in Genesis 2–3; and for many Christians, including (but not restricted to) Reformed ones, Augustine's reading of Paul is in a similar way constitutive for their theological thinking on these themes.

Therefore, although John Walton may be right that "Augustine pushes beyond what Paul says, and Paul has moved beyond what Genesis says,"[75] I

---

71. Abraham van de Beek borders on this view in his "Evolution, Original Sin, and Death," *Journal of Reformed Theology* 5 (2011): 206-20: "Sin had come into human consciousness as an inerasable reality, for it belonged to the genetic code of humankind" (209).

72. Paul Ricoeur, *The Symbolism of Evil* (Boston: Beacon, 1967), 251 (cf. 233).

73. Cf. N. P. Williams, *The Ideas*, 118-22.

74. Southgate, *The Groaning of Creation*, 146, who adds that Paul's understanding of the Fall can be seen as such a lens through which one may read Gen. 3 (although Southgate himself does not do so).

75. Walton, *Lost World of Adam and Eve*, 155.

consider Augustine an appropriate hermeneutical lens for reading both Paul and Genesis. That is not to say that all of Augustine's views on these issues are to be considered sacrosanct. For example, his portrayal of paradise as a state of perfection where no desires were involved in sexual intercourse is, as all contemporary Christians will agree, open to criticism. His basic intuition, however, that sin came into the world through a wrong act of the human will and that this act had disastrous consequences has been constitutive for Reformed theological thought on sin and grace ever since the sixteenth century—and for many strands of Catholic thinking even longer.[76] If evolutionary theory forces us to abandon this belief, it is understandable that Reformed Christians will continue to relate in antagonistic ways with evolutionary theory.

But does it? Or is it possible to retain the notion of an "evental" Fall as sketched here, given what we found in §6.2 on the evolutionary background and development of our species? That is the question to which we now turn.

### 6.5 The Fall and Original Sin in an Evolutionary Context

It is often claimed that what we know about our evolutionary backgrounds does not allow for the possibility of a Fall as sketched in the previous section. Yet, during the past couple of years several scholars have come up with proposals for reimagining the biblical Fall within an evolutionary context. Some of these proposals are more plausible than others. The most vulnerable ones are those that try to harmonize biblical and scientific sets of data while ignoring the distinctive character of theology as having to do first and foremost with our human relationship to God. Other proposals, however, are more nuanced and sophisticated in that they try to respect the integrity of both the sciences and theology but still attempt to relate the two in meaningful ways.

To start with, in line with the prehistorical approach of Genesis 2–3 as discussed in §6.3, several authors have suggested that the first humans who accord-

---

76. It is important to add that in other strands of the Christian tradition, too, this Augustinian line of thinking continues to be influential. A fine recent example from a Roman Catholic perspective is Matthew Levering, *Engaging the Doctrine of Creation: Cosmos, Creatures, and the Wise and Good Creator* (Grand Rapids: Baker Academic, 2017). Levering argues that central theological tenets of the Augustinian interpretation of Gen. 2–3—in particular the notions that human nature was created good, that the first sin was "entirely free," and that this sin has "a corruptive impact on all humans"—remain indispensable to the Christian faith (250).

ing to the Genesis narrative were addressed by God need not be equated with the first members of *Homo sapiens*. As Derek Kidner argued quite some time ago already: "The intelligent beings of a remote past, whose bodily and cultural remains give them the clear status of 'modern man' to the anthropologist, may yet have been decisively below the plane of life which was established in the creation of Adam. Nothing requires that the creature into which God breathed human life should not have been of a species prepared in every way for humanity."[77]

Thus, God bestowed his image on two people out of many—presumably Neolithic farmers around ten thousand years ago—who then received the calling to serve and worship him, just like Abram and Sarai would later receive a special calling from among their contemporaries.[78] It was at this stage of human history that the Fall took place, when the same Neolithic couple that had received God's image started to sin. We may wonder, however, whether this way of bringing together science and theology is convincing, since the Genesis narrative is not just about some arbitrary human beings but is a story of origins.[79] It would definitely go against the grain of Genesis 2–3 to suppose that these chapters relate events that took place some "30,000 years *after* the archaeological appearance of religion and human culture."[80]

Alternatively, therefore, it is sometimes proposed that the bestowal of the divine image on certain hominids more or less coincided with the emergence of *Homo sapiens*. Given that exact dates and periods of time in the study of human origins are subject to ongoing (and probably never-ending) debate, we might rather refrain from speculating about them, but as we have seen, the usual estimation of when this happened is somewhere around two hundred

---

77. Derek Kidner, *Genesis* (Leicester: IVP, 1967), 28; Kidner's approach has been adopted in a "white paper" by Tim Keller; see his "Creation, Evolution, and Christian Laypeople," *BioLogos*, February 23, 2012, 10–12, https://biologos.org/uploads/projects/Keller_white_paper.pdf. It is also, basically, Denis Alexander's view; cf. his *Creation or Evolution*, 289–94, 300–304.

78. For this view see also Joshua M. Moritz, "Evolution, the End of Human Uniqueness, and the Election of *Imago Dei*," *Theology and Science* 9 (2011): 307–39.

79. Cf. §6.3 above.

80. James P. Hurd, "Hominids in the Garden?," in *Perspectives on an Evolving Creation*, ed. Keith B. Miller (Grand Rapids: Eerdmans, 2003), 224, who adds: "If Adam lived at the time of the Neolithic, how should we classify these pre-Adamic forms so abundant in the fossil record? If they walked like humans, worked like humans, and worshipped like humans, were they not human?" It seems clear that these creatures, especially those who "worshipped like humans," must indeed be classified as humans, bearing the image of God.

thousand years ago.[81] It seems that this is far too early for the Fall to have taken place, since at that time human consciousness was yet at a very low level. Humans were still unable to communicate in high-level conscious ways with God and each other, and it is hard to see how the concept of guilt might be applicable at this stage. As Celia Deane-Drummond rightly observes, "on purely historical grounds it seems that *Homo sapiens* existed as a biological species long before the cultural revolution and any religious belief was in evidence."[82] Therefore, if we want to relate the Genesis story to some particular phase in human history, we might best associate both the bestowal of the divine image and the fall into sin with the aforementioned "dramatic so-called cultural big bang, the Upper Paleolithic revolution and the explosive growth of human creativity around 45,000 years ago."[83] This, as we have seen, was also the time period in which the first signs of religious awareness have been detected.

Therefore, in making sense of the Genesis story, we may consider it as dealing with these first "behaviorally modern" human beings, and doing so in the garments of the later cultural and linguistic framework of the Neolithic transition toward a sedentary and agricultural lifestyle. Arguably, this later framework reflects the views of both the authors of Genesis 2-3 and their first readers. Using this framework, the story relates what must have happened when the first "real" human beings arrived on the scene. That Adam and Eve were not supposed to be the only people around at that time is already reflected in the Genesis text itself (see especially Gen. 4:14, 17). Moreover, as we observed above, it is not necessary—nor even credible—that Adam and Eve belonged to the first group of *Homo sapiens*. John Walton remarks that their story is not about material human origins but about sin's origins.[84] Closely related to this, however, it is also about the origin of members of the species *Homo sapiens* who came to bear the image of God—that is (among other things), who became *personally addressable and accountable* human beings. Genesis 2-3, then, is a story of origins in that Adam and Eve are the representatives of this first group of human beings that could be addressed by God in personal communication—and I know of no better label for this group than

---

81. Other sources (e.g., Hurd, "Hominids in the Garden?," 218) mention earlier dates, up to one hundred thousand years ago.

82. Celia Deane-Drummond, "Response: *Homo Divinus*—Myth or Reality?," in *Darwinism and Natural Theology: Evolving Perspectives*, ed. Andrew Robinson (Newcastle upon Tyne: Cambridge Scholars, 2013), 42.

83. Van Huyssteen, *Alone in the World?*, 66.

84. Walton, *Lost World of Adam and Eve*, 103.

*Homo divinus*.[85] It is the "people of God" that is created here. Adam and Eve may somehow have been their tribal leaders, but what is theologically important is that they function as their *federal heads* (to put it in typically Reformed phraseology)—and as *ours* as well, since we belong to this same group. As Paul argues in Romans 5, in this way Adam mirrors Christ, being "a type of the one who was to come" and whose "act of righteousness leads to justification and life for all" (5:14, 18).[86]

Interestingly, this representative role of Adam corresponds to the ambivalence in the Hebrew text of Genesis as to whether "Adam" should be read as a proper name or as a general noun ("the human being").[87] As Karl Rahner points out, even if we accept polygenism, that is, the view that modern humans do not stem from a single first couple but have a plurality of first human ancestors, the first human population was a biological-historical unity. They shared both the same biotope and the same divine destination.[88] Indeed, as we have seen in §6.2, the standard view on human origins among contemporary paleoanthropologists is the "out of Africa" hypothesis, which suggests that we humans have a unified origin rather than stemming from various lineages that developed independently from each other at various places.[89] Let us suppose, then, that at some point in time there was a first population of modern human beings who received the image of God.[90]

---

85. Deane-Drummond, "Response," emphatically rejects this label because "it sets up a divide that splits humans apart from their evolutionary history" (42), but it is not clear why this should follow. In God's making or selecting our species for its special role, no degradation of our evolutionary background (or of other species) need be implied.

86. The similarities and differences between Adam and Christ currently receive new attention in light of human evolution; see, e.g., Celia Deane-Drummond, "In Adam All Die? Questions at the Boundary of Niche Construction, Community Evolution, and Original Sin," in Cavanaugh and Smith, *Evolution and the Fall*, 23-47.

87. Of course, it is also consonant with the intimations in Gen. 2-4 that more people than Adam and Eve were around at the dawn of human civilization, such as Cain's wife (Gen. 4:17) and "all those" of whom Cain was afraid (Gen. 4:14), not to mention those for whom he built a city (Gen. 4:17).

88. Karl Rahner, "Evolution and Original Sin," *Concilium: International Review of Theology* 6, no. 3 (1967): 30-35.

89. Such "polyphyletism" would imply new theological challenges, which I will not discuss here; I leave this to future theologians in case it might turn out that polyphyletism rather than monophyletism is true after all.

90. John H. Walton, "A Historical Adam: Archetypal Creation View," in Barrett

If we ponder the enormous number of species that preceded us, it is of course bewildering that in such marvelous ways we came to be what we are: bearers of God's image. As Graeme Finlay asks:

> Is it feasible that a lineage of ape-like creatures progressively losing its ability to make vitamin C, its hair, and its sense of smell, and sustaining the random invasion of myriad retrotransposons, could be ancestral to *Homo divinus*? Could such inauspicious beginnings precede the creature which would reconstruct its evolutionary past, reflect on its future, and respond to its creator?[91]

Here, a notion that has taken pride of place in many Reformed theologies may be illuminating: divine election. This notion (which is by no means *exclusively* or typically Reformed) has been distilled from the biblical pattern that God often chooses unexpected and even relatively insignificant individuals to work out great things. Such "elected" people are not better or more worthy than others. It is only by God's grace that they have been chosen for their specific roles and tasks. Nor are they chosen for their own sake, but for the sake of others and to the honor of God. In this way, God chose Abram out of countless pagans (Gen. 12:1-3), and God chose Israel as a small and insignificant people ("A wandering Aramean was my ancestor," Deut. 26:5). He chooses the poor and weak and despised to form his church (1 Cor. 1:27-38), and he even chose a rejected and crucified human being to become "both Lord and Messiah" (Acts 2:36).

Seen in this theological perspective, it is not at all strange or out of sync with what we know from the Bible that God at some point in time chose one of the many ape-like creatures that had evolved through time and equipped it to bear his image. As Finlay explains:

> Our very creation is an act of sheer grace. In an initiative of unconditioned love, God conferred his likeness upon a member of the ape family and brought into being *Homo divinus*, the ape-in-the-image-of-God, with the unique capacity to know, love and serve its creator. There is no room for *hubris* here.

---

and Caneday, *Four Views on the Historical Adam*, 114, tentatively links this population to "the moment that geneticists refer to as the bottleneck when humanity nearly became extinct," but theologically there is no reason to do so and biologically it seems that this bottleneck came too early (i.e., at a stage in which personal awareness of God may not yet have been possible).

91. Graeme Finlay, "*Homo Divinus*: The Ape That Bears God's Image," *Science and Christian Belief* 15, no. 1 (2003): 38.

Our biological roots remind us that we are human not because of any inherent or necessary superiority to the rest of the animal kingdom, but in creaturely dependence on God's goodness.[92]

Rather than playing the concepts of *election* and *selection* against each other, as has sometimes been done, we might better align them to one another, viewing natural selection as one way in which God chose to materialize his elective purposes.[93]

At some moment in time, however, this elected population was confronted with the dilemma of either responding to God's calling or rejecting it by continuing to behave in accordance with the aggressive inclinations it had inherited from its animal and hominin ancestors. At that crucial juncture of history, they took the wrong track. Keith Ward colorfully envisages the course of events as follows:

> The first human beings[94] had a responsible choice between their lustful, aggressive dispositions and the more altruistic, co-operative dispositions that would have led them to grow in the knowledge and love of God. . . . From a religious viewpoint, the deepest purpose of human existence is the free development of a relationship of joyful obedience to the will of God, within a community of justice, peace and love. It is that purpose which was rejected when the fateful choice was made of a path of autonomy, of rational self-will, which placed the descendants of those first humans in bondage to self and its consequent conflict and suffering.[95]

---

92. Finlay, "*Homo Divinus,*" 38; cf. Moritz, "Evolution, the End of Human Uniqueness, and the Election of *Imago Dei.*" Speaking of "the ape family" and especially "the ape-in-the-image-of-God" is slightly misleading, though, since the hominins were branched off from the line that brought forth the apes long before *Homo sapiens* came into existence.

93. Van der Meer points to various Reformed scholars in the past who "saw an analogy between natural selection and divine election"—an analogy that for them "made selection acceptable as a form of divine providence." See Jitse M. van der Meer, "European Calvinists and the Study of Nature," in *Calvinism and the Making of the European Mind,* ed. Gijsbert van den Brink and Harro M. Höpfl (Leiden: Brill, 2014), 120–21. Next to the scholars mentioned by Van der Meer, Herman Bavinck is another case in point; see his *Reformed Dogmatics,* vol. 2, *God and Creation* (Grand Rapids: Baker Academic, 2004), 399.

94. This should be read as referring to the first "modern" human beings, not the first anatomical humans.

95. Keith Ward, *God, Faith, and the New Millennium* (Oxford: Oneworld, 1998), 133.

I suggest that we may appropriately call this dramatic event the *Fall*. It seems reasonable to think of this event as a gradual process rather than a split-second event.[96] Even so, this process must have started somewhere and at some time. The first human beings fell, not from a state of spiritual perfection but surely from a state of *innocence*, since they had not been morally responsible beings before the enormous widening of their consciousness and their emerging awareness of the divine presence and will. In that sense, and in that sense only, "God made human beings straightforward" (Eccles. 7:29).[97] This is not to deny that they used to kill and deceive, were sexually promiscuous, and did other things that were bequeathed to them by their hominin and animal ancestors. Such acts, however, while sinful for us, were not sinful *for them*. It is only when they became aware of God's calling them to another way of life that the possibility of sin entered the scene. As Paul says, when no law applies—and that is the case as long as there is no relationship within which a moral law can be established—no sin can be imputed (Rom. 5:13). It is the fact that we humans can act deliberately and intentionally that makes the difference.[98]

---

96. Cf. John Polkinghorne, "Scripture and an Evolving Creation," *Science and Christian Belief* 21 (2009): 166: "That human turning inwards was the Fall—not a single disastrous ancestral event, but a process that was an attempt to claim human autonomy and to refuse heteronomous dependence upon divine grace, a deeply mistaken move of which we are now all the heirs." See also Smith, "What Stands on the Fall?," 63: "the Fall is still historical, temporal, and even 'evental,' though it is something like an episode-in-process."

97. Cf. Marguerite Shuster, *The Fall and Sin: What We Have Become as Sinners* (Grand Rapids: Eerdmans, 2005), 15, according to whom the notion of a "state of righteousness" does not require that we have an exalted view of the first humans' capacities. "All it requires is that God had indeed made them in his image, with awareness of a relationship to him, of his command, and of right and wrong as related to that command." See also C. John Collins, "Adam and Eve in the Old Testament," in Madueme and Reeves, *Adam, the Fall, and Original Sin*, who argues that in Gen. 2-3 "the humans were created morally innocent ('innocence' is not naïveté or moral neutrality), but not necessarily perfect" (20).

98. Cf. Walton, *Lost World of Adam and Eve*, 155. Here I part company with the otherwise instructive paper of Georg Etzelmüller, "The Evolution of Sin," *Religion and Theology* 21 (2014): 107-24, who applies sin-talk to forms of prehuman behavior in evolutionary history, especially to "cruel" forms of predation: "While the fact that life lives at the expense of life must be attributed to the shadow side of creation, the cruelty that may result from this can be understood as an expression of sin." It seems to me that concepts like "sin" and "cruelty" can be properly applied only to human behavior.

Rather than preserving this state of guiltlessness, however, the first humans chose to spoil it by acting against God's will for them. They refused to find their fulfillment in God and to operate as faithful stewards of God's creation. One may object that given their evolutionary background, it would have been very difficult for them to avoid the aggressive acts of their ancestors; but then we must imagine that their awareness of God and God's aims with them "was particularly clear, uncluttered by the spiritual darkness that eventually clouded the minds of the human race because of its turning away from God."[99] It was this turning away from God's fellowship rather than their behavior as such that made the first human beings guilty. Moreover, they had also inherited from their ancestors dispositions to the types of socially constructive and cooperative behavior that (as we saw in §4.6) can still be observed in many primates today. However, they used their newly acquired capacities of high-level consciousness (i.e., a level that includes moral awareness) and free choice in the wrong way. By refusing to trust and obey God, they turned their gifts into a means of self-assertion. As natural as that may have been for their ancestors, it was unnatural for them, since it went against the grain of their evolved moral consciousness.[100]

Can we also continue to say that the Fall had disastrous consequences for all subsequent life? In any case, we may suppose that although death and decay had been part of the natural world all along, the situation was aggravated after the first human beings gave in to sin. From then on they more and more turned toward violence, as not only the Genesis record testifies, but as is also witnessed (in what may be seen as a remarkable instance of "consonance") in archaeological findings about the transition from relatively peaceful hunter-gatherer societies toward more aggressive agricultural ones.[101] Humans also started to threaten the well-being of the natural world, since they had chosen the path of autonomy instead of taking care of the earth in responsible ways. Nature would never be safe from human encroachment again. More generally, the tendency toward

---

99. Robin Collins, "Evolution and Original Sin," in Miller, *Perspectives on an Evolving Creation*, 470.

100. Raymund Schwager, SJ, *Banished from Eden: Original Sin and Evolutionary Theory in the Drama of Salvation* (Leominster, UK: Gracewing, 2006), 52. For a critical analysis of Schwager's book, see Jonathan Chappel, "Raymund Schwager: Integrating the Fall and Original Sin with Evolutionary Theory," *Theology and Science* 10 (2012): 179-98.

101. This is called the agricultural or Neolithic revolution, some 10,000 years ago. According to some scholars, this transition led to much higher levels of violence and deprivation among humans. See, e.g., Van Schaik and Michel, *Good Book*, 29-66.

self-assertion at the expense of obedience to God became deeply ingrained in human nature—as the doctrine of original sin testifies. Arguably, this tendency was handed down through the generations not only (epi)genetically but also by social factors.[102] Thus, the classical locus of the "consequences of sin" can and should still be upheld within an evolutionary context. Something precious broke down when humans started to disobey God's will—something that could not easily be repaired but, on the contrary, turned out to have dire consequences. Henceforth, our salvation could only be brought about by God (Gen. 3:15).

In this way, then, it seems to me that we can make sense of the classical notions of the Fall and original sin within an evolutionary history as it has been disclosed by contemporary science. In doing so, we are once again not following a concordist strategy by harmonizing every element of the Genesis accounts with contemporary science, as if the results of modern science are somehow hidden in the Bible. Nor are we yielding theological ground to the sciences, retreating to an ever-diminishing and ever more ethereal "realm of faith." Rather, we attempt to show how the biblical narrative of creation, fall, and redemption continues to provide a compelling theological understanding of our common history as human beings, given all that we know (or seem to know) about that history as well-educated twenty-first-century citizens.[103] Rather than belittling the disastrous role of sin and evil—a tendency especially Reformed theology has always sought to resist—this theological understanding enables us to take human evil with full seriousness. Here, Christian theology continues to point beyond metaphysical accounts according to which evil is an ontological ingredient of our natural existence.[104]

    102. See, e.g., Benno van den Toren, "Original Sin and the Coevolution of Nature and Culture," in *Finding Ourselves after Darwin: Conversations on the Image of God, Original Sin, and the Problem of Evil*, ed. Stanley P. Rosenberg (Grand Rapids: Baker Academic, 2018), 173–86. For a more elaborate attempt at linking the doctrine of original sin to evolutionary theory than I can offer here, see my paper in the same volume, "Questions, Challenges, and Concerns for Original Sin," 117–29.

    103. Here I am in line with Alister McGrath's project of developing a Christian natural theology, that is, a Christian account of the natural world that accommodates evolutionary theory (among other things) but makes better sense of it than rival interpretations. For as McGrath, *Darwinism and the Divine: Evolutionary Thought and Natural Theology* (Oxford: Wiley-Blackwell, 2011), rightly warns: "The failure of a worldview to gain significant traction with the empirical worlds of human observation and experience raises serious questions about both its intellectual validity and existential relevance" (283).

    104. This point is rightly emphasized by Smith, "What Stands on the Fall?," and

## 6.6 Human Death as the Wages of Sin

If we can indeed recontextualize the notion of the Fall within the framework of our evolutionary history along these lines, using the contemporary scientific worldview to express a fundamental tenet of Christian belief, then how should we think about human *death*? We have discussed the issue of animal death in chapter 4. For those who want to take the Bible seriously (as Reformed Christians do), however, human death is a more complicated and sensitive issue. According to the narrative of Genesis 2-3, it belongs to the consequences of human sin, along with a more troublesome relationship to the thorny ground and stronger pangs in giving birth for women. Indeed, these phenomena continue to characterize human life as it is portrayed in the entire Old Testament.[105] Whereas the narrative of Genesis 2-3 already suggests that human death entered the world as a divine punishment for the sins of our first ancestors, Paul makes the connection even more explicit. Both in Romans 5:12 and 6:23 and in 1 Corinthians 15:21, he argues that death entered the world as a result of human sin. Now clearly, neither the authors of Genesis 2-3 nor Paul had in mind the death of animals in this connection (in Rom. 5:12 the restriction to *human* death is even explicit). As we saw, there is no biblical warrant for the view that animals have to die because of human sin. But both Genesis and Paul suggest that before the human race fell into sin, not a single human being had to die.

Can this view be retained within an evolutionary framework? It does not at all seem so. Like animals, humans must have been subject to processes of death and decay right from their earliest beginnings. Therefore, it is a common strategy to explain all biblical references to the sin-death connection as pertaining to what is called our *spiritual* death. The idea here is that humans became alienated from God and lost their intimate relationship with him because they had started to sin. From a religious or spiritual perspective, not having a personal relationship with God but being alienated from God amounts to being dead—even though physically one is still alive.[106]

---

elaborated by Conradie, *Redeeming Sin?*, 61-105. It is also highlighted by Levering, *Engaging the Doctrine of Creation*, 249-50.

105. The view that the Fall narrative of Gen. 3 has only an isolated and marginal position in the Old Testament has been refuted by T. Stordalen, *Echoes of Eden: Genesis 2-3 and Symbolism of the Eden Garden in Biblical Hebrew Literature* (Leuven: Peeters, 2000).

106. Terence E. Fretheim, *God and World in the Old Testament: A Relational The-*

Indeed, in both the Old and the New Testament "death" is a layered concept. For example, the author of Revelation discerns a "second death" next to our physical death (Rev. 2:11; 20:14), thus referring to an eternal death from which no redemption is possible. And Paul could write to the Ephesians that they were "dead through [their] trespasses" (Eph. 2:5), obviously not referring to either physical or eternal death but to their spiritual death from which they were now saved by God's grace. It has been argued that in Genesis 2-3 as well, this spiritual death is what the author(s) must have had in mind. For example, it is remarkable that despite the warning in Genesis 2:17, the man and his wife do not physically die "in the day" they ate from the forbidden tree. Apparently it is another kind of death that enters their life at that moment—spiritual death.[107]

Similarly, given that references to death in texts such as Romans 6:4, 6:21, and 8:10 clearly do not apply to physical death, it has been suggested that even in the most famous "proof texts" for the connection between sin and death, Paul had in mind spiritual rather than physical death. In this way, the threatening conflict between evolutionary biology and biblical interpretation could be neatly solved: science tells us about our physical states, religion about our spiritual conditions. It seems to me, however, that this is too easy a way out. A close reading of Romans 5 hardly allows for the interpretation that Paul is discussing the cause of our spiritual death here. As Mark Harris points out:

> It is true that Paul was certainly capable of using "death" as a metaphor for the separation between God and humankind . . . , and this appears in various places in the New Testament (e.g. Lk. 15:32; Rom. 6:2-11; Eph. 2:1,5; Col. 2:13). But it is always clear from the context when "death" is meant as a metaphor, and this is emphatically not the case in Romans 5, since Paul introduces "death" by referring to Christ's very literal death (vv. 8-10).[108]

---

*ology of Creation* (Nashville: Abingdon, 2005), 77, refers to "the experience of death within life" in this connection.

107. C. John Collins, *Genesis 1-4*, 161, 175; Alexander, *Creation or Evolution*, 324, even writes: "Genesis 3 provides for us one of the most powerful descriptions of spiritual death in the whole of Scripture." One may compare the way in which the complementary notion of human *birth* often stands for a spiritual (re)birth in the Bible (e.g., John 3:3).

108. Mark Harris, *The Nature of Creation: Examining the Bible and Science* (Durham, UK: Acumen, 2013), 141. Likewise, in 1 Cor. 15:20-26 it is the physical nature of Christ's death and resurrection that requires us to interpret the human death and future resurrection in this passage as physical as well (Harris, 141).

Thus, Paul sees a direct causal relation here (as well as in 1 Cor. 15:21–22) between human sin and human physical death. As to Genesis 2–3, whereas humans were created mortal ("from the dust of the ground," 2:7), only after they had sinned, and because of that fact, did physical death become inevitable (3:19).

Given an evolutionary framework, could we still make sense of the strong suggestion of Genesis and the clear statement of Paul that our physical death is somehow inextricably linked with God's judgment about sin? This negative assessment of death corresponds to our intuitive response to, for example, the sudden death of a younger person whom we have known closely. Like sin, such a death is usually experienced as "not the way it's supposed to be" instead of as something that is all in the game.[109] In line with this, we might suggest that according to the Bible we humans, as opposed to other species, are being exposed to "the full reality of death."[110] Although every living organism quite naturally has to cease to exist after some time, this is not the whole story for humans. The meaning of human death can be fathomed only if we recognize the element of divine disapproval inherent in it—a disapproval of the way we post-Fall humans live our lives in estrangement from God.[111]

We can extrapolate this line of thought by pointing out that in the narrative plot of Genesis 2–3 the first humans would somehow have received immortality, or the capacity to keep themselves alive endlessly, if they had not sinned. This is clearly suggested by Genesis 3:22, where the human being is expelled from the garden lest "he might reach out his hand and take also from the tree of life, and eat, and live forever."[112] Although it is impossible for us to

---

109. In my experience, even nonbelievers are inclined to ask questions like "what did I do wrong?" when they see themselves confronted with the possible prospect of a relatively early death.

110. Bimson, "Doctrines of the Fall," 114 (quoting Terence Fretheim).

111. Needless to say, I hope, such disapproval is not specifically connected to an *early* death; it is the collective human experience of having-to-die-at-some-time and of being cut off from one's relevant relationships that is pertinent here. Schwager, *Banished from Eden*, 18–19, suggests that whereas the first *Homo sapiens* were subject to physical death, they were free from *violent* death, which entered their lives only after the rise of human sin and violence (Gen. 4)—but this is implausible given what we know about the role of predation in evolutionary history. Levering, *Engaging the Doctrine*, 250, holds that "human death as we now experience it" is the result of sin, but it is not entirely clear what he means by this phrase.

112. According to Mettinger, *The Eden Narrative*, the tree of life functions as "the symbol of the potential reward" (123; cf. 130–31). James Barr, *The Garden of Eden and the*

*Evolution and Covenantal Theology*

spell out *how* humankind could have received immortality given our biological constraints, such a scenario is not ruled out by what we know about human origins (just as the occurrence of miracles is not ruled out by evolutionary science, or by any science).[113] For those who believe that God will eschatologically give imperishable life to his children (cf. 1 Cor. 15:42), it is quite natural to believe that God could have done so right from the beginning "if Adam had remained upright."[114] Since Adam did not remain upright, we humans had to die just like the animals—death in our case being the "wages of sin" (Rom. 6:23). Here, it seems, we have a way of thinking about the origin and nature of human physical death that on the one hand continues to do justice to the biblical view of the causal connection between sin and death,[115] while on the other hand not being made obsolete by the findings of evolutionary biology.

Now, how does the covenant narrative continue, and what further issues does that raise?

### 6.7 The Scope of the Christian Message of Salvation

The covenantal history that is related in the biblical narrative runs from the creation of the human being through its fall into sin to God's mighty acts of salvation. Thanks to God's free and sovereign grace as revealed in the history of Israel and as culminating in the person and work of Jesus Christ, human beings can again be accepted by God and receive his salvation, even though they did not deserve it. The New Testament uses various words and metaphors to express the nature of this gracious salvation or redemption, each illuminating one of its manifold aspects: atonement, reconciliation, justification, adoption, eternal life, etc. It is this message of salvation that, according to the common

---

*Hope of Immortality* (London: SCM, 1992), even argues that "how human immortality was almost gained, but in fact was lost" (4) is the story's central theme, although the way in which he pits this against the traditional view of Gen. 2-3 as a story on "the origins of sin and evil" is unconvincing.

113. Cf. Alexander, *Creation or Evolution*, 486n279.

114. John Calvin, *Institutes of the Christian Religion*, ed. John T. McNeill, trans. Ford L. Battles (Philadelphia: Westminster, 1960), 1.2.1. That this is what Calvin actually thought is clear from his commentary on Gen. 2:17: "His [= Adam's in his prefallen state] earthly life truly would have been temporal; yet he would have passed into heaven without death, without injury."

115. This connection is also emphasized and clarified in, e.g., Wolfhart Pannenberg, *Systematic Theology*, vol. 3 (Grand Rapids: Eerdmans, 1998), 556-62.

witness of the church, forms the heart of the Christian gospel. Taking its inspiration especially from Saint Paul and Augustine, Reformed theology has found reason to emphasize the merciful, unmerited, and all-decisive nature of God's saving acts even more than some other Christian traditions. In line with this, the free offer of God's saving grace to sinful human beings has become the focus of many Reformed and evangelical sermons up till today.

Suppose that the standard version of evolutionary theory as expounded in chapter 2 is true, what would that mean for this central Christian message of redemption? Christians who are skeptical toward evolutionary theory often fear that "everything is at stake" in accepting evolution, including the very heart of the gospel. According to them, we can no longer uphold the message of God redeeming us from sin and death through the work of Jesus Christ if we ascribe an evolutionary history to humankind. For, as a famous creationist motto goes: "No Adam, no Fall; no Fall, no Atonement; no Atonement, no Savior."[116] If this were true, the heart of the gospel would indeed be jeopardized. Clearly, however, it is not true. First, the nature and necessity of Christ's atoning work do not logically depend on the way in which we became sinners, but on the fact that we *are* sinners. And neither evolutionary theory in general nor the specific picture of human descent given above rules out this fact. Second, we have already seen that accepting evolutionary theory does not annul the possibility of belief in a human fall into sin—and by consequence, it does not invalidate belief in the atonement.

We may even go one step further. For the seriousness of our sinfulness (both as a state and as a series of acts) and thus our need for redemption is even *underlined* by our evolutionary background, as is sometimes acknowledged by theologians who take this background seriously. Christopher Southgate, for example, claims that liberal models of the atonement as only a subjective experience in human beings provoked by Christ's example of self-giving love won't do the job. Rather, "the impact of the Christ-event must be an objective one."[117] To be sure, our understanding of what salvation actually amounts to will be

---

116. Cf. George McCready Price, *Back to the Bible* (Washington, DC: Review & Herald, 1920), 124. For a recent version of what is essentially the same claim, see David Anderson, "Creation, Redemption and Eschatology," in *Should Christians Embrace Evolution? Biblical and Scientific Responses*, ed. Norman C. Nevin (Nottingham: Inter-Varsity Press, 2009), 91.

117. Southgate, *The Groaning of Creation*, 76. Domning and Hellwig, *Original Selfishness*, 152, are more inconsistent when they conceptualize salvation in terms of the life of Jesus being offered to us by God as a pattern we should imitate, because this largely Pelagian view fits ill with their emphasis on original sin as handed on by propagation rather than imitation (145).

affected by the way we view our evolutionary origins, since it becomes difficult to unpack this notion in terms of a return to a glorious past state. It is false, however, to suggest that the very structure of Christian soteriology collapses as a result of this. As George Murphy contends: "The Christian claim is that a savior is needed because all people are sinners. It is that simple. *Why* all people are sinners is an important question but an answer to it is not required in order to recognize the need for salvation."[118]

Note, however, that for us humans to be sinners implies that collectively we must somehow *have become* sinners. If we necessarily commit sinful acts because we are built to do so, then presumably our acts would not be sinful at all since we are not accountable for them. But even the doctrine of original sin does not deny our personal responsibility for our sinful behaviors. Here the Christian faith differs structurally from the tragic view of life found in the Oedipus tragedy and other Greek tragedies, as well as in various Eastern religions, which view humans primarily as victims. That is why we maintained the notion of a primordial human fall into sin. In this way we safeguarded that doing evil is not inherent to what it means to be human, since *that* idea would indeed fundamentally change the Christian story of redemption. As C. John Collins argues:

> Theologically, if we say that being prone to sin is inherent in being human with a free will, then we must say the Bible writers were wrong in describing atonement the way they did; and we must say that Jesus was wrong to describe his own death in these terms (e.g. Mark 10:45). Further, this approach makes nonsense of the joyful expectation of Christians that they will one day live in a glorified world from which sin and death have been banished (Rev. 21:1-8).[119]

Although we may hold that an almighty God could also liberate humans from evil if they had been its victims from the very beginning, Collins is right in arguing that by naturalizing sin we end up "telling a different story from the one we find in the Bible."[120] For clearly, key biblical concepts such as atonement, reconciliation, and justification would no longer apply in that case.

---

118. George Murphy, "Roads to Paradise and Perdition: Christ, Evolution, and Original Sin," *Perspectives on Science and Christian Faith* 58 (2006): 110.

119. C. John Collins, *Did Adam and Eve Really Exist?*, 47-48; unfortunately, Collins unhelpfully mixes up the existence of sin and the existence of suffering and pain in the world that God has made—as we have seen, these are different things that should be dealt with each on its own terms.

120. C. John Collins, *Did Adam and Eve Really Exist?*, 48. For a similar view from

This scenario does not emerge, however, when we recontextualize the notions of the Fall, human death, and original sin within an evolutionary context, as we have attempted to do in the preceding sections. To the contrary, if we assume an evolutionary history of the human race, the Christian doctrine of redemption only becomes more relevant and encompassing. For if we accept that the world God created was not by all means a perfect world right from the outset, we may recognize that the extent or scope of Christ's saving work goes far beyond "putting right what had gone wrong." It is through Christ and through the Spirit that God has intended from all eternity to finish and perfect his work of creation. Christ became incarnate and the Spirit was poured out not just in order to redeem us from sin (which is very significant already!); they were also meant to deliver the creation from its bondage to decay—that is: to make an end to the suffering, pain, and death that "until now" (Rom. 8:22) are pervasive in it. In more technical theological terminology, we should conceive of the significance of the salvific work of Christ and the Spirit not only in infralapsarian but also in supralapsarian terms: though the reversal of the negative effects of the Fall remains key to our salvation, divine redemption goes far beyond that. That makes the work of Christ and the impact of the Spirit only greater, and gives us all the more reason for awe and adoration.

That this is not just a modern adaptation of the biblical story inspired by the constraints of evolutionary science may become clear when we look at an often neglected passage in 1 Corinthians 15. At the end of this chapter, Paul compares two forms of bodily existence with each other. The first is called natural or physical and is characterized by being perishable, vulnerable, and weak (i.e., susceptible to illness). The second one is called spiritual, and this spiritual body is imperishable, glorious, and strong (vv. 42–44). Paul does not say that the first form of bodily existence was a divine punishment for sin. Instead, he argues that God endowed us with this weak, vulnerable, and perishable body when he created us. Indeed, it is remarkable that Paul does not refer to Genesis 3 in this connection but goes even further back, to the creation story of Genesis 2 (v. 45). Thus, according to Paul, *we have been created* with a body that won't last forever and is vulnerable to weakness and decay. It is only thanks to Christ, who has become "a life-giving spirit," that this will change. Clearly, in Paul's view redemption and salvation are much richer and include

---

a Roman Catholic perspective, see Levering, *Engaging the Doctrine*, 249: "If . . . sin and alienation were conceived as intrinsic to the human condition, there might need to be a healing, but one could hardly speak of a need for *reconciliation*, since the creator God would be responsible for humans' original disharmony" (emphasis added).

much more than just a return to a lost stage of creation. Rather, they bring us to a higher stage of existence—and not just us but the entire created world. For through Christ "God was pleased to reconcile to himself all things" (Col. 1:20). In brief, we do not need evolutionary theory to believe such things—we could have found them in the Bible, if only we had eyes to discern them.

Of course, it is not our business to find out *why* God has decided to work in this way. We are not in a position to fathom why this has pleased God. Like Job, we have to learn to let God be God. But yet, Paul discovers a pattern in God's creative activities: "It is not the spiritual that is first, but the physical, and then the spiritual" (1 Cor. 15:46). Apparently, he perceived a certain logic in the way in which God moves from more simple forms of life toward more complex ones. Similarly, in his many paraenetical appeals to lead a Christian life, Paul often contrasts life in the Spirit (which is an anticipation of the new life that is to come) with life in the flesh as driven by all kinds of old desires and inclinations that we now know have an evolutionary background.[121] Thus, here as well, there is a remarkable consonance between classical faith and contemporary science. This is not to suggest that Paul was aware of evolutionary theory. It is to suggest that, far from detracting from the gospel story, evolutionary theory may help us discover wonderful aspects of it that have for long remained hidden in relative obscurity.

Does this "discovery" imply that we have to leave the Augustinian (and Reformed) tradition behind us? Not necessarily, because Augustine already famously made a distinction between the proton and the eschaton, that is, between the situation before the Fall and the state of our future consummation. Whereas after being created, humans found themselves in a position of "being able not to sin" (*posse non peccare*), in the eschatological future the children of God will have moved toward a state of "not being able to sin" (*non posse peccare*).[122] As a result, a new fall into sin will be impossible. Thinking along such lines, we may extend the supralapsarian overtones in Augustine's soteriology by recognizing other aspects of creation—such as our creaturely vulnerability to death and decay—that will also be remedied and perfected when God finally reaches his goal with creation in the eschatological future.

---

121. See, e.g., Gal. 5:16-21; cf. Etzelmüller, "The Evolution of Sin."

122. See Augustine, *Enchiridion* 118 on the "fourfold state" of humanity—a theme that was adopted and elaborated in Reformed theology. Cf., e.g., Thomas Boston, *Human Nature in Its Fourfold State* (Edinburgh: Banner of Truth Trust, 1964), originally published in 1720.

CHAPTER 7

# Natural Selection and Divine Providence

> If you accept God already, it is still very much open to you to think of God as great inasmuch as He has created this wonderful world. . . . In the spirit of Baden Powell, one might think that God's magnificence is confirmed as one realizes that He does so much with so simple a mechanism as natural selection.
>
> —Michael Ruse[1]

## 7.1 Introduction

In the preceding chapters we have discussed the theological challenges posed by the first two layers of the neo-Darwinian theory of evolution set out in chapter 2: the appearance of various forms of life over vast periods of time on the geological timescale and the common background of all those forms of life in ultimately "a few forms or only one."[2] It is now time to move on to the

---

1. Michael Ruse, *Can a Darwinian Be a Christian? The Relationship Between Science and Religion* (Cambridge: Cambridge University Press, 2000), 114–15. The reference is to Baden Powell, *Essays on the Spirit of the Inductive Philosophy* (London: Longman, Brown, Green & Longman, 1855), 272: "Precisely in proportion as a fabric manufactured by machinery affords a higher proof of intellect than one produced by hand; so a world evolved by a long train of orderly disposed physical causes is a higher proof of Supreme intelligence than one in whose structure we can trace no indications of such progressive action." (Note that this quote precedes Darwin's *Origin* by a couple of years.)

2. Cf. the famous final sentence of Darwin's *Origin of Species*: "There is grandeur in this view of life, with its several powers, having been originally breathed into a few forms or into one"; Charles Darwin, *On the Origin of Species by Means of Natural Selection* (London: John Murray, 1859), 490 (quoted from www.darwin-online.org.uk).

third layer of evolutionary theory: the theory of natural selection operating on random mutations. Or, to formulate this a bit more precisely: the idea that natural selection operating on random mutations is the dominant mechanism behind the evolutionary process. Although hardly anybody denies that natural selection plays a role in biological evolution, as we saw in chapter 2, evolutionary biologists and others (including philosophers) strongly disagree about the extent to which this mechanism can be held responsible for the entire evolutionary process. Whereas "orthodox Darwinians" among the scientists and philosophers still hold on to this conviction, others are more skeptical and argue that one or more other mechanisms, not being reducible to natural selection, play a crucial part as well.

We will not pursue this scientific debate further, but for the sake of argument we will take the pivotal relevance of natural selection as the principal motor of evolution for granted. So suppose that natural selection working on random genetic mutations is key to all evolutionary processes, what concerns would that raise for Christian—and, more specifically, for Reformed—theology? Just as we have been able to specify two theological concerns raised by the other two layers of evolutionary theory, here again two specific issues come to mind. Interestingly, the first is one of the oldest problems Christians have had with evolutionary theory, and the second is one of the newest.

First, if the mutations (or "transmutations," as Darwin had it) that are selected by nature because of their fitness-enhancing character are *random*, how can this be combined with one of the most fundamental Christian convictions: the doctrine of God's sovereignty and providence over all that goes on in God's creation (that is, in the entire universe)? Apparently, if randomness rules, there is no proper place for purpose, guidance, directedness, and so forth. In that case even God has to wait and see how things develop. Second, the theory of natural selection has itself gone through a remarkable evolution in that it became applied not only to biological but also to *cultural* evolution. That is, according to many observers, cultural phenomena such as morality and religion (to mention only two of the most remarkable ones) evolved and spread through human societies because of their significance in the struggle for reproductive success.[3] Somehow, nature must have favored human (or

---

3. Cf., e.g., Michael Ruse and Edward O. Wilson, "The Evolution of Ethics: Is Our Belief in Morality Merely an Adaptation Put in Place to Further Our Reproductive Ends?" *New Scientist* 108 (October 17, 1985): 50-52; Pascal Boyer, *Religion Explained: The Evolutionary Origins of Religious Thought* (New York: Basic Books, 2001); David Sloan Wilson, *Darwin's Cathedral: Evolution, Religion, and the Nature of Society* (Chicago:

prehuman) groups and individuals that were moral or religious over groups and individuals that were not. But what would that mean for the reliability of moral and religious truth claims? If such claims are just a function of evolution, could they possibly still be true, and if so, could we know that? In brief: Can we still know about good and evil, and know God?

It is to this second set of questions that we will turn in chapter 8, focusing on theories about the evolutionary rise of morality and especially of religion. Let us first, however, discuss the way in which the theory of natural selection may backfire on the doctrine of divine providence.

## 7.2  High Stakes: Divine Guidance or a Purposeless Universe?

In investigating the impact of the theory of evolution on the doctrine of divine providence, we will not focus here on the enormous amount of suffering, predation, starvation, and extinction that pervades the natural world. Of course, these phenomena as well might cast doubt on the belief that God is in providential control of anything that happens. We discussed the problem of evolutionary evil in chapter 4, however, because it is a corollary of the very first layer of evolutionary theory: gradualism. Here, we will deal with another important question, namely, whether evolution can be harmonized with God's guidance and providence at all (i.e., apart from the suffering and pain that accompany it). Adherents of the so-called independence model of science and religion may argue that there is no problem here, since evolution is a scientific theory whereas the doctrine of providence constitutes a confession of faith: an expression of ultimate trust in God.[4] However, most Christians, including Reformed ones, will argue that such trust should somehow correspond to the nature of reality. If the natural world is replete with processes characterized by sheer randomness and coincidence, and if even humankind is the product of

---

University of Chicago Press, 2002); Michael Tomasello, *A Natural History of Human Morality* (Cambridge, MA: Harvard University Press, 2016).

4. According to this independence model, science and religion each have their own domains, aims, methods, etc., which do not overlap. It became especially popular as a result of Stephen Jay Gould's defense of the so-called NOMA principle (religion and science as *non-overlapping magisteria*) in his *Rocks of Ages: Science and Religion in the Fullness of Life* (New York: Ballantine Books, 1996). See also Ian G. Barbour's famous typology of ways of modeling the science-religion interface, e.g., in his *Religion and Science: Historical and Contemporary Issues* (New York: HarperCollins, 1997), 77-105.

blind forces, it is hard to see how God might steer all things to their destination through his providential control. Therefore, it is impossible to detach faith in God's providence from what actually happens in our mundane reality.[5]

This means that we must ask whether there is a necessary connection between evolution and dysteleology: Does the stress of the neo-Darwinian theory of evolution on the role of chance not make any overarching divine goal and guidance of the evolutionary process impossible? If so, the prospects for recontextualizing the classical doctrine of providence in an evolutionary setting (as we recontextualized the doctrine of the Fall in the preceding chapter) are dim. For clearly, if chance and randomness reign, the very heart of what is expressed in the doctrine of divine providence—namely, that *God* reigns over the universe as well as over our personal lives—is at stake. This is especially the case when we specify that this God is the almighty Father of Jesus Christ, who has shown us that he is not at all indifferent to our fate but who "so loved the world that he gave his only Son" (John 3:16). Such a belief is incompatible with the idea that our lives are from beginning to end the products of blind and arbitrary forces. Therefore, do Darwinian evolution and a Christian account of divine providence simply exclude each other? Must we conclude that the evolutionary process rules out any belief in a God who guides his creation, through all that happens, to the purposes he has set?

Given the centrality of the notions of God's sovereignty and providence in Reformed theology,[6] it is perhaps little wonder that, in the past, especially representatives of the Reformed tradition have found reason to reject evolutionary theory with an appeal to God's providence. In fact, evolutionary theory's purported denial of God's overarching control over history was one of the first and most important objections brought forward against it

---

5. Recent introductions to the doctrine of divine providence from a Reformed perspective include Benjamin W. Farley, *The Providence of God* (Grand Rapids: Baker Books, 1988); Paul Helm, *The Providence of God* (Leicester: Inter-Varsity Press, 1993); Charles M. Wood, *The Question of Providence* (Louisville: Westminster John Knox, 2008); David Fergusson, *The Providence of God: A Polyphonic Approach* (Cambridge: Cambridge University Press, 2018); Philip Ziegler and Francesca Aran Murphy, eds., *The Providence of God: Deus Habet Consilium* (London: T&T Clark, 2009); Mark W. Elliott, *Providence Perceived: Divine Action from a Human Point of View* (Berlin: de Gruyter, 2015). See also Cornelis van der Kooi and Gijsbert van den Brink, *Christian Dogmatics: An Introduction* (Grand Rapids: Eerdmans, 2017), 233–44.

6. Cf. §1.4 above; Alister E. McGrath, *Darwinism and the Divine: Evolutionary Thought and Natural Theology* (Oxford: Wiley-Blackwell, 2011), 214n9, also points out that "the role of providence is particularly significant for Reformed writers."

by Christian critics. If Darwinian evolution is incompatible with notions of divine purpose and design, then evolutionary theory is indeed bad news not only for Reformed Christianity but also for religion much more generally conceived. For, as Roman Catholic scientist and theologian John Haught has argued, "although theology can accommodate many different scientific ideas, it cannot get along with the notion of an inherently purposeless universe."[7] If William Paley's famous watchmaker turns out to be blind, then there is not much that we can hope for.[8] In what follows, we will briefly examine what we can learn from the past about this issue (§7.3) before turning to the present (§7.4). As we will see, much depends on how the relevant concepts are defined and interpreted.

## 7.3 Evolution, Chance, and Teleology in the Past

The discussion about the relationship between evolution and Christian faith has shifted back and forth over time. The interpretation of the Bible has been the main concern, or the issue of anthropology ("we do not stem from the animals, do we?"), or the question of theodicy. During the first decades of the Christian reception of the theory of evolution, however, the struggle was almost totally confined to the issue of divine providence.[9] Both within Christian circles and beyond, the answers given to the question whether Darwinian evolution could be reconciled with any notion of teleology were quite varied. And in fact, this has not changed over time. The topic became so complicated because from the outset both Christians and naturalists introduced worldview considerations into the debate, arguing either in favor of or against the randomness of evolution because of their specific view of life. As a result, it

---

7. John F. Haught, *God after Darwin: A Theology of Evolution* (Boulder, CO: Westview, 2000), 26.

8. Cf. Richard Dawkins, *The Blind Watchmaker: Why the Evidence of Evolution Reveals a Universe without Design* (New York: Norton, 1985).

9. David Livingstone in particular has shown how Protestant responses to evolutionary theory (and esp. to the issue of its compatibility with providence) differed from place to place and were highly influenced by local perceptions and discussions. See his *Dealing with Darwin: Place, Politics, and Rhetoric in Religious Engagements with Evolution* (Baltimore: Johns Hopkins University Press, 2014). For the way in which the debates developed in Princeton, New Jersey, see Bradley J. Gundlach, *Process and Providence: The Evolution Question at Princeton, 1845–1929* (Grand Rapids: Eerdmans, 2013).

often was—and sometimes still is—difficult to distinguish between scientific and (anti)religious considerations.

The first reaction came from those keen to use the Darwinian hypothesis to deal a final blow to the Christian faith in God and providence. At long last, these notions were no longer necessary! As Richard Dawkins would later famously put it: "Darwin made it possible to be an intellectually fulfilled atheist."[10] Indeed, we should not underestimate the liberating force of the discovery of the pivotal role of natural selection in adaptation and speciation. The highly sophisticated ways in which the physical parts and organs of plants and animals are "fine-tuned" to their natural environment needed to no longer be explained with reference to the supernatural intelligence of a divine Creator. Instead, their origin and development could be conceived of along immanent lines. All sorts of specific details—the smell of honeysuckle, the long neck of the giraffe, the brain volume of human beings, and many other amazing things—that had thus far been regarded as the product of a special divine design could now much more simply be explained with an appeal to natural causes.

Many added to this that, apparently, teleological explanations had turned out to be just false: if natural selection is responsible for what happens in the natural world, then evidently there is no God who oversees and controls everything. This conclusion was most loudly proclaimed by epigones of Darwin who went further than the master himself in connecting Darwinian evolution with a monistic and often materialistic worldview.[11] One may think of Ernst Haeckel (1834-1919)[12] in Germany and also of Thomas Huxley (1825-1895) in England, who was famously called "Darwin's bulldog." Although Huxley acknowledged "a wider teleology which is not touched by the doctrine of Evolution," he envisaged this teleology in terms of a deterministic process governed

---

10. Dawkins, *The Blind Watchmaker*, 6. For a thoughtful response to this quote, see Denis O. Lamoureux, "Toward an Intellectually Fulfilled Christian Theism," *Faith and Thought* 55 (2013): 2-17, and cf. its sequel in *Faith and Thought* 57 (2014): 3-20.

11. Darwin's own religious views continued to waver between agnosticism and theism until the end of his life. Cf., e.g., Denis O. Lamoureux, "Theological Insights from Darwin," *Perspectives on Science and Christian Faith* 56, no. 1 (2004): 2-12.

12. Cf. Robert J. Richards, "Ernst Haeckel and the Struggles over Evolution and Religion," *Annals of the History and Philosophy of Biology* 10 (2005): 90: "The antagonism between conservative religion and evolutionary theory, brought to incandescence at the turn of the century and burning still brightly in our own time, can be attributed, in large part, to Haeckel's fierce broadsides launched against orthodoxy in his popular books and lectures."

by definite laws and leading from primordial matter to the world's current complexity.[13] As to "teleology as commonly understood," that is, the teleological view according to which the goal or *telos* of processes in the natural world is set by God, Huxley declared that it had "received its deathblow at Mr. Darwin's hands."[14]

Clearly, however, one cannot conclude from the fact (assuming for the moment that it is a fact) that this teleological view is no longer *needed*, that it must therefore be *false*. The latter just does not follow from the former. We need more assumptions to make this argument work, such as that God cannot use natural selection to attain his goals with creation. Darwin's atheist followers, however, ignored this point and directly inferred from Darwin's theory that all attempts to connect God and nature had failed.[15] The rhetorical force used to this very day by atheist Darwinians when suggesting this inference and when defining Darwin's theory in terms of a materialistic worldview made many Christians wary of Darwinism. If Darwin's theory does indeed put an end to the belief that God is involved in the course of nature by leading it to its ultimate goal, then they could only be opposed to it. They also feared that when God's supervision became replaced by natural processes, the foundations of morality were compromised. Does not a world without purpose imply human lives without meaning or purpose?[16]

Prominent Reformed theologians such as Abraham Kuyper in Amsterdam and Charles Hodge in Princeton rejected Darwinian evolution because of its putative incompatibility with divine providence. According to Kuyper, the "real motive" behind the appeal to natural selection in Darwin's theory of evolution

---

13. "Professor Huxley on the Reception of the 'Origin of Species,'" in *The Life and Letters of Charles Darwin*, ed. Francis Darwin (London: John Murray, 1887), 2:201, http://darwin-online.org.uk/content/frameset?keywords=wider%20teleology&pageseq=217&itemID=F1452.2&viewtype=text, accessed May 13, 2019. Cf. Michael Ruse, *Darwin and Design: Does Evolution Have a Purpose?* (Cambridge, MA: Harvard University Press, 2003), 141–42. For a different interpretation, see McGrath, *Darwinism and the Divine*, 186–87.

14. Thomas Huxley, "Criticisms on 'The Origin of Species'" [1864], in his *Darwiniana: Essays* (New York: Appleton and Co., 1896), 82.

15. Cf. Jon H. Roberts, "Myth 18: That Darwin Destroyed Natural Theology," in *Galileo Goes to Jail and Other Myths about Science and Religion*, ed. Ronald L. Numbers (Cambridge, MA: Harvard University Press, 2009), 161–69.

16. Cf. Ernst Mayr, *The Growth of Biological Thought: Diversity, Evolution, and Inheritance* (Cambridge, MA: Belknap Press of Harvard University Press, 1982), 514–17 (on "reasons for the strength of resistance against selection").

was "the desire of the human heart to rid itself of God."[17] Using his characteristically baroque idiom, he declared that the Darwinian theory of evolution "is the 'form of unity' that currently unites all the priests of modern science in their secularized temple."[18] In his famous 1900 inaugural address "Evolution," Kuyper expanded on his main objection, strongly criticizing evolutionary theory's "sworn enmity against every presupposition of a previously established goal toward which the development of living organisms would be impelled, either by means of an indwelling principle, or through divine power working from without."[19] Kuyper relied, in particular, on Haeckel for his perception of natural selection. His language was once again very straightforward: "The Christian religion and the theory of evolution are two mutually exclusive systems."[20] Kuyper's problem was not with the theory of common ancestry, or, in Kuyper's words, the "spontaneous unfolding of the species in organic life from the cytode or the nuclear cell," since that theory could be interpreted in theistic terms.[21] Such an interpretation, however, was different from Darwinism, since "the pre-established purpose [*Zweck*] would then not have been banished but would have been all-controlling, and then the world would not have constructed itself mechanistically, but God would have constructed it by the use of elements that He himself had prepared."[22]

A quarter of a century earlier, Charles Hodge had voiced a similar opinion in the United States. In his well-informed study *What Is Darwinism?* he did not leave the reader in any uncertainty how the question of this title had to be answered. Whether Darwin himself had to be branded an atheist was another matter, but Hodge had no doubt about his theory: "It is atheism."[23] The reason

17. Abraham Kuyper, "The Blurring of the Boundaries," in *Abraham Kuyper: A Centennial Reader*, ed. James D. Bratt (Grand Rapids: Eerdmans, 1998), 377.

18. Kuyper, "Blurring of the Boundaries," 378; cf., however, 376n68, where Bratt shows that at this stage already (1892) Kuyper had a "dual attitude" toward evolutionary theory, at the same time praising it for the "vast gains for our knowledge" it had yielded and for the fact that it had restored the "unity of nature."

19. Abraham Kuyper, "Evolution," *Calvin Theological Journal* 31 (1996): 20-21.

20. Kuyper, "Evolution," 15.

21. Kuyper, "Evolution," 47.

22. Kuyper, "Evolution," 46-47. Cf. in more detail, Gijsbert van den Brink, "Evolution as a Bone of Contention between Church and Academy: How Abraham Kuyper Can Help Us Bridge the Gap," *Kuyper Center Review* 5 (2015): 92-103. Over against atheistic uses (or abuses) of the theory of evolution, Kuyper's words are still to the point.

23. Charles Hodge, *What Is Darwinism?* (New York: Scribner, Armstrong and Co., 1874), 83.

for this was the lack of any notion of a divine design or divine guidance. "The grand and fatal objection to Darwinism is this exclusion of design in the origin of species."[24] For Hodge this was nothing less than the defining characteristic of Darwinism. Other theories of evolution might perhaps be compatible with the Christian faith, even if they operated with the principle of natural selection, but not so with Darwinism, since according to Hodge Darwin emphatically denied any teleological element in his theory. Hodge even included in the book "Proof of Darwin's Denial of Design from His Own Writings" and also referred to the materialistic interpretation of Darwin's theory by, among others, Huxley and Haeckel. As with Kuyper, other considerations, such as the exegesis of the first chapters of Genesis, do not seem to have played a major role in Hodge's rejection of evolution. What counted was the denial of God's providence. For "this universal and constant control of God is not only one of the most patent and pervading doctrines of the Bible, but it is one of the fundamental principles of even natural religion."[25]

Whereas Hodge was quite right in this latter statement, his rejection of Darwinism because of the antiteleological bias in Darwin's take on natural selection was vulnerable to criticism even from his fellow Christians at the time. For example, the Scottish philosopher and president of Princeton University James McCosh (1811-1894) countered that it "could be easily shown that the doctrine of development [= theory of evolution], properly understood, and kept within inductive limits, is not inconsistent with final cause." We just had to await a new Paley who would offer proofs for design in the Darwinian scheme. Hodge might have been right that Darwin rejected teleology, "but his facts are teleological whether he acknowledges it or no[t]."[26] Similarly, in his review of Hodge's *What Is Darwinism?* the distinguished Harvard botanist Asa Gray endorsed the possibility of a teleological version of Darwinism (suggesting that God might provide favorable mutations in the selection process). In his view, Hodge's obsession with the antiteleological definition of natural selection he had found in the works of Darwin counterproductively induced a gradual drifting apart of the community of naturalists (i.e., what we would now call natural scientists) from their Christian background.[27]

---

24. Hodge, *What Is Darwinism?*, 80.
25. Hodge, *What Is Darwinism?*, 21.
26. James McCosh, *Ideas in Nature Overlooked by Dr. Tyndall* (New York: Robert Carter and Brothers, 1875), 33; quoted from Livingstone, *Dealing with Darwin*, 168 (see 160-74 for McCosh's attitude toward evolution).
27. Livingstone, *Dealing with Darwin*, 173-74; Gray's review appeared in the *Na-*

## 7.4 Evolution, Chance, and Teleology Today

In present times the discussion about Christian faith and evolutionary theory has broadened to include domains such as those examined elsewhere in this book. The debate about natural selection's compatibility or incompatibility with providence, however, has never ended. In the previous century, it received a new impetus when the French biochemist Jacques Monod famously claimed that the evolutionary process was a matter of "pure chance, absolutely free but blind."[28] By then George Simpson had already phrased his equally well-known statement: "Man is the result of a purposeless and natural process that did not have him in mind. He was not planned."[29] And forty years later, Edward O. Wilson especially credited the mechanism of natural selection in this connection: "Evolution in a pure Darwinian world has no goal or purpose: the exclusive driving force is random mutations sorted out by natural selection from one generation to the next. Many . . . , however, on religious grounds, [cannot] accept the operation of blind chance and the absence of divine purpose implicit in natural selection."[30] Again, the general thrust is that natural selection excludes any divine goal or guidance. Others, however, protested that this view imposes an atheistic view of life on the sober scientific facts.

So who is right? It all depends on what exactly the notion of randomness as it figures in "random mutations" means. What kind of chance is involved?[31] We will attempt to answer this question using a recent exchange between Al-

---

tion, May 28, 1874, 348–51. On Gray, see the biography by A. Hunter Dupree, *Asa Gray: American Botanist, Friend of Darwin* (Baltimore: Johns Hopkins University Press, 1988).

28. Jacques Monod, *Chance and Necessity: An Essay on the Natural Philosophy of Modern Biology* (New York: Knopf, 1971), 112–13.

29. George G. Simpson, *The Meaning of Evolution: A Study of the History of Life and of Its Significance for Man*, rev. ed. (New Haven: Yale University Press, 1967), 345.

30. E. O. Wilson, introduction to *From So Simple a Beginning: The Four Great Books of Charles Darwin*, ed. E. O. Wilson (New York: Norton, 2005), 12. Mayr, *Growth of Biological Thought*, 515, as well, is quite confident: "it has been rightly said that natural theology as a viable concept died on November 24, 1859" (i.e., the day when Darwin's *Origin of Species* came out).

31. For the various meanings of words like "chance" and "randomness," see René van Woudenberg, "Chance, Design, Defeat," *European Journal for Philosophy of Religion* 5 (2013): 31–41. Sometimes a distinction is made between randomness (as sheer arbitrariness) and chance (as referring to a number of possible outcomes with assignable probabilities). See for more details on this issue, James Lennox, "Darwinism," *Stanford*

vin Plantinga and Herman Philipse in which the controversy is elaborated with considerable acumen and precision.[32] In the first chapters of his *Where the Conflict Really Lies*, Plantinga argues that natural selection on the basis of random mutations does not exclude the possibility of divine providence.[33] He points out that the randomness of random mutations does not mean that these mutations have no cause or are just a matter of chance. Contemporary biologists construe the notion much more precisely as conveying that there is no causal connection between the occurrence of genetic mutations in an organism and any adaptational needs that organism may have.[34] In other words, the mutations do not occur because they have adaptive value in the struggle for survival, but they occur "at random"—and many mutations do, indeed, not have such value at all. Plantinga refers to a statement by the prominent philosopher of biology Elliott Sober, who puts this point as follows: "There is no *physical mechanism* (either inside organisms or outside of them) that detects which mutations would be beneficial and causes those mutations to occur."[35] In *that* sense the mutations are random. Sober explicitly acknowledges that in a stronger, metaphysical sense the mutations need *not* be random, so that "theistic evolution is a logically consistent position."[36]

Thus, evolutionary biologists indicate that the term "random" is used with a very specific, technical meaning, which does not exclude the idea that genetic mutations may somehow be caused by God and occur under his prov-

---

*Encyclopedia of Philosophy*, last revised May 26, 2015, §3.1, https://plato.stanford.edu/entries/darwinism/; I owe this reference to Daan Oostveen.

32. Much more so, e.g., than in Philip Kitcher, *Living with Darwin: Evolution, Design, and the Future of Faith* (Oxford: Oxford University Press, 2007). Here Kitcher argues that Darwinian evolution rules out "providentialist religion," which involves "belief that the universe has been created by a Being who has a great design, a Being who cares for his creatures, who observes the fall of every sparrow and who is especially concerned with humanity" (122-23). In making his case, however, he just points to two well-known theological riddles (the problem of evil and religious pluralism), which obviously predate the Darwinian theory of evolution.

33. Alvin Plantinga, *Where the Conflict Really Lies: Science, Religion, and Naturalism* (Oxford: Oxford University Press, 2011), 3-63.

34. Plantinga, *Where the Conflict Really Lies*, 11.

35. Elliott Sober, "Evolution without Naturalism," in *Oxford Studies in Philosophy of Religion*, vol. 3, ed. Jonathan L. Kvanvig (Oxford: Oxford University Press, 2011), 193; I have changed the title and added the page reference, since Plantinga (*Where the Conflict*, 11) quotes a prepublication version of this article that had a slightly different title.

36. Sober, "Evolution without Naturalism," 190.

idence. Now what objections does Philipse have to this view? First, Philipse points to the testimony of biologists and other evolutionary specialists who deny the possibility of mutations that are caused by the Creator—with Darwin (who had contradicted Asa Gray on this point) as the first in line. Philipse even argues that this is the "consensus view."[37] However, the claim of Sober that we just quoted suggests otherwise. And Sober is not the only one. Ernst Mayr (also cited by Plantinga) may be somewhat less explicit, but he also limits his definition to the technical aspects, steering clear of metaphysical overtones: "When it is said that mutation or variation is random, the statement simply means that there is no *correlation* between the production of new genotypes and the adaptational need of an organism in a given environment."[38] In a later exchange, Plantinga mentioned even more definitions of this type by evolutionary biologists, most of them atheists (and thus not having an interest in reconciling theism and Darwinian evolution).[39] Philipse is probably right that a majority of evolutionary biologists deny any notion of teleology and, by extension, of divine providence. The question is, however, whether they do so *qua* scientists, or whether this just reflects their view of life (which they then mix up with their science). If we want to know this, we have to explore their arguments.

Second, here is the argument brought forward by Philipse: given the evolutionary meaning of the term, *randomness* is not "logically compatible with being guided by God, if at least 'guided' means that mutations are somehow directed towards new adaptations, or to the development of new species. And what else should it mean? Clearly, then, the thesis that evolution is unguided is an integral part of the Modern Evolutionary Synthesis."[40]

Over against Plantinga, Philipse argues that *causing* some entity or pro-

---

37. Herman Philipse, "The Real Conflict between Science and Religion: Alvin Plantinga's *Ignoratio Elenchi*," *European Journal for Philosophy of Religion* 5 (2013): 87-110 (92).

38. Ernst Mayr, *Towards a New Philosophy of Biology: Observations of an Evolutionist* (Cambridge, MA: Harvard University Press, 1988), 98. Elsewhere, however, Mayr claims that one can be a Darwinist only if one explains the origin and diversity of life exclusively by means of natural causes; see his *One Long Argument: Charles Darwin and the Genesis of Modern Evolutionary Thought* (Cambridge, MA: Harvard University Press, 1991), 98-100, and cf. n. 30 above.

39. Alvin Plantinga, "Seeking an Official Definition of 'Randomness': A Reply to Jay Richards," *Evolution News and Science Today*, April 3, 2012, http://www.evolutionnews.org/2012/04/seeking_an_offi058161.htm.

40. Philipse, "Real Conflict," 92.

cess is different from *guiding* it.[41] Philipse concedes that random mutations may be caused by God, but in his view this does not mean that they can also be guided by God, in the sense of having been given a particular purpose by him. For this, Philipse maintains, would contradict the fact that mutations do *not* have a purpose or goal. It can be questioned, however, whether this distinction holds water, since it seems that to guide some event is to cause it to happen in a particular way. For example, if I guide or steer car X in the direction of Amsterdam, I see to it that it moves toward Amsterdam, thus causing the event "car X moves toward Amsterdam." Let us suppose, however, that Philipse can easily rephrase his argument, for example, by stipulating that God can cause some events (e.g., mutations coming into existence) but not others (e.g., mutations having particular outcomes).

It is not easy to see why this would be the case. But even if it is true, we may wonder whether, as Philipse suggests, "guidance" can be understood only in terms of directing specific mutations to a particular goal. Is it not also possible that God oversees the general system without steering all of its specific details? The entire process of natural selection on the basis of random mutations does in any case have an unambiguous effect, as it results in a stunning increase in adaptivity and complexity of life-forms. This effect is often seen as a kind of immanent "goal" of the process.[42] To achieve this goal, however, it is not necessary that every single mutation has a fitness-increasing value. It will do if only a certain percentage of the enormous number of mutations that occur have a fitness-enhancing effect. And if many mutations of all varieties occur uninterruptedly, as is the case, there is a good chance that some are helpful in the struggle for survival. Thus, even though each of its specific mutations may be unguided and even though it has no ultimate goal "in mind" toward which it strives, the process of natural selection may still not be as blind as has often been suggested, since it is clearly directed toward the enhancement of the inclusive fitness of a population.[43]

---

41. Philipse, "Real Conflict," 92 (where Plantinga is accused of confusing both concepts, 92n20).

42. McGrath, *Darwinism and the Divine*, 187-91, points out that well-known contemporary biologists such as Ernst Mayr and Francisco Ayala, despite their fear of being misunderstood or abused by those who smuggle theological or metaphysical theories into scientific accounts, no longer shy away from using teleological language in biological explanation. "Natural selection itself should be considered a teleological process, in that it is directed to the goal of increasing reproductive efficiency" (191).

43. This is rightly pointed out (against such popular notions as Richard Dawkins's "blind watchmaker") by Ernan McMullin, "Could Natural Selection Be Purposive?," in *Divine Action and Natural Selection: Science, Faith, and Evolution*, ed. Joseph Seckbach

Why, however, would it be impossible that through this process also an external goal—either set by a providential God or otherwise—could be achieved? Of course, this would be a claim on a metaphysical or theological level, which cannot be empirically demonstrated or serve as an explanation on the same level as scientific explanations. The question is, however, whether this idea is *incompatible* with the evolutionary process, and it is not easy to see why that should be the case.[44] Thomas Aquinas already perceived that God's providential guidance may include events and processes that would be defined as "chance" in another setting.[45] Aquinas uses an example that may clarify this for us, even today. Suppose a lord sends two of his servants, independently from each other, to the market with different errands, and that he plans this in such a way that they arrive at the same time. For the servants it would appear that they meet at the market "by chance." One was, for instance, going to the blacksmith and the other had to draw water from the central well. It seems totally superfluous to appeal to other factors than time and chance for their simultaneous presence at the market square. Everyone who becomes aware of the meeting of the two servants will be satisfied with an explanation that is based on chance. However, in reality this certainly is, at another (maybe one might say, higher) level, a clear instance of intentionality and an all-encompassing sense of purpose.[46]

---

and Richard Gordon (London: World Scientific, 2009), 114-25. McMullin makes an enlightening distinction between two types of teleology (117): a "Platonic" one in which there is "a consciously entertained goal, an intentionally pursued purpose," and an "Aristotelian" one in which the sole issue is internal functionality ("the regularities of nature perceived as furthering the finalities of nature").

44. Cf. Mikael Stenmark, *Scientism: Science, Ethics, and Religion* (Aldershot, UK: Ashgate, 2001), 115-23, who rightly concludes that "it's not true that science (or evolutionary biology) itself [i.e., apart from an extrascientific worldview] undermines the religious belief that there is a purpose or meaning to the existence of the universe and to human life in particular. Science cannot establish that the universe and humans are here for no reason" (122).

45. Thomas Aquinas, *Summa contra Gentiles* 3.74.

46. I found this example—unfortunately without an indication of its source—in Stephen M. Barr, "The Concept of Randomness in Science and Divine Providence," in Seckbach and Gordon, *Divine Action and Natural Selection*, 473-74. With reference to Eccles. 9:11 (but we might also think of, e.g., 1 Kings 22:34 and Luke 10:31 in this connection), Barr points out that the Bible recognizes the role of chance. Aquinas may rather have derived his example from Aristotle, *Physica* 2.4-6, though (I am grateful to Jeroen de Ridder for this reference).

Aquinas's example may seem to convey a deistic view, according to which God is involved only in the beginning of a certain history; but of course, it should not be stretched beyond its point. And this point is a very important insight that is also recognized in the Christian doctrine of providence, namely, that God and immanent actors may both be involved as causal factors that can explain the occurrence of a certain event but *at a different level*. Acting as the *causa prima*, God does not directly interfere with the *causae secundae* but still is involved as the one who enables these to operate and who may arrange them in certain ways. In Reformed scholastic theology, this way of seeing things has been captured in the notion of divine concurrence (or *concursus*): "Concurrence or cooperation is the operation of God by which He co-operates directly with the second causes as depending upon Him alike in their essence as in their operation, so as to . . . operate along with them in a way suitable to a first cause and adjusted to the nature of second causes."[47]

Admittedly, however, the example offered by Aquinas involves no randomness in the statistical sense, as is the case with natural selection. This difference may be relevant, because, however significant the chance that a specific mutation will occur at some moment, in statistical processes within finite populations that chance can never be 100 percent. It remains possible that the way the coin lands will always show heads and never tails. Consequently, although the emergence of complex species may have been "guided," perhaps also of such complex species that resemble humans, the emergence of the specific human species as we now know it cannot be predetermined, for that would require very specific mutations. We may compare this with the way in which an ovum is fertilized: the fact that this, under normal circumstances, will happen quite frequently is reasonably predictable given the great number of sperms that move toward the ovum; but it cannot be predicted which sperm will win the race and which person will as a result be born.[48]

Should we therefore accept that God's providence may concern the bigger picture but not the finer details? If the coming into existence of the human species is seen as such a detail, this would be incompatible with the Christian

---

47. Heinrich Heppe, *Reformed Dogmatics: Set Out and Illustrated from the Sources*, rev. and ed. Ernst Bizer (Grand Rapids: Baker Books, 1978), 258 (the definition is from J. H. Heidegger, *Medulla Theologiae Christianae* [Zürich: D. Gessner, 1696]). Reformed accounts tend to stress the active involvement of God more than most other ones, but they as well are nonetheless keen to acknowledge the relative independence of the secondary causes.

48. Cf. Barr, "Concept of Randomness," 475.

faith. For, according to the Christian faith, God not only wanted the human species to exist but even every human individual (cf. Ps. 139). What is more, according to the classic doctrine of providence, even the smallest details are subject to divine guidance, since "even the hairs of your head are all counted" (Matt. 10:30).[49] Thus, even though Philipse and others do not succeed in refuting all notions of divine guidance by appealing to the randomness of genetic variation, their arguments might still pose a problem to traditional (and especially Reformed) accounts of divine providence.

In order to move forward here, let us take another look at the encompassing process of natural selection on the basis of random mutations. Theoretically, a complete description of this process is feasible. For even though debates on how exactly genetic mutations come about continue unabated, all participants agree that these mutations must be caused. If they are caused, however, it is possible in theory to describe the exact cause-and-effect relationships. As human beings, we do not know all relevant factors, and most likely we will never know them. But suppose that someone does know all these factors. And suppose further that this person is also the Creator of heaven and earth, and therefore possesses a great number of capacities. Nothing in the neo-Darwinian scenario excludes this possibility. But then it can neither be excluded that such an omniscient and powerful God also *wanted* this entire complex of cause-and-effect relationships. As Keith Ward argues, even if we assume that only natural causal factors are at work in the process, "yet the whole causal structure may itself depend on the existence of a creator God."[50] Since in the statistically accidental mutations and the connected processes of natural selection some factors play a role that we cannot know or measure—either as a matter of principle or as the de facto present situation—it can even be imagined that, as Asa Gray thought, God *influences* such processes, guiding them in certain directions, such as (for example) toward the emergence of

---

49. This all-encompassing scope of God's providence has been especially emphasized (and intensified) in Reformed accounts of the doctrine of divine providence. See, e.g., Heidelberg Catechism, Q&A 27; John Calvin, *Institutes of the Christian Religion*, ed. John T. McNeill, trans. Ford L. Battles (Philadelphia: Westminster, 1960), 1.16.2-3; Heppe, *Reformed Dogmatics*, XII, 19: "Divine providence thus covers the whole of the world and every living element and member of the world, appear it to be ever so small, consciousless and intelligent creatures, angels and men, elect and non-elect, their good and bad actions, the contingent and the necessary."

50. Keith Ward, *The Big Questions in Science and Religion* (West Conshohocken, PA: Templeton, 2008), 68.

human life.⁵¹ Contemporary philosophers indeed often acknowledge the logical possibility that processes of chance of a statistical nature may be part of a complex whole that as such (and in some of its outcomes) is intended, and that in this way the evolutionary process is compatible with divine providence.⁵²

Interestingly, at some point in his argument Herman Philipse concedes the validity of this consideration:

> Because our evidence is limited in principle, it is always logically possible that there are hidden variables, which are still undetected or even undetectable. For example, our available evidence concerning the Cretaceous-Paleogene extinction event does not contradict the hypothesis that God caused it by steering an asteroid towards Earth. . . . Finally, . . . in a sense there is no contradiction between neo-Darwinism and the theistic doctrine of guided evolution.⁵³

Thus, despite the robust language with which he begins his argument, Philipse appears in the end to disagree with Plantinga only about the *relevance* of the compatibility of the neo-Darwinian and theistic explanations, not about this

---

51. Cf. Alexander Pruss, "A New Way to Reconcile Creation with Current Evolutionary Biology," *Proceedings of the American Catholic Philosophical Association* 85 (2011): 213-22. Pruss argues that if "random" in "random variation" just means that there is no correlation between fitness and variation, this is compatible with the claim that a number of variations were produced by God. Pruss rightly distinguishes this view from intelligent design theories in that he denies that there is empirical evidence for such divine intervention.

52. See (apart from the paper mentioned in the previous footnote), e.g., Del Ratzsch, "Design, Chance, and Theistic Evolution," in *Mere Creation: Science, Faith, and Intelligent Design*, ed. William A. Dembski (Downers Grove, IL: InterVarsity Press, 1998), 289-312; Michael Ruse, *Can a Darwinian Be a Christian? The Relationship between Science and Religion* (Cambridge: Cambridge University Press, 2000), 114-15; Peter van Inwagen, "The Compatibility of Darwinism and Design," in *God and Design: The Teleological Argument and Modern Science*, ed. Neil A. Manson (London: Routledge, 2003), 347-62; René van Woudenberg, "Darwinian and Teleological Explanations," in *Evolution and Ethics: Human Morality in Biological and Religious Perspective*, ed. Philip Clayton and Jeffrey Schloss (Grand Rapids: Eerdmans, 2004), 182-83.

53. Philipse, "Real Conflict," 93. This recognition is in tension with Philipse's earlier statement that "clearly, then, the thesis that evolution is unguided is an integral part of the Modern Evolutionary Synthesis" (92).

compatibility itself. According to Philipse, it is only "trivially true" that there is no contradiction between neo-Darwinism and the theistic doctrine of guided evolution, since the hypothesis that God causes certain things is just "gratuitous speculation." But this, of course, is changing the playing field, since Plantinga was not at all taking stock of the evidence for theistic evolution or theism in general (he insistently argues in other places that the theist need not do so), but only argued for its compatibility with neo-Darwinism.[54] Contrary to what Philipse suggests, if there is such compatibility, this is not at all trivial, for it refutes the standard rhetoric of those who argue that Darwinism implies atheism. Presumably being part of the "culture wars" around science and religion, such authors jump to conclusions by smuggling atheism into their scientific account of neo-Darwinism.[55]

In any case, Plantinga's main point that natural selection and divine providence are compatible is accepted by Philipse. Indeed, it does not follow from the sober scientific theory of natural selection operating on random mutations that any form of divine providence is ruled out.[56] That theory simply does not allow for any conclusions about what is beyond the reach of empirical science, since that is of another, metaphysical order. It is quite understandable that we are deeply impressed—and at times even overwhelmed—by the whimsical and unpredictable nature of the ongoing evolutionary process. The famous line of Shakespeare's Macbeth about life being "a poor player that struts and frets his hour upon the stage and then is heard no more"[57] seems to apply to entire species. We may especially stagger at the enormous number of natural selection's casualties—all those living organisms that fell prey to suffering and

---

54. Strangely enough, Philipse makes this move in a section called "Logical Conflicts" (89). In fact, he concedes that there is no logical conflict, in order to then subtly shift the debate to the evidential problem.

55. Cf., e.g., Mark Ridley, *Evolution*, 3rd ed. (Oxford: Blackwell, 2004), 256: "It is one of the most fundamental claims in the Darwinian theory of evolution that natural selection is the only explanation for adaptation." The word "only" seems over the top here, since it attempts to decide the issue whether or not natural selection is compatible with theistic explanations of biodiversity with an authoritative statement.

56. René van Woudenberg and Joëlle Rothuizen-van der Steen, "Both Random and Guided," *Ratio* 28 (2015): 332-48, rightly argue that the claim that "the evolutionary process is unguided by God" is *not* part of evolutionary theory, since it does not explain any phenomenon that is not yet explained by other parts of evolutionary theory. So it "is inessential to E[volutionary] T[heory]. It does no work. It is an add-on, a philosophical gloss on ET" (344).

57. William Shakespeare, *Macbeth*, act 5, scene 5.

death long before coming to their full potential.[58] But the evolutionary process as a whole need not be as chaotic as it seems, and it does not follow from the facts that it is meaningless and without any goal or direction. It seems, then, that Kuyper and Hodge were wrong when they insisted on the incompatibility of Darwinism and divine providence—although they were right in pointing out that many contemporary Darwinians interpreted the evolutionary process along materialistic lines that did not allow for teleology. Today, we should not let ourselves be misled by contemporary Darwinists who interpret the evolutionary process along materialist lines and therefore leave no room for any form of directionality, goal, or guidance. For the conclusion that must be drawn from our analysis is clear: Darwinian evolution and divine providence are not mutually exclusive.

## 7.5 Beyond Compatibility? Convergence and Consonance

We saw how, ever since the publication of Darwin's *Origin of Species*, there has been a sharp divide among scholars about whether or not natural selection is compatible with some form of guidance or purpose in the biosphere. Many years after 1859 the debate still continues unabated. We have tried to dismantle the ideological and metaphysical assumptions that often play a role in these discussions by concentrating on natural selection as a sober scientific theory. A brief survey of the arguments led us to conclude that the "randomness" of the genetic mutations sifted through by natural selection does not rule out the possibility that the process as a whole is guided by God. In other words, natural selection and divine providence are logically compatible.

Can we go one step further? Could it, in particular, be argued that there is not just compatibility but also *consonance* between the two, that is, that the two make a good fit, so that it is far from strange to believe in divine guidance when pondering the nature of the evolutionary process? Once again, opinions differ. On the one hand, there are theists who think there is no reason to assume that a divine purpose steers the evolutionary process at all. God is just the creator of the natural laws out of which the universe with all that is in it emerged. Or, alternatively, rather than causing or steering it, God is

---

58. We will not pick up the problem of evolutionary evil in this connection, since we have discussed this in chapter 4. Whereas the two issues are often conflated, in the current chapter I exclusively focus on the problem that was still unaddressed, viz., the relationship between evolutionary randomness and divine providential guidance.

pulling and "luring" the world toward its ultimate future by arousing it toward self-creativity in an ongoing process of creation.[59] On the other hand, there are theists who perceive phenomena that in their view unambiguously point to a divine purpose or design. The first group clearly falls short of the strong notions of divine providence that we have encountered in Reformed theology.[60] The second group, as we saw in chapter 2, runs the risk of situating God's purpose-giving activity in phenomena that we do not yet fully understand, thus invoking a "God of the gaps."

There are, however, some interesting scientific developments that suggest directionality as seen *from within* the evolutionary process. Jeffrey Schloss has explored a couple of "credible, and testable, theoretical options that make directionality plausible, even likely," focusing on directional evolutionary "trends."[61] One such development (only mentioned in passing by Schloss here) is the discovery of evolutionary convergence. This development is indissolubly linked with the name of one paleobiologist (who happens to be a Christian): Simon Conway Morris. Like Stephen Gould, Conway Morris studied the famous fossils of the Burgess Shale fauna in Canada, which display all kinds of odd forms of life—dating from the so-called Cambrian explosion—that have disappeared long ago. The two of them came to radically different conclusions, however.

Gould had become convinced of the radical contingency of the evolutionary process. In his view, sudden and unpredictable events such as tectonic "accidents" (or the meteorite impact that presumably eradicated the last of the dinosaurs) decisively influenced the course of evolutionary history, which otherwise most probably would have developed in totally different directions—

---

59. Cf. John F. Haught, *Deeper Than Darwin: The Prospect of Religion in the Age of Evolution* (Boulder, CO: Westview, 2003), 130. For similar views on evolution inspired by process theism (as well as some countervoices), see John B. Cobb, ed., *Back to Darwin: A Richer Account of Evolution* (Grand Rapids: Eerdmans, 2008). In this volume, Haught, "Darwinism, Design, and Cosmic Purpose," takes purpose "to mean not design but, following [process philosopher] Alfred North Whitehead, an overall aim toward the actualizing of value" (325).

60. Harris, *The Nature of Creation*, 33, warns that those who draw theologically on the scientific concept of chance (as process theists do) rather than counterbalancing it by a notion of preestablished purpose should not be seen as "anti-Christian." True as that may be, for our purposes this option will not do, since it is at odds with the notions of divine guidance and sovereignty that are so close to the heart of Reformed theology.

61. Jeffrey P. Schloss, "Divine Providence and the Question of Evolutionary Directionality," in Cobb, *Back to Darwin*, 330-50 (331-32).

for example, not at all leading to the emergence of human beings or beings with a similar level of intelligence. This inspired Gould to use the vivid metaphor of rewinding the tape of life in what came to be a famous and oft-quoted statement: "Wind back the tape of life to the early days of the Burgess Shale; let it play again from an identical starting point, and the chance becomes vanishingly small that anything like human intelligence would grace the replay."[62] Some years later, Gould intimated that "we are, whatever our glories and accomplishments, a momentary cosmic accident that would never rise again" under similar conditions.[63]

In stark opposition to such interpretations, Conway Morris pointed to the phenomenon of evolutionary *convergence*, which he defines as "the recurrent tendency of biological organization to arrive at the same solution to a particular need."[64] Conway Morris does not at all deny the role of chance events such as random genetic mutations, but in a fascinating survey he assembled hundreds of examples from the literature of how natural selection has led to the same sort of "solutions" to evolutionary challenges that various species and groups had to face at different times and places. One of the most well-known examples is the highly sophisticated so-called camera eye, which evolved quite a number of times independently of each other, especially in cases of mobile and predatory animals who "needed" sharp eyes (ranging from types of squid and jellyfish to humans and other vertebrates).[65] Other examples include the habit of many lizards and snakes to keep the egg inside the mother until their fully developed young is born—a technique that seems to have evolved separately about one hundred times—and the many similarities between marsupials that have evolved independently from each other in Australia and South America, filling similar ecological niches. Apparently, life has only a limited number of "building blocks," and through the intricate interplay of chance

---

62. Stephen Jay Gould, *Wonderful Life: The Burgess Shale and Natural History* (New York: Norton, 1989), 14; cf. 233: "we [humans] would probably never arise again even if life's tape could be replayed a thousand times."

63. Stephen Jay Gould, *Full House: The Spread of Excellence from Plato to Darwin* (New York: Harmony Books, 1996), 18.

64. Simon Conway Morris, *Life's Solution: Inevitable Humans in a Lonely Universe* (Cambridge: Cambridge University Press, 2003), xii. Interestingly, atheist philosopher Daniel Dennett had already recognized the occurrence of convergence at a relatively early stage; see his *Darwin's Dangerous Idea: Evolution and the Meanings of Life* (New York: Simon & Schuster, 1995), 308: "Replay the tape [of life] a thousand times, and the Good Tricks will be found again and again."

65. Conway Morris, *Life's Solution*, 147-73.

and necessity evolutionary history develops increased complexity along more or less the same lines.

In light of the many constraints that nature imposes on the number of solutions (in terms of bodily organs, structures, and other properties) to evolutionary challenges and on the pathways that in theory could be followed to reach them, it becomes much more plausible that somehow intelligent beings such as humans *should* emerge at some point in time. Indeed, this is exactly what Conway Morris claims. Referring to Gould's tape metaphor, he contends that it is "widely thought that the history of life is little more than a contingent muddle punctuated by disastrous mass extinctions. . . . Rerun the tape of the history of life, as S. J. Gould would have us believe, and the end result will be an utterly different biosphere. Most notably there will be nothing remotely like a human."[66] Countering this view, Conway Morris then goes on: "Yet, what we know of evolution suggests the exact reverse: convergence is ubiquitous and the constraints of life make the emergence of the various biological properties very probable, if not inevitable."[67] It is this process that led to the phenomenon of self-conscious intelligence as it finally emerged in human beings. Given the law-like patterns that underlie the evolutionary process, we humans might definitely have looked different (e.g., having six instead of five digits on each hand and foot), but "something like ourselves is an evolutionary inevitability."[68] Due to observations such as those sampled by Conway Morris, Gould's earlier interpretations of the Burgess Shale fossils are now often seen as superseded.[69]

Of course, the empirical phenomenon of evolutionary convergence (or convergent evolution) does not demonstrate that life on earth falls under the providential control of God or is guided by God toward a future fulfillment. Still, it is very much open to such an interpretation. It is, at the very least, not strange to interpret this empirical state of affairs in terms of direction-

---

66. Conway Morris, *Life's Solution*, 283.

67. Conway Morris, *Life's Solution*, 283-84; cf. Simon Conway Morris, "Evolution and the Inevitability of Intelligent Life," in *The Cambridge Companion to Science and Religion*, ed. Peter Harrison (Cambridge: Cambridge University Press, 2010), 148-72.

68. Conway Morris, *Life's Solution*, xv. Conway Morris adds to this an observation that is significant in light of what we discussed in chapter 5: "Contrary to popular belief, the science of evolution does not belittle us [humans]."

69. There are still many unresolved and disputed questions in this area, though. For critical discussions of convergence, see the theme issue "Are There Limits to Evolution?," *Interface Focus* 5, no. 6 (2015), http://rsfs.royalsocietypublishing.org/content/5/6.

ality and purposefulness. Using labels from the debate on the relationship between science and religion, we might see *consonance* here, or, in older language, complementarity: there is no logical connection between convergence and providence, but the two fit in with each other very well.[70] Consonance and complementarity are stronger notions than compatibility, since clearly propositions can be compatible without fitting in with each other very well. In that sense, convergence is consonant with a plan-like theistic view of evolutionary history: it enhances the probability of design in nature. It seems that, despite all appearances to the contrary, we live in an ordered biosphere, which is by no means entirely random. Of course, the human being is still seen by many as a "momentary cosmic accident," as Gould had it. "But perhaps now it is the time to take some of the implications of evolution and the world in which we find ourselves a little more seriously."[71] Indeed, the convergences found in evolutionary history evoke a sense of marvel and make us wonder about the possibility of "a providential account for the overall meaning of biological diversity, including ourselves, in which God has intentions and purposes for the created order, and render less plausible claims . . . that evolutionary history is a totally random walk."[72]

This type of argument is different from the common arguments for intelligent design in that it does not start from what we do not know but from what we (seem to) know. Thus, it is not a "God of the gaps" argument to which I am appealing here. Even if we came to know the underlying law-like patterns (which might enable us to predict future evolutionary paths), the ubiquity of evolutionary convergence will still challenge the view of evolution as an

---

70. The term "consonance" was introduced in the science-and-religion debate by Ernan McMullin, "How Should Cosmology Relate to Theology?," in *The Sciences and Theology in the Twentieth Century*, ed. Arthur Peacocke (Notre Dame: University of Notre Dame Press, 1981), 17-57. It was elaborated in, for example, Ted Peters, ed., *Science and Theology: The New Consonance* (Boulder, CO: Westview, 1998). The concept of complementarity dates back to Donald M. MacKay, "'Complementarity' in Scientific and Theological Thinking," *Zygon* 9 (1974): 225-44. From the perspective of Reformed theology, the metaphor of the two books we discussed in chapter 1 comes to mind in this connection.

71. Conway Morris, *Life's Solution*, xvi. Conway Morris reflects more explicitly on the significance of his work for the relationship between science and the Christian faith in "Creation and Evolutionary Convergence," in *The Blackwell Companion to Science and Christianity*, ed. J. B. Stump and Alan G. Padgett (Oxford: Blackwell, 2012), 258-69.

72. Alexander, *Creation or Evolution*, 432.

entirely open-ended process.⁷³ And even if convergence turns out to be less ubiquitous than it now seems, the compatibility between evolution and providence that we found in the previous section is not at stake. Thus, we need not build our faith on the phenomenon of convergence. It is enough to know that the neo-Darwinian synthesis at the very least does not discredit faith in a God who leads the world through all its vicissitudes to its final destination.

Having discussed the objection of some Christians that evolution is a chance process that is incompatible with the providential God of the Bible, Denis Alexander rightly concludes as follows:

> So as Christians we can perceive the evolutionary process as the way that God has chosen to bring biological diversity into being, including us. That is the way things are and our task . . . is to describe the way things are—what God has done in bringing this vast array of biological diversity into being. Of one thing we can be sure: the evolutionary process provides no grounds for thinking that the universe is a chance process in any ultimate metaphysical sense. In fact, quite the reverse.⁷⁴

Similarly, after having reviewed Conway Morris's research, Jacob Klapwijk concludes that "there appear to be interesting reasons to mark the development of the cosmos and evolution on earth as events that are not entirely devoid of purpose and meaning."⁷⁵

On the other hand, Christians should be prepared to acknowledge that, rather than controlling every single detail of natural history, God may make use of probabilistic (or stochastic or statistical) processes, such as random mutations, to achieve his ends. Though God has counted all the hairs of our

---

73. Cf. McGrath, *Darwinism and the Divine*, 193.

74. Alexander, *Creation or Evolution*, 159; Alexander distinguishes three kinds of chance in this connection: chance that is the result of our ignorance (since we don't know all causal relations and conditions), quantum chance (which is due to ontic rather than epistemic conditions), and what he calls "metaphysical chance," that is, the idea that the universe came into being at random. Alexander rightly points out that the first two kinds of chance are compatible with an overall purposefulness.

75. Jacob Klapwijk, *Purpose in the Living World? Purpose and Emergent Evolution* (Cambridge: Cambridge University Press, 2012), 181; especially, there are phenomena that "seem to anticipate the human presence" (181). Klapwijk creatively connects his view of emergent evolution with the philosophical thought of Herman Dooyeweerd.

head (Matt. 10:30), he may not have arranged every single mutation that pops up in nature. Especially for Reformed Christians, there is perhaps some work to do here, since "there is a strong theological tradition (Calvinist in sympathies) which will not countenance chance in any shape or form."[76] If there is a price to be paid by Christians for accepting the mechanism of natural selection working on random mutations as evolution's most dominant driving force, it is this. It is not too high a price when we start to see what we have seen in this chapter, namely, that, as David Bartholomew puts it, "chance does not, in reality, exclude purpose."[77] Indeed, it might increase our awe for God's creative wisdom when we acknowledge that he might even use the subtleties of random processes to reach his goals.

## 7.6 Conclusion: Consonance instead of Conflict

We may conclude that the randomness of the genetic mutations on which natural selection operates does not exclude the possibility of divine guidance. Of course, the question may come up why God does not "do a better job" in guiding the natural world, allowing so much suffering and death, but we discussed the problem of evolutionary evil and suffering in chapter 4. The theory of natural selection does not offer us a reason to return to this issue. In any case, God may use the entire evolutionary process (including its many chance events) for his plans. He may also, as Asa Gray argued, direct certain specific mutations toward his goal.[78] We need not even shy away from the idea that God may incorporate processes of "deep chance" in his plans.

Could it be that the process of natural selection operating on random mutations is not only compatible but even *consonant* with belief in God's providential role? Simon Conway Morris's work on evolutionary convergence gave us reason not to dismiss this suggestion out of hand. In any case, the evolutionary process seems to have an *internal* purposefulness in the marvelous ways

---

76. David J. Bartholomew, "God and Probability," in *Creation: Law and Probability*, ed. Fraser Watts (Minneapolis: Fortress, 2008), 139-53 (151); in this connection Bartholomew (who has written extensively on God and chance over the years) critically points to a book of Presbyterian theologian R. C. Sproul: *Not a Chance: The Myth of Chance in Modern Science and Cosmology* (Grand Rapids: Baker Books, 1994).

77. Bartholomew, "God and Probability," 151.

78. For a defense of this latter possibility, see John S. Wilkins, "Could God Create Darwinian Accidents?" *Zygon* 47 (2012): 30-42.

in which it strives toward ever greater complexity. Although this is different from the external purpose theists ascribe to the universe, it is not strange or inappropriate to link the two to each other, especially in light of what we now know about the amazing cosmological constraints that enabled the emergence of life on earth (the so-called anthropic principle). If this is correct, then, in spite of so many voices to the contrary, we may posit consonance instead of conflict between natural selection and divine providence—or, if you wish, between Darwin and God. For in that case both science and religion suggest that we and our world are not the products of mere chance.

CHAPTER 8

# Morality, the Cognitive Science of Religion, and Revelation

> In the cognitive science and religion discussion, any claim to objectivity is illusory. Everyone who discusses this topic has an acknowledged or unacknowledged interest in it, whether one is using cognitive science to undermine religion, defend religion, or simply understand it better. No one can legitimately wrap themselves in the mantle of science and feign objectivity.
> —James W. Jones[1]

## 8.1 Biological, Social, and Cultural Evolution

In addition to the reasons discussed in the previous chapters, there is at least one other reason why many Christians—and others—are wary of evolutionary theory: they think that accepting Darwinian evolution would radically change their view of social life, including politics, economics, morality, and religion. In this chapter, we will examine to what extent this concern is valid. Indeed, from its very beginning, Darwinian evolution has been associated with theories about the ideal arrangement of society. Contemporaries of Darwin like the philosophers Herbert Spencer (1820-1903) and Ernst Haeckel (1834-1919) started to extrapolate the principles of Darwinian evolution to societal processes. Here as well, they argued, we find natural selection in place, and Spencer coined the expression "survival of the fittest" to indicate what natural selection amounted to: "the preservation of favoured races in the struggle for life."[2] In society as well as in nature, progress was achieved

---

1. James W. Jones, *Can Science Explain Religion? The Cognitive Science Debate* (Oxford: Oxford University Press, 2016), 4.
2. Herbert Spencer, *Principles of Biology* (London: Williams & Norgate, 1864),

when stronger individuals—and in Spencer's economic theories, these were the *richer* ones in particular—succeeded in securing resources for survival, thus enhancing their chances to pass on their traits to later generations.[3] Therefore, strictly speaking, charity was counterproductive, and governments should not take measures to protect the poor but should stimulate free-enterprise capitalism.

In this way the biological mechanism of natural selection was applied to society, economics, and politics in what came to be called (usually in a pejorative sense) "social Darwinism."[4] The most notorious part of this development was the rise of *eugenics*, that is, the belief—eventually coupled with a set of laws and practices—that the genetic quality of the human race should gradually be improved by intentional action. The term was coined by a cousin of Darwin, Francis Galton (1822–1911), who indeed tried to stimulate "eugenic marriages" between people with extraordinary capacities and to discourage "dysgenic marriages." In the first decades of the twentieth century, the eugenics movement became very popular in the United States, where it received the support of many biologists and led to compulsory sterilization programs and legal restrictions on marriages of mentally disabled people.[5] These developments culminated in Nazi Germany, where sterilization laws were first widely applied and then replaced by a euthanasia program for the mentally disabled, with an appeal to the scientific law of natural selection that secured

---

445 (quoting Darwin). In the fifth edition of the *Origin of Species* (1869), Darwin credited Spencer for proposing this "more accurate" term as a substitute for the notion of natural selection that, Darwin conceded, unduly personified nature (as if nature consciously "selects" certain phenomena). Darwin was unhappy with the ideological overtones in Spencer's discourse, however.

3. Both Darwin and Spencer read "fittest" as "most well adapted," but of course, the term was easily understood in the more popular sense of "strongest."

4. See for a short sketch, Naomi Beck, "Social Darwinism," in *The Cambridge Encyclopedia to Darwin and Evolutionary Thought*, ed. Michael Ruse (Cambridge: Cambridge University Press, 2013), 195–201. In fact, Spencer's evolutionary philosophy was dependent on Lamarck's version of evolutionary theory (allowing for the heredity of acquired character traits) much more than on Darwin's theory of natural selection; still, since many came to confuse Darwinism with evolutionism *tout court*, "social Darwinism" became the catchword. Cf. Peter J. Bowler, *The Eclipse of Darwinism: Anti-Darwinian Evolution Theories in the Decades around 1900* (Baltimore: Johns Hopkins University Press, 1983), 69–71.

5. Cf. Daniel J. Kevles, *In the Name of Eugenics: Genetics and the Use of Human Heredity* (Cambridge, MA: Harvard University Press, 1985).

the survival of the fittest. This program was then extended to other groups, in particular the Jews, thus ending in the Holocaust.[6]

In addition to the reasons discussed in the previous chapters, it is because of this fateful trajectory from Darwin through Spencer, Galton, and others to Hitler that many Christians—including Reformed ones—have become very suspicious of the biological theory of evolution. In response to this, however, we must first of all realize that the reasons for the Holocaust are complex and diverse and cannot simply be reduced to the rise of social Darwinism. Second, logically speaking, all these social extensions do not follow from the sober Darwinian theory of biological evolution. Any attempt to deduce some ideal arrangement of society from evolutionary theory suffers from what has been called, with a term introduced by British philosopher G. E. Moore (1873–1958), the naturalistic fallacy: one concludes from how things (apparently) *are* to how they *should* be. This is plainly wrong, however, since we cannot (as Hume already argued) derive an "ought" from an "is."[7] For example, we cannot conclude from the fact that cannibalism, rape, and sexual promiscuity are common among various animal species that such behaviors are morally acceptable to us humans as well. And if selfishness is the key characteristic promoted by the evolutionary process (a view that is, as we will see, highly debatable), it does not follow that we humans should act in selfish ways as well. What we need here is an independent normative perspective from which we can infer which types of behavior are morally acceptable and which are not; nature does not in and of itself teach us what is morally good.

Denis Alexander rightly points out that social Darwinism and eugenics belong to the "ideological transformations of biology": after the theory of natural selection had become successful as a scientific theory, specific interest groups started to hijack it in order to use it for the advancement of their specific causes. In this way, evolutionary theory has been used to lend support to various isms, such as capitalism, Marxism, racism, colonialism,

---

6. For short summaries of these events, see the essays of Mark A. Largent, "Darwinism in the United States, 1859–1930," and Richard J. Roberts, "The German Reception of Darwin's Theory, 1860–1945," in Ruse, *The Cambridge Encyclopedia of Darwin and Evolutionary Thought*, 226–34 and 235–42 (esp. 241–42). See also question 34 ("How Is Darwinism an Ideology?") in Kenneth D. Keathley and Mark F. Rooker, *40 Questions about Creation and Evolution* (Grand Rapids: Kregel, 2014), 335–38.

7. Some philosophers have argued that there are exceptions to this rule; thus (most famously perhaps) John Searle, "How to Derive 'Ought' from 'Is,'" *Philosophical Review* 73 (1964): 43–58; this debate does not affect the point made here, however.

militarism, and (as we saw in the previous chapter) atheism.[8] As many observers—including most biologists—today recognize, however, "none of these ideologies can be rationally derived from the biological theory of evolution; all are parasitic upon the theory, attempting to extract plausibility by repeated association of each ideology with the theory itself."[9] Therefore, Christians and others who reject evolutionary theory because of social Darwinism commit the logical fallacy of "guilt by association": although there is no logical connection between the two, the first is rejected because it is associated with the disreputable second.

In fact, already in the aftermath of the Second World War social Darwinism and the concomitant eugenics fell into disrepute among the general public. What remained, however, was the intuition that the shaping of social behavior is to a large extent constrained by the same mechanism that had been detected as the main engine behind biological evolution: natural selection leading to the survival of evolutionarily advantageous traits and dispositions. This intuition was forcefully elaborated in what came to be called "sociobiology." Humans were emphatically included in this study of the evolutionary backgrounds of social behavior: characteristics of human social behavior such as aggression, communication, sexual urges, and parental care were considered to have evolved from the animal world in similar ways as our biological traits, all of them being selected because of their contribution to the survival of our genes (or group, or species).[10] Since sociobiology continued to be associated with social Darwinism, from the mid-1980s it became relabeled as "evolutionary psychology" (the subtle differences between the two need not bother us here). It is important to see how evolutionary psychology and sociobiology on the one hand differ from social Darwinism on the other: whereas social Darwinism presents the natural selection of the fittest as something to support and emulate, evolutionary psychology restricts itself to the claim that

8. Denis Alexander, *Creation or Evolution: Do We Have to Choose?* 2nd rev. ed. (Oxford: Monarch Books, 2014), 205; cf. 206-11 for brief descriptions of each of these movements.

9. Alexander, *Creation or Evolution*, 211. Often, of course, these associations are not entirely arbitrary but based on *extrapolations* from Darwinian evolution—extrapolations that are unwarranted since they cross the border between is and ought, science and ideology.

10. See esp. E. O. Wilson, *Sociobiology: The New Synthesis* (Cambridge, MA: Belknap Press of Harvard University Press, 1975). Wilson's book provoked a wide-ranging debate on "biological determinism": the view that nature rather than nurture determines most if not all of our doings.

our human behaviors and social life can to a large extent be understood along the lines of Darwinian evolution. In particular, it attempts to explain human psychological and behavioral traits as adaptations evolved in the struggle for survival and reproductive success. In other words, whereas social Darwinism was a normative view, even an ideology, evolutionary psychology is a descriptive and explanatory field of study.

Yet, this is not the whole story. For the claim of many evolutionary psychologists that our psychological and social behaviors can largely be explained along lines of Darwinian evolution has far-reaching consequences and is not at all uncontested. As evolutionary psychologist Robert Wright has stated, evolutionary psychology easily transcends the level of empirical science since it becomes a new way of seeing everyday life. Unlike quantum mechanics, for example, "it can entirely alter one's perception of social reality."[11] Social processes are different from what they appear to be if in the end they are driven by the same "selfish" mechanisms as biological processes. But are they? Many observers argue—rightly in my view—that evolutionary psychology is misguided in that it does insufficient justice to *cultural* factors that explain our behavior next to biological ones (let alone that it cannot account for phenomena such as free will, responsibility, etc.). They point to what is called "cultural evolution" as deeply affected by, but at the same time in crucial respects different from, biological evolution.[12]

Indeed, the sheer variety of human forms of behavior cannot be fully explained with an appeal to Darwinian processes of variation, competition, and heredity, since humans have the capacity to set their own courses in often unpredictable ways. A well-known example is the decline in birthrate from five to two children on average among Italian women in the course of the nineteenth century. Scholars agree that it is implausible to see this development as a result of natural selection, since at the time mothers having more than two children offered greater "fitness benefit" than those giving birth to only two babies. It seems that forms of cultural transmission have been pertinent here that cannot simply be reduced to biological or genetic transmission. For example, it may be that women who for some reason had more social prestige started to give birth to fewer children, a habit that gradually came to be imi-

---

11. Robert Wright, *The Moral Animal: Why We Are the Way We Are—the New Science of Evolutionary Psychology* (New York: Vintage Books, 1994), 5; Wright speaks of "a new worldview" in this connection (4).

12. Cf., e.g., Peter J. Richerson and Robert Boyd, *Not by Genes Alone: How Culture Transformed Human Evolution* (Chicago: University of Chicago Press, 2005).

tated by others.[13] This is not to deny that many of our psychological and social traits can at least partially be explained along the lines of natural selection (as evolutionary psychologists contend), but it is to point out that human behavior is far more complex than to be reducible to biological factors.

In this connection, Richard Dawkins's popular claim that culturally transmitted ideas and techniques ("memes") replicate themselves, jumping as it were from one mind and generation to another and trying to perpetuate themselves just as genes do in biological transmission, however popular it may be, "has not become well established in scientific circles."[14] It appears that cultural forms of evolution cannot be linked so strongly to biological ones. Moreover, one of the problems of evolutionary psychology is the difficulty it has in formulating testable hypotheses. It is quite easy to propose Darwinian explanations for virtually any human behavioral inclination—from our fondness of sweets through our involvement in gossip to the existence of morality and religion—but it is much more difficult to specify how such explanations might be tested. As a result, many of them have a high "just so story" level: they may ring true, but as long as we have no way of sorting out whether they are indeed true, we just don't know.[15] There is a relevant difference here with biological evolution, where we can actually see the working of natural selection, for example, in groups of bacteria that gradually develop resistance to antibiotics.

Even if we believe that cultural factors are much more relevant than biological ones in explaining our social behaviors, it should not be denied that, deep down, biological influences are operative as well. Indeed, it seems that at least to some extent our behavioral dispositions (such as our sexual desires) are determined by evolutionary mechanisms related to natural selection. There-

---

13. The classic study here is L. L. Cavalli-Sforza and M. W. Feldman, *Cultural Transmission and Evolution: A Quantitative Approach* (Princeton: Princeton University Press, 1981). See Tim Lewens, "Cultural Evolution," in *Stanford Encyclopedia of Philosophy*, last revised May 1, 2018, http://plato.stanford.edu/entries/evolution-cultural/. Of course, the tendency to imitate group members with a high social status is usually seen as having evolutionary roots; but this is not decisive here, since it leaves unresolved why the outcome (collectively giving birth to fewer children) goes against the grain of Darwinian evolution.

14. Lewens, "Cultural Evolution," §4; cf. Richard Dawkins, *The Selfish Gene* (Oxford: Oxford University Press, 1976). In 2005, the *Journal of Memetics*, then in its ninth volume, was even discontinued.

15. See David Stove, *Darwinian Fairytales: Selfish Genes, Errors of Heredity, and Other Fables of Evolution* (New York: Encounter Books, 1995).

fore, in this chapter we will focus on the possible implications of biological Darwinian evolution for—or extensions of Darwinian evolution to—aspects of human sociality. How do Darwinian explanations of human social behavior affect our theological outlook, and how could they be addressed from a (Reformed) Christian perspective? It may seem that, even if the arguments put forward in the previous chapters have allayed other concerns about evolutionary theory, here finally a concern emerges that cannot be so easily removed. For here we may suspect that, as Daniel Dennett has so eloquently put it, natural selection turns out to be "a universal acid" that "eats through just about every traditional concept, and leaves in its wake a revolutionized worldview, with most of the old landmarks still recognizable, but transformed in fundamental ways."[16]

Presumably, two of the most venerable of these traditional phenomena that may come to look entirely different when approached from the perspective of Darwinian evolution are *morality* and *religion*. Because of their theological sensitivity, I will concentrate in what follows on these phenomena, surveying various evolutionary explanations that have recently been proposed for each of them. First, I briefly examine the issue of morality (§8.2), and then I explore current theories on the emergence and persistence of religion (§§8.3 and 8.4). In doing so, however, I have to greatly restrict myself and can sketch only some lines of contemporary research, given that an enormous amount of work has been done in both fields over the past couple of years. Nevertheless, limited though they are, our findings may enable us to get a glimpse of the theological challenges that come with this type of research, and also to offer some provisional suggestions as to how these challenges might be adequately met (§8.5).

## 8.2 The Evolution of Morality and Theological Ethics

The evolution of morality has been the focus of intensive scientific research over the past decades. This research has led to some remarkable discoveries.[17] For a long time biologists had wondered about "altruistic" types of behav-

---

16. Daniel C. Dennett, *Darwin's Dangerous Idea: Evolution and the Meanings of Life* (New York: Simon & Schuster, 1995), 63.
17. For what follows I have been helped by an insightful essay of the Dutch primatologist Jan van Hooff in *Darwin en het hedendaagse mensbeeld*, ed. Luc Braeckmans et al. (Antwerp: Antwerp University Press, 2009), 17-43.

ior among animals. How can, for example, meerkats, vampire bats, ants, and honeybees with particular roles in their groups sacrifice their lives for other members of their groups? Clearly, organisms with an inclination to self-sacrificing behavior reduce their reproductive success, so that one would expect natural selection to eradicate altruistic behavior in the long (or more probably not-so-long) run.

Drawing on earlier work of, among others, J. B. S. Haldane,[18] W. D. Hamilton in 1964 proposed a strategy that might be operative here and even formalized it in what came to be known as "Hamilton's rule." Hamilton assumed that genes are the units of selection. Thus the strategy behind "kin selection" or "kin altruism" is this: since one's genes are to some extent also represented in one's family members, the closer individuals are related to each other, the more rewarding (in the evolutionary sense) it will become for them to invest in each other's survival chances. Thus, under specific conditions, risking one's own life for the benefit of other members of one's kin group (as, for example, meerkats do) may be an effective strategy for passing on considerable parts of one's genetic material to the next generation. Hamilton's rule captures this strategy in a simple mathematical formula: $rB > C$, where $r$ is the degree of an altruistic organism's relatedness to the beneficiary (e.g., statistically 50 percent for siblings, 25 percent for grandchildren, etc.), B stands for the benefit the recipient gains by the altruistic act, and C for the cost paid by the altruistic actor (in terms of reproductive success). The formula states that an altruistic act is effective—and that therefore the underlying "altruistic gene" can survive—if the product of the degree of relatedness and its benefit for the recipient is greater than the costs for the actor.[19] As Haldane famously summarized its pivotal idea: "I would gladly lay down my life for two brothers or eight cousins."[20]

---

18. Cf. J. B. S. Haldane, *The Causes of Evolution* (London: Longmans, Green, 1932); Haldane, a scientific materialist, reportedly served as the model for the villainous Professor Weston in C. S. Lewis's space trilogy, *Out of the Silent Planet*, *Perelandra*, and *That Hideous Strength* (1938–1945).

19. W. D. Hamilton, "The Genetical Evolution of Social Behaviour," *Journal of Theoretical Biology* 7 (1964): 1–52. The problem that it is hard to see how, for example, social insects can recognize their close relatives and distinguish a brother from a cousin was solved by Hamilton (1) by pointing to the evolutionary importance of inbreeding avoidance, which requires some sort of kin recognition, and (2) by assuming that the relevant groups usually display low rates of dispersal, so that relatives stay closely together over time within the group. In such a situation, there is no need for more precise forms of kin recognition.

20. As passed on by his student John Maynard Smith, Haldane came to this in-

In this way, Hamilton introduced what he called the theory of "inclusive fitness": our drive to pass on our genes may not only be expressed in our production of offspring (thus enhancing our "direct fitness") but it may also extend to our support of relatives who are thus enabled to replicate, among other ones, the genes we share with them (thus enhancing our "indirect fitness").[21] Although some empirical studies have provided support for Hamilton's rule, it is still contested because it is difficult to properly calculate the costs and benefits of specific altruistic behaviors, given the numerous nonquantifiable factors that are involved.[22] Still, it continues to be one of the two main mechanisms appealed to by evolutionary biologists to explain the rise of altruistic behavior.[23] The other mechanism, called "reciprocal altruism," accounts for forms of altruistic behavior that extend beyond one's kin group. It points to the win-win situations that emerge when recipients do not just profit from their donors (as so-called freeriders do) but reciprocate the favor extended to them at a later stage.[24] Since in this way the donors advance their own chances of reproductive success (next to those of their recipients), what we have here is another instance of inclusive fitness—this time unrelated to kin selection.

As far as concern for others is at the heart of morality, altruism among humans—loosely defined as showing concern for other individuals (*alteri*)—arguably signifies a very clear case of moral behavior. Yet, altruism does not by definition express or presuppose morality. Altruistic behavior as observed in ants and bees, for example, is best explained without attributing concern

---

sight already in the fifties, while talking to some of his graduate students in a London pub, but it was Hamilton who would later follow it up. Cf. Georg Breuer, *Sociobiology and the Human Dimension* (Cambridge: Cambridge University Press, 1982), 33-34.

21. Thus, an organism's "inclusive fitness" is the sum of its direct and indirect fitness (= chance of reproductive success).

22. For criticism along such lines, see Martin A. Nowak and Roger Highfield, *Super Cooperators: Beyond the Survival of the Fittest; Why Cooperation, Not Competition, Is the Key to Life* (New York: Free Press, 2011), 95-112.

23. A third proposed mechanism is group selection—nature's favoring of groups characterized by relatively high degrees of cooperation and altruism—but since this requires more stable groups than we observe among most primates, its working is often considered to be limited at best. For a recent defense, though, see Nowak and Highfield, *Super Cooperators*, 81-94.

24. The classic paper here is Robert Trivers, "The Evolution of Reciprocal Altruism," *Quarterly Review of Biology* 46 (1971): 35-57. Elaborations of this theory are to be found in Robert Axelrod, *The Evolution of Cooperation* (London: Penguin Books, 1984), and more recently in Nowak and Highfield, *Super Cooperators*, 21-69.

for others in the sense of feelings of empathy. For this reason scholars usually call types of animal behavior that ants and bees display "pro-social" rather than "altruistic." Still, it is a hotly debated question to what extent especially higher primates exhibit moral or protomoral behavior. It seems reasonable to assume that empathy and similar feelings (concern for another's well-being, a sense of fairness, etc.) belong to the constitutive "building blocks" of morality.[25] In any case, it is not just actual behavior that constitutes morality but also—and perhaps even more so—certain underlying capacities, such as the capacity to show concern, to have empathy, and to sympathize with others. From a series of fascinating experiments, the Dutch American primatologist Frans de Waal has concluded that these capacities and feelings are to be found extensively among nonhuman advanced primates.[26] Especially chimps and bonobos, but also other apes and monkeys, are able to show empathy, to spontaneously help each other, to console mourning conspecifics, to strive for reconciliation after a conflict, and even to display "moral" indignation.[27] If we combine these findings with the pro-social types of behavior among "lower" animals that we mentioned already, it does not seem an overstatement to conclude that animals "occupy several floors of the tower of morality."[28] Whereas nonhuman primates lack more advanced capacities required for morality such as moral reasoning and morally evaluating situations in a disinterested and impartial way, they share some of its lower constitutive compartments with us humans.[29]

From a Christian theological perspective, this conclusion does not seem to be in any way problematic. On the contrary, it helpfully qualifies the troublesome image of nature as "red in tooth and claw" (Alfred Tennyson) that

---

25. Frans de Waal, *Primates and Philosophers: How Morality Evolved*, ed. Stephen Macedo and Josiah Ober (Princeton: Princeton University Press, 2006), 20.

26. See, e.g., Frans de Waal, *Good Natured: The Origins of Right and Wrong in Humans and Other Animals* (Cambridge, MA: Harvard University Press, 1996).

27. Cf., e.g., Frans de Waal and Angeline van Roosmalen, "Reconciliation and Consolation among Chimpanzees," *Behavioral Ecology and Sociobiology* 5 (1979): 55-66; S. F. Brosnan, C. Freeman, and F. B. M. de Waal, "Equitable Behavior, Not Reward Distributions, Affect Capuchin Monkey's (*Cebus apella*) Reactions in a Cooperative Task," *American Journal of Primatology* 68 (2006): 713-24. It should be added, however, that not all of De Waal's experiments could be replicated.

28. De Waal, *Primates and Philosophers*, 181.

29. Some of De Waal's interlocutors in *Primates and Philosophers*, such as Philip Kitcher (120-39) and Peter Singer (140-58), emphasize these differences—a point De Waal concedes (175) without dropping his main thesis that there is continuity here between animals and humans.

bothered us in chapter 4. Apart from death and decay, nature displays many things that are good and even delightful, thus, as believers will be inclined to say, mirroring the mind of its Maker. What happens, however, when we add to this the theory of common descent? First, we may readily agree with De Waal that from an evolutionary perspective there is continuity between animals and humans in the emergence of morality. Indeed, we may intuitively sense such continuity. For example, we recognize the inclination, which is ours as well, to prioritize relatives and friends (one's "in-group") in everyday life. We might even refer to the Bible in this connection, where Jesus already recognizes and criticizes this tendency: "If you love those who love you, what credit is that to you? For even sinners love those who love them. If you do good to those who do good to you, what credit is that to you? For even sinners do the same."[30] Second, however, as is in fact already clear from the biblical appeal to go beyond in-group thinking, there is a clear discontinuity between humans and nonhuman primates as well when it comes to morality. If we define moral behavior as behavior that is motivated by a concern for the good (i.e., for what is perceived as objectively good), it is clear that only humans are able to display such behavior. Presumably, it is for this reason that, of all species, we hold only human beings accountable for their (intentional) behaviors.

Still, evolutionary explanations of human morality raise a couple of serious theological questions—especially pertaining to the realm of theological ethics. If at the most fundamental level we hold our moral norms and values because they somehow enhance our inclusive fitness, can we really act in altruistic ways or are we necessarily selfish all the time? Can we even be sure that our norms and values are actually correct, or should we rather conclude that, as Michael Ruse and E. O. Wilson once put it, "in an important sense, ethics as we understand it is an illusion fobbed off on us by our genes to get us cooperate"?[31] Indeed, from an evolutionary perspective, it is enough that we

---

30. Luke 6:32-33 (these are words of Jesus in the Sermon on the Mount). Arguably, these words do not deny that we have special obligations to friends and family members (which makes it normal and good that we sometimes prioritize them), but they criticize our tendency to do good only to those who do good to us.

31. Michael Ruse and Edward O. Wilson, "The Evolution of Ethics: Is Our Belief in Morality Merely an Adaptation Put in Place to Further Our Reproductive Ends?" *New Scientist* 108 (October 17, 1985): 52 (50-52); the subtitle-question is answered affirmatively in the same passage (51). The paper is also available as Ruse and Wilson, "The Evolution of Ethics," in *Religion and the Natural Sciences: The Range of Engagement*, ed. James E. Huchingson (Orlando: Harcourt Brace, 1993), 308-12.

*Morality, the Cognitive Science of Religion, and Revelation*

humans hold on to certain moral beliefs, not that these beliefs are actually true. But how about the theological conviction that the basis of morality lies in what is objectively good as expressed in God's will?[32] Does biology show us that there is no such "external grounding"?[33]

These are questions that are absolutely critical from a theological point of view. I will not engage them here, however, since their answers closely resemble the answers we supply to the even more sensitive questions raised by work on the purportedly evolutionary origins of *religion*. Recent theories in this field extend the explanatory scope of natural selection to the field of religious beliefs and practices, claiming that not only morality but also religion can be completely explained as a natural (as opposed to a supernaturally based) phenomenon. In this way, these theories seem to strongly discredit the classical Christian—and Reformed—account of divine revelation, according to which people are religious because God somehow has *revealed* Godself to humanity. It is to this issue that we shall turn now.

### 8.3 The Cognitive Science of Religion

It has been established that 85 to 90 percent of the world's current population believe in God or gods.[34] Many scholars wonder how this strong persistence of religious beliefs and practices in our science-imbued and secular age should be explained. Atheist observers in particular are puzzled by how it is possible that so many people remain religious while there is so little "proof" for their

---

32. I circumvent the so-called Euthyphro dilemma here by (in line with a long-standing Christian tradition) identifying the Good and God's will; God's moral will is usually seen as expressed both in the conscience of all or most human beings (Rom. 2:14-15) and in God's special revelation (e.g., in the Ten Commandments).

33. Ruse and Wilson, "The Evolution of Ethics," 52: "It is without external grounding. Ethics is produced by evolution but not justified by it, because . . . it serves a powerful purpose without existing in substance." See also Michael Ruse, "Evolutionary Ethics: A Phoenix Arisen," in *Issues in Evolutionary Ethics*, ed. Paul Thompson (New York: State University of New York Press, 1995), 235: "What is really important to the evolutionist's case is the claim that ethics is illusory inasmuch as it persuades us that it has an objective reference. This is the crux of the biological position."

34. P. Zuckerberg, "Atheism: Contemporary Numbers and Patterns," in *The Cambridge Companion to Atheism*, ed. Michael Martin (Cambridge: Cambridge University Press, 2007), 47-65.

religious claims.[35] The generally accepted assumption of methodological naturalism (which limits scientific research to the study of natural causes) forbids them to refer to God's existence and revelation in this connection. Therefore, they have to explain the near universality of religion by appealing to natural causes. But what kind of natural factors can be involved here? Classical theories of religion like that of Émile Durkheim considered religions as being inextricably bound up with the formation of human culture. In fact, religions were seen as the necessary cement that held cultures together. Such theories came under pressure, however, when in the slipstream of sociobiology, evolutionary accounts of cultural phenomena came to the fore. Isn't our propensity for religion embedded much more deeply than in our cultural constructs, namely, in the naturally evolved structure of our brains? Thus, theory formation on natural processes that might explain the human aptness for religion moved on to other fields, such as the neurosciences. Research in this field led to what was called the "God spot" in the human brain and, following on this, to the development of what came to be labeled "neurotheology."[36] Today, however, theorizing about the rise and persistence of religious experiences, beliefs, and practices finds much more fecund soil in the burgeoning area of the *cognitive science of religion*—or CSR, for short. Nancey Murphy may well be right when she claims that "CSR has replaced Freud, Marx and others as the current best explanation of religion."[37]

In what follows I will first sketch the contours of this new field of research, and then in the next section (§8.4) I will offer a survey of some of the most widely discussed theories that have been proposed in recent years in this field. Next, I will try to assess the status of these theories and—in particular—clarify the relationship of the discussed natural explanations of morality and religion to the theological doctrine of revelation (§8.5). Are we here dealing with theories that necessarily discredit the classical Christian—and

---

35. Thus (along with many others), e.g., Kai Nielsen, *Naturalism and Religion* (Amherst, NY: Prometheus, 2001), 35.

36. See especially the work of Andrew Newberg, in, e.g., Andrew Newberg, Eugene d'Aquili, and Vince Rause, *Why God Won't Go Away: Brain Science and the Biology of Belief* (New York: Random House, 2001); Andrew B. Newberg, *Principles of Neurotheology* (Farnham, UK: Ashgate, 2010).

37. Nancey Murphy, "Cognitive Science and the Evolution of Religion," in *The Believing Primate: Scientific, Philosophical, and Theological Reflections on the Origin of Religion*, ed. Jeffrey Schloss and Michael J. Murray (Oxford: Oxford University Press, 2009), 266.

## Morality, the Cognitive Science of Religion, and Revelation

Reformed—view that in the final instance religion finds its source in God's self-revelation?[38] Or are they, in fact, fully compatible with this view?

*Cognitive science* focuses on the study of mental phenomena and processes in humans and animals. It is not a specific discipline but rather an interdisciplinary area of research of, among others, cognitive psychologists, neuroscientists, philosophers, and linguists. With regard to humans, questions that have a central place in it include: What, in fact, is the human mind, and how does it function? How do people actually think—how do they observe, interiorize, and reason, and what mental mechanisms do they use to solve particular problems? For a long time such questions were studied (if at all) by psychologists and so-called philosophers of mind in nonempirical, purely conceptual ways. Gradually, however, a more empirical branch split off from these disciplines, first in the direction of psychology ("cognitive psychology") but later also in what is now more broadly referred to as *cognitive science*. A simple definition—but adequate for our purpose—is the one that has been proposed by CSR scholar Justin Barrett: "Cognitive science is the interdisciplinary area of scholarship that considers what the human mind is and how it functions."[39] It is called "cognitive" since its focus is on *cognitions*: mental activities—both conscious and unconscious, both rational and experiential—involved in the processing of information and the acquisition of knowledge and understanding. It is believed that humans are equipped with a mental toolkit that comprises diverse evolved tools or capacities—sometimes (but controversially) seen as isolated "modules"—which help them to successfully navigate their way through the world.

The cognitive science of *religion* studies mental processes that concern religious phenomena, such as religious experiences, beliefs, practices, and dispositions. Those who work in the field of CSR are especially interested in

---

38. Cf. John Calvin, *Institutes of the Christian Religion*, ed. John T. McNeill, trans. Ford L. Battles (Philadelphia: Westminster, 1960), 1.3.1: "There is within the human mind, and indeed by natural instinct, an awareness of divinity [*sensus divinitatis*]. This we take to be beyond controversy. To prevent anyone from taking refuge in the pretense of ignorance, God himself has implanted in all men a certain understanding of his divine majesty." The words "This we take to be beyond controversy" reflect the fact that Calvin voices the common opinion of the preceding Christian tradition here.

39. Justin L. Barrett, *Cognitive Science of Religion and Theology: From Human Minds to Divine Minds* (West Conshohocken, PA: Templeton, 2011), 5. For more on cognitive science, cf. Barbara Von Eckardt, *What Is Cognitive Science?* (Cambridge, MA: MIT Press, 1993; fifth printing 1996), and, for an introduction that is both succinct and accessible, Jones, *Can Science Explain Religion?*, 14–24.

cross-cultural constants, that is, in patterns that recur in all sorts of different cultures, because such patterns may teach us something about the human brain in general, that is, regardless of cultural influences. For the same reason, they are interested in dispositions that emerge at a very early stage in the development of children, since these as well may point to brain functions that are universally human. Universally human brain functions, in turn, are seen to be largely determined by our evolutionary past. Thus, in brief, CSR "is the scientific study of religion as a natural, evolved product of human thinking."[40] Its beginnings are usually dated in the early 1990s, when a few publications, more or less independently from each other, started to point in the same direction: if we want to understand religiosity in a natural way, we must study the cognitive processes that (according to these authors) constitute its basis.[41] From the turn of the millennium, the work being done within this new "paradigm" grew exponentially. That CSR has in the meantime developed into a flourishing discipline is clear from the academic centers that have been established for its study in a range of locations,[42] as well as from the establishment in 2006 of the International Association for the Cognitive Science of Religion (IACSR). This body organizes scientific conferences, supports other initiatives in this area of research, and biannually publishes the *Journal for the Cognitive Science of Religion* (since 2013).[43]

From a Christian point of view, one might be inclined to adopt a skeptical attitude toward CSR. Isn't it in a sense peculiar to see atheistic scholars worrying about the continuing prevalence of religion and coming up with one

---

40. Helen De Cruz and Johan De Smedt, *A Natural History of Natural Theology: The Cognitive Science of Theology and Philosophy of Religion* (Cambridge, MA: MIT Press, 2015), 13.

41. Pascal Boyer, "Explaining Religious Ideas: Outline of a Cognitive Approach," *Numen* 39 (1992): 27–57, is often seen as the publication that marks its beginning, next to Thomas E. Lawson and Robert McCauley, *Rethinking Religion: Connecting Cognition and Culture* (Cambridge: Cambridge University Press, 1990). An important precursor of CRS is a paper by the anthropologist Steven Guthrie, "A Cognitive Theory of Religion," *Current Anthropology* 21 (1980): 181–203.

42. Dimitris Xygalatas, "Cognitive Science of Religion," in *Encyclopedia of Psychology and Religion*, ed. David A. Leeming, 2nd ed. (New York: Springer, 2014), 344, enumerates academic centers in Belfast, Aarhus, Oxford, London, Paris, Brno, Vancouver, and Boston; however, some of these centers study the relationship between cognition and culture in a broader sense, rather than focusing on religion.

43. See http://www.iacsr.com/iacsr/Home.html and https://journals.equinoxpub.com/index.php/JCSR.

theory after another as to how this phenomenon might be explained? Isn't this precisely because they don't want to consider the claim that God might have revealed, and might continue to reveal, himself to humanity, as we heard Calvin claim—a revelation that is then arguably tweaked and twisted in as many different directions as there are religions? If in the end religion and religiosity could be scientifically explained on the basis of evolutionary processes, that would probably come as a great relief to atheistic scholars. At long last, it would become demonstrable that religion is based on an illusion—something that older theories such as those of Freud, Marx, and Durkheim were unable to show, since these could not be empirically confirmed. Thus, it may be suspected that the very intention behind CSR is ideologically driven. The quote of Jim Jones that serves as epigraph for this chapter clearly voices this concern.

Even though we should keep the possibility of ideological distortions in mind, I suggest that a more open attitude to the rapidly growing field of CSR is due. For it is hardly surprising that many scholars are eager to contribute to an increase of our knowledge, when in a certain field of research significant progress appears to be possible through a combination of methods. Viewed from this perspective, no atheistic conspiracy needs to be surmised. We may assume that this development is prompted by a purely scientific curiosity and a desire to widen the perimeters of our knowledge. In this connection, it is interesting that also some religious scientists are involved in CSR—some of them even playing a very prominent role. A clear example is Justin Barrett, a Christian psychologist working at Fuller Theological Seminary (Pasadena), who has, next to his theoretical work in CSR, developed a constructive view as to how CSR relates to the Christian faith.[44] In any case, it is important to take the developments in this field seriously from a theological perspective, if only because the ideas formed here gradually settle in the heads and hearts of many people. As Barrett argues: "Cognitive science is rapidly gaining prominence in shaping how people think about themselves and the world, and the theologian who ignores it voluntarily surrenders a useful tool for her . . . vocation, and risks limiting her relevance to the contemporary world."[45]

So, what theories are currently being developed by CSR scholars? Because of limitations of space, I can only point to the most prominent ones here, and

---

44. See, e.g., Justin L. Barrett, "Cognitive Science, Religion, and Theology," in Schloss and Murray, *The Believing Primate*, 76-99.

45. Barrett, *Cognitive Science*, 146; with reference to Michael Heller, *Creative Tension: Essays on Science and Religion* (West Conshohocken, PA: Templeton Foundation, 2003), 28.

we will have to restrict ourselves to the pivotal ideas on which they hinge. Further, given the focus of this book, I will pay special attention to those theories that draw on the role of processes of natural selection in the course of human and cultural evolution.[46]

### 8.4 A Panoply of Theories: Religion as By-Product or Adaptation

Various recent experiments have led cognitive psychologists to conclude that children have the spontaneous inclination to form (proto)religious intuitions.[47] For example, children often think natural phenomena have intentions and purposes (the sun exists to give warmth, etc.), and they tend to see an intentional designer at work behind them. Similarly, young children display a natural propensity to believe in life after death.[48] This also applies to children with a secular upbringing who had no prior knowledge of concepts like "God," "heaven," and "hereafter." One of the researchers in this field, Deborah Kelemen, suggests that children are "intuitive theists."[49] Indeed, though the ascription of intentions to natural entities and even belief in an afterlife are not necessarily religious phenomena, it seems that religion has a natural background. From this point of view, it is not strange to suggest that some of the most basic religious beliefs—such as the belief that a god exists—may be innate.[50] This is a very remarkable conclusion from recent empirical research. Philosophers and theologians have long speculated about "the naturalness of religion," but for the first time now this naturalness might be demonstrated.

46. We will therefore not discuss recent statistical research about the effect of prayer, or the aforementioned neuroscientific search for areas of the brain that react when certain mystical/religious experiences take place; usually, these types of research are not included in CSR, since they have no direct link with cognitive processes.

47. For surveys of such research, see, e.g., Jesse Bering, *The God Instinct: The Psychology of Souls, Destiny, and the Meaning of Life* (London: Nicholas Brealey, 2011); Justin L. Barrett, *Born Believers: The Science of Children's Religious Belief* (New York: Free Press, 2012).

48. Jesse Bering, "The Cognitive Psychology of Belief in the Supernatural," *American Scientist* 94 (2006): 142-49.

49. Deborah Kelemen, "Are Children 'Intuitive Theists'? Reasoning about Purpose and Design in Nature," *Psychological Science* 15, no. 5 (2004): 299.

50. De Cruz and De Smedt, *A Natural History of Natural Theology*, argue in this connection that the classic "proofs" for God's existence derive their appealing force from the fact that they correspond with spontaneous religious intuitions we have.

This means, among other things, that those parents who offer religious education to their children do not by definition brainwash or indoctrinate them, as some new atheists have argued, but rather link up with their children's deepest intuitions. The brainwashing is rather being propagated by atheists who want us to suppress any religious inclinations in children.

How can this remarkable outcome of empirical research be explained? CSR scholars usually look for evolutionary mechanisms bound up with natural selection that might serve as an explanation. It is here that theory formation comes in. We can loosely divide the existing theories into two groups. Some theories regard religion as an evolutionarily useless (i.e., not fitness-enhancing) by-product of certain useful cognitive capacities that emerged in the course of evolution—we might say: as an "evolutionary accident."[51] In other theories, religion *itself* confers an adaptive value on human evolution. These two groups of theories do not necessarily exclude each other, since it is possible to think that important building blocks of religion arose incidentally but then received a new function once religion was in place.[52] Let us discuss both groups in more detail.[53]

### Religion as a By-Product

To the first category belong some intriguing theories that have recently drawn much attention. Most of these are based on the view that certain cognitive functions (or "modules") led to the rise of religious beliefs because they were, as it were, tuned too sharply. The fact that they were tuned so sharply, however, was on average advantageous to those who possessed these traits. There-

---

51. Thus Paul Bloom, "Religious Belief as an Evolutionary Accident," in Schloss and Murray, *The Believing Primate*, 118-27.

52. De Cruz and De Smedt, *Natural History*, 22, even consider this to be a third option: the "redeployment view" or "recycling view," according to which older cognitive capacities (which came into being either as adaptations or as by-products) may be reused to deal with newly surfaced cultural challenges. It seems more to the point to see this possibility as a combination of the adaptationist and the by-product view, though.

53. For what follows I am indebted to Hans Van Eyghen, "Scientific Theories of Religion as Naturalistic Challenges," *Studies in Science and Theology* 15 (2015-2016): 93-109. For the complexity of the current CSR scene, cf. Léon Turner, "Introduction: Pluralism and Complexity in the Evolutionary Cognitive Science of Religion," in *Evolution, Religion, and Cognitive Science: Critical and Constructive Essays*, ed. Fraser Watts and Léon Turner (Oxford: Oxford University Press, 2014), 1-20.

fore, thanks to natural selection, these sharply tuned modules could spread widely, taking religiosity along in their wake. The two most well-known theories in this category are the following ones.[54]

Justin Barrett suggests the best-known example of such a mechanism that, supposedly, was involved.[55] He points out that whenever our ancestors at some stage in evolutionary history heard a sound or saw a movement somewhere in the forest, it may have been to their advantage to always suspect the presence of an intentional actor. For clearly, the thought that the sound might have been "accidental," for example, caused by the wind, would not as easily lead to making preparations in the face of possible danger. Therefore, one who thinks this way will more easily become a victim when there is a real danger. Thus, it increased one's chances of survival to postulate "actors" behind everything that one heard and saw in nature. This means that those human individuals who did so more consistently than others were eventually to become dominant. But people with such a *hyperactive agency detection device* (HADD) were also the ones who saw actors at work behind events that were *not* caused by humans or animals—and therefore started to postulate invisible actors, or higher powers, or, in the end, gods. Barrett does not go so far as to claim that this mechanism explains all the different varieties in which religion presents itself, but he holds that it does clarify at the very least one of its important *building blocks*.

The American developmental psychologist Jesse Bering—who thinks that the findings of CSR "debunk" religious belief, that is, show it to be illusory—advanced another theory in this category. Bering uses the so-called *theory of mind* (or ToM, for short). This is not a theory about the nature of the mind but about the notion developed by most of us in early childhood that other people *have* minds, that is, intentions, desires, feelings, etc., just as we have them ourselves. This notion enables us to make sense of the behavior of other people: we intuitively interpret such behavior in light of the desires, intentions, etc., we think they have. From an evolutionary point of view, ToM helped us tremendously to cooperate with other people (and to fight possible ene-

---

54. We leave aside here by-product theories that rely on supposed nonevolutionary mechanisms, such as the so-called theory of minimally counterintuitive concepts; see on this, e.g., Pascal Boyer, *Religion Explained: The Human Instincts That Fashion Gods, Spirits, and Ancestors* (London: Vintage Books, 2002).

55. Cf. Justin Barrett, "Exploring the Natural Foundations of Religion," *Trends in Cognitive Sciences* 4, no. 1 (2000): 29-34; Justin Barrett, *Why Would Anyone Believe in God?* (Walnut Creek, CA: AltaMira, 2004), chap. 3.

mies), so that it had a clear survival value.⁵⁶ Bering argues that this capacity helps us to give meaning not only to forms of human conduct but also to life events that strike us as meaningful.⁵⁷ Being used to give meaning to the way people act by positing a "mind" behind their conduct, we have been prompted to posit a mind behind everything for which we have no explanation, even including our very existence—thus developing an *existential theory of mind* (or EToM). This resulted from the fact that our ToM was overly active (which is why children typically also attribute minds to their pet dogs, etc.). "So much has human evolution invested in the theory of mind system, that the drive to attribute mental states has expanded into corridors of human cognition that probably had nothing whatever to do with the system's initial selection."⁵⁸ Thus, for example, people have felt tempted to interpret a particular cloud formation as a token of an invisible intentional being (a "mind") who supposedly provided it with some meaning. In particular, we humans tend to intuitively ascribe events that strike us as meaningful, such as a beautiful sunset, to an invisible higher being.

Like HADD, this mechanism cannot explain all aspects of religion(s), but—or so the argument goes—it does offer an interpretation of how some of its constitutive factors may have emerged, especially the belief in supernatural beings with mental states and intentional actions. Given a scenario in which several of these constitutive factors come together (including, for example, teleological thinking, or the belief in an afterlife), we can imagine that at some point all elements were in place and religiosity came naturally to people. In any case, according to this group of theories, religious belief must have evolved as the accidental by-product (or "spandrel") of some cognitive trait or traits that turned out to be beneficial to humans in their evolutionary history and therefore became widespread as a result of natural selection.⁵⁹

56. ToM has received special attention in recent research because it is suspected that this mental capacity may be poorly developed in people who suffer from autism, as a result of which they are less capable of empathizing with others.

57. Jesse Bering, "The Existential Theory of Mind," *Review of General Psychology* 6 (2002): 3–24, later elaborated in his *The God Instinct* (which was published in the United States under the title *The Belief Instinct* [New York, 2011]).

58. Bering, "Existential Theory of Mind," 10.

59. The term "spandrel" is used in architecture for a space or feature of a building that came into being for some structural reason but could be put to some unexpected use, such as being a space for art. The term was applied to evolutionary biology by Stephen J. Gould and Richard C. Lewontin, "The Spandrels of San Marco and the Panglossian Paradigm: A Critique of the Adaptationist Programme," *Proceedings of the*

### Religion as an Evolutionary Adaptation

A second group of theories suggests that religiosity itself is (or has been) somehow evolutionarily useful. Whereas until recently by-product theories were favored, at the moment adaptationist theories are gaining popularity. Most of them follow from Émile Durkheim's (in)famous reduction of religion to its social functions. I will briefly introduce the three of them that probably draw most attention in current literature.

First, according to the *supernatural punishment theory* (or SPT), belief in a powerful supernatural being facilitated pro-social behavior among humans and thus strengthened the cohesion of the groups in which they lived. As a result, such groups survived and procreated more easily than groups that did not believe in such supernatural beings. This was especially the case when the supernatural beings in question were supposed to have a moral concern for human behavior, rewarding those who did good and punishing those who transgressed the norms. Especially the latter part is important: Azim Shariff and Ara Norenzayan have found a connection between belief in gods that punish and relatively low levels of cheating, whereas such a connection could not be established for belief in more loving gods.[60] Dominic Johnson has aptly adumbrated the mechanism that seems at work here in the title of his book-length defense of SPT.[61] Some adherents of SPT, however, deny that this theory can explain the rise of religious belief in general. They argue that belief in "big gods" gained traction only when in the course of cultural evolution human groups became so large that it became impossible to know each and every member of one's group. To make sure it was safe to deal with each of them, one had to know that they could be supposed to follow certain moral norms. It is precisely belief in a "big god" (i.e., a god who is morally good, omniscient, and omnipotent) that made this possible: it prevented people from becoming

---

*Royal Society of London*, ser. B, 205 (1979): 581–98. Jones, *Can Science Explain Religion?*, 30, offers the example of the space below a staircase that can be used as a storage room.

60. Azim F. Shariff and Ara Norenzayan, "Mean Gods Make Good People: Different Views of God Predict Cheating Behavior," *International Journal for the Psychology of Religion* 21 (2011): 85–96. Cf. Ara Norenzayan, *Big Gods: How Religion Transformed Cooperation and Conflict* (Princeton: Princeton University Press, 2013).

61. Dominic P. Johnson, *God Is Watching You: How the Fear of God Makes Us Human* (New York: Oxford University Press, 2015). Cf. Dominic Johnson and Jesse Bering, "Hand of God, Mind of Man: Punishment and Cognition in the Evolution of Cooperation," in Schloss and Murray, *The Believing Primate*, 26–43.

"freeriders" and stimulated them to cooperate as morally trustworthy group members. Thus, it paved the way toward modern large-scale civilizations.[62]

Second, some scholars have argued that a similar mechanism might work *within* groups. In particular, religion may contribute to the social cohesion of a group on a much smaller scale by providing what is called *socially strategic information* (SSI). Human beings are, much more than any other species, dependent on cooperation with conspecifics. Relying on others for such cooperation is absolutely necessary for their survival. However, which members of their group may be trusted and which may not? To assess this one needs information on which to base one's strategy for social interaction. For example, one needs to know whether a possible candidate to cooperate with is likely to behave in morally praiseworthy ways or, by contrast, is inclined to lying and cheating. If this candidate believes that a powerful and righteous supernatural being sees everything she does, she will, generally speaking, behave in a more socially desirable way than if she believes in the law of the jungle. Indeed, psychological experiments have made clear that people who think they are being watched display more pro-social behavior than those who think they act anonymously.[63] As a consequence, a believing person will be considered by most people as more trustworthy and will easily be regarded as a friend instead of an enemy. Seen from this angle, religiosity would have an obvious evolutionary advantage not only on the level of one's group but also on an individual level.[64]

Third, the *costly signaling theory* (or CST) also hinges on the notion of socially strategic information but does not appeal to belief in a supernatural being who is watching all human beings. Instead, it starts off from the

---

62. Azim F. Shariff, Ara Norenzayan, and Joseph Henrich, "The Birth of High Gods: How the Cultural Evolution of Supernatural Policing Agents Influenced the Emergence of Complex, Cooperative Human Societies, Paving the Way for Civilization," in *Evolution, Culture, and the Human Mind*, ed. Mark Schaller et al. (New York: Psychology, 2009), 119-36.

63. Cf. De Cruz and De Smedt, *Natural History*, 123-24 (also for further references). This will probably be the reason why local authorities in the Netherlands have recently placed billboards with stringent-looking human faces at places that are susceptible to theft (esp. bicycle garages at railway stations).

64. Several researchers support this theory, such as Boyer, *Religion Explained* (see 152 for a definition of SSI); Scott Atran, *In Gods We Trust: The Evolutionary Landscape of Religion* (Oxford: Oxford University Press, 2002); David Sloan Wilson, *Darwin's Cathedral: Evolution, Religion, and the Nature of Society* (Chicago: University of Chicago Press, 2002).

biological phenomenon that some attributes or attitudes which at first sight are useless and even disadvantageous to their possessors may turn out to be evolutionarily beneficial in a more crucial way. A well-known example is the very ostentatious tail of the male peacock. This tail makes the peacock more vulnerable to its predators, but at the same time it has been established that it makes him more sexually attractive to female peacocks. Overall, having this tail has turned out to be beneficial. This phenomenon may offer a clue to the explanation of religious practices that require self-denial and asceticism by believers—such as fasting, celibacy, circumcision, strict dress codes, bringing sacrifices, giving generous gifts, and so forth. By participating in these sorts of practices, dysfunctional though they may seem, believers emit a signal: religion may cost us something (hence the term "costly signaling"). This demonstrates the sincerity of their faith, which in turn signals that they would also adhere to the concomitant social rules, such as being reliable. As a result, this allowed individuals to be trusted to a greater degree in their own group (and, as a result, live longer on average) and also helped the human species to solve one of the biggest problems in its evolutionary history: how to form groups that are keen to cooperate out of individuals who tend to be deeply egoistic. Clearly, without the formation of such groups humans would never have been able to survive.[65]

In conclusion, all three theories agree that religious practices and rituals (even those which at first sight seem ineffective and costly) played an important role in deterring freeriders and strengthening the internal cohesion of groups, thus offering these groups a decisive evolutionary advantage.

## 8.5 Revelation Debunked?

When Christian believers assess this panoply of theories—HADD, ToM, SPT, SSI, CST (and, as said, this is just a selection consisting of some of the most salient and prominent ones)—they may find it difficult to suppress a feeling of

---

65. Cf. Richard Sosis, "Why Aren't We All Hutterites? Costly Signaling Theory and Religious Behavior," *Human Nature* 14 (2003): 91–127; Michiel van Elk, *De gelovige geest. Op zoek naar de biologische en psychologische wortels van religie* [The believing mind: In search of the biological and psychological roots of religion] (Amsterdam: Bert Bakker, 2012), 178–81. Van Elk points to the Amish and to certain orthodox Reformed communities as groups in which such costly signaling still seems essential for their persistence.

## Morality, the Cognitive Science of Religion, and Revelation

amazement. Is this really what religion is all about? It may even be tempting to conclude that apparently this is where we end up when we start giving the theory of evolution the benefit of the doubt. Should we not thoroughly rethink matters before we take this route? Indeed, the theories we have surveyed are often used by "new atheists" to argue that religion (just like morality, as we saw in §8.2) is illusory: now that we are able to give a natural explanation (linked to our evolutionary past) of religions and their continuing appeal to many people, it is clear that a supernatural explanation is out of the question. We may be enticed to think that there is a God out there who created the world, serves as our ultimate judge, etc., but in fact we are just tricked into believing such irrational things by our genes. Matthew Alper, for example, intimates: "If belief in God is produced by a genetically inherited trait, if the human species is 'hardwired' to believe in a spirit world, this could suggest that God does not exist as something 'out there,' beyond and independent of us. . . . If true, this would imply that there is no actual spiritual reality, no God or gods, no soul, or afterlife."[66]

Let us see, however, whether this conclusion does indeed follow. In spite of Alper's claim, CSR theories do not first of all challenge belief in God's existence; they challenge belief in God's *revelation*. For clearly, even if religious belief could be fully explained by the fact that we are drawn into it by our genes, God (or a god, or gods) might still exist. In that case there just would not be a direct causal connection between our religious beliefs and this God. Our beliefs would not somehow be caused or triggered by God's making himself known to us, since they were sparked off by our genes. Therefore, I take it that it is the *doctrine of divine revelation* rather than God's existence that is at stake here. For Christians, Reformed ones included, the doctrine of revelation articulates one of the most basic tenets of their faith: God "has not left himself without a witness" (cf. Acts 14:17) but has spoken to humans in a myriad of ways (Heb. 1:1). It is only because of this that we humans can come to know God. For example, God has shown his eternal power and divine majesty through the things he has made in creation (Rom. 1:19-20). We already saw how John Calvin interpreted these words in terms of a universal awareness of God (*sensus divinitatis*) implanted in the human mind. As Richard Muller explains, "even in its false forms, it [religion] presumes the fundamental *sensus divinitatis* and is grounded in the objective reality of the God who must be

---

66. Matthew Alper, *The "God" Part of the Brain: A Scientific Interpretation of Human Spirituality and God* (Naperville, IL: Sourcebooks, 2006), 79.

worshiped."[67] Thus, we typically find a "pairing of revelation and religion" in the prolegomena of Reformed orthodox expositions of doctrine.[68] Many of these expositions even open with a locus on divine revelation, thus testifying to its vital importance as a starting point for theology.[69]

Now suppose that CSR provides a complete explanation of religious belief (which I take here in its broadest possible sense, that is, including religious experiences, practices, attitudes, etc.), that is, an explanation that takes all causal factors into account. Then indeed humans do not believe in God because of Calvin's *sensus divinitatis*, Holy Scripture, the inner working of the Holy Spirit, the testimony of the church, the experience of being addressed by God through a sermon, or any other means of revelation, but because of the way in which a propensity toward religion is hardwired in their brains. But do CSR theories such as those sampled above provide a complete explanation of our religiosity? Do they even provide a sound explanation—or how should we assess these theories? To begin with, we do well to point to the scientific problems attached to the theories just summarized. Whereas some theories on the evolution of morality, such as that of kin selection (cf. §8.2), are fairly well established today, evolutionary explanations of the emergence and evolution of religion are not.

First, these theories are very diverse in nature. Each of them may serve as an explanation for one or two particular aspects of religion, but none of them is equipped to explain religion in all its complexity. For example, one cannot easily extrapolate from childish personifications of clouds to Karl Barth's *Church Dogmatics*, or to a random Christian sermon, for that matter. Second, some theories seem to fit better with some religious traditions (e.g., animism) than with others (e.g., contemporary Christianity). Even cumulatively, the

---

67. Richard A. Muller, *Post Reformation Reformed Dogmatics*, vol. 1, *Prolegomena to Theology*, 2nd ed. (Grand Rapids: Baker Academic, 2003), 167.

68. Muller, *Post Reformation Reformed Dogmatics*, 171.

69. Cf. Heinrich Heppe, *Reformed Dogmatics: Set Out and Illustrated from the Sources*, rev. and ed. Ernst Bizer (Grand Rapids: Baker Books, 1978), 1–11. For contemporary defenses and articulations of the Christian doctrine of revelation, see, e.g., Richard Swinburne, *Revelation: From Metaphor to Analogy*, 2nd ed. (Oxford: Clarendon, 2007 [1992]); Colin E. Gunton, *A Brief Theology of Revelation* (London: T&T Clark, 1995); Matthew Levering, *Engaging the Doctrine of Revelation: The Mediation of the Gospel through Church and Scripture* (Grand Rapids: Baker Academic, 2014); Mats Wahlberg, *Revelation as Testimony: A Philosophical-Theological Study* (Grand Rapids: Eerdmans, 2014). Its most vigorous reinterpretation in the twentieth century no doubt came from Reformed theologian Karl Barth.

theories don't offer an encompassing explanation of "religion."[70] Indeed, some of them are mutually exclusive.[71] Third, it remains highly questionable whether the mechanisms that supposedly impact the rise and persistence of religion do indeed function in the ways that are proposed by the supporters of the respective theories. For example, it seems that SPT won't work well anymore when people find out that the supernatural being they believe in usually does *not* (immediately) punish morally wrong behavior.[72] And CST suffers from so many weaknesses, pertaining both to its coherence and to its explanatory potential, that some actually call it a "nonstarter."[73] Indeed, important objections have been raised to most if not all CSR theories. Therefore, a lot of theoretical work remains to be done before a serious candidate for a comprehensive natural explanation of religion can be proposed.

But even so, there is a fourth problem that is no less serious: so far there is hardly any empirical confirmation for these various theories. And without such confirmation, they remain "just so stories": interesting ideas, but nothing more than that. Barrett is therefore correct when he states that CSR has no future if its supporters do not attempt to find empirical confirmation, or falsification, for their claims.[74] At the moment, however, this project is still

---

70. For an attempt to combine various cognitive and cultural-evolutionary processes in order to explain religion's pro-social effects (a much more modest goal than "explaining religion"), see Scott Atran and Joseph Henrich, "The Evolution of Religion: How Cognitive By-Products, Adaptive Learning Heuristics, Ritual Displays, and Group Competition Generate Deep Commitments to Prosocial Religions," *Biological Theory* 5, no. 1 (2010): 18–30. For the incompleteness of current CSR explanations, see Jones, *Can Science Explain Religion?*, 60–74.

71. This is rightly observed by Van Eyghen, "Scientific Theories of Religion," 105.

72. Cf. Jeffrey P. Schloss and Michael J. Murray, "Evolutionary Accounts of Belief in Supernatural Punishment: A Critical Review," *Religion, Brain, and Behavior* 1 (2011): 46–99 (one may counter, though, that this is why belief in a final judgment and an afterlife became important—a point that was brought to my attention by Herman Philipse).

73. As does Justin Barrett, "Cognitive Science of Religion: Looking Back, Looking Forward," *Journal for the Scientific Study of Religion* 50 (2011): 232–33; for the objections, see, e.g., Michael J. Murray and Lyn Moore, "Costly Signaling and the Origin of Religion," *Journal of Cognition and Culture* 9 (2009): 225–45. Murray and Moore, for example, point out that, especially in cases of relatively minor costs (e.g., participating in rituals), "signals will quickly become dishonest, and will soon cease to confer any adaptive advantage. The logic of signaling seems, from an evolutionary standpoint, self-undermining" (226).

74. Barrett, "Cognitive Science of Religion," 232. Note that the notions of em-

in its infancy, and theologians might therefore argue that they do not need to bother about these theories as long as there is no hard evidence for them (thus repeating the response of Kuyper, Bavinck, and others to Darwin's theory of evolution—a response that was not unreasonable during the scientific low tide of Darwinian evolution around 1900).

We are, however, ill advised to go this way, clinging as it were to the "gaps" in current science in order to hold on to a *super*natural origin of (our) religious belief. All too often it has turned out that such gaps in our knowledge could be closed after all by further scientific research; and even though it is hard to see how we might find decisive empirical evidence for a particular theory about the rise of religion (since we cannot do experiments and hardly know anything about the time when this happened), we should not rule out this possibility in advance. Therefore, let us assume that in the future we will see such an empirical undergirding or even confirmation of one or more of the theories referred to above (or of some other theories). Would this then mean that—at long last—it has been proven—or at least made plausible—that religious faith is indeed an illusion, since we have found a satisfactory immanent explanation for its origin? It is vitally important to see that, despite all kinds of naturalistic claims to the contrary, this is by no means the case. As James Jones forcefully argues:

> The crusaders or debunkers . . . claim that cognitive science heralds the demise of religion, all religion. They interpret, in their own particular way, the findings of cognitive science that suggest that religious beliefs utilize common cognitive processes like pattern recognition, attributions of causality, perception of others' minds, etc. . . . It is almost universally asserted that such findings prove that religion is a purely natural phenomenon. Clearly these findings do no such thing. All they suggest is that natural, human processes are at work in religion, something virtually no one denies. That does not logically entail that only natural processes are present. . . . The debunkers seem to be assuming that if natural processes are at work, nothing else can be. But no argument is offered to support that assumption.[75]

---

pirical confirmation and falsification are already complex in themselves, given the underdetermination of scientific theories by the data; cf. Kyle Stanford, "Underdetermination of Scientific Theory," in *Stanford Encyclopedia of Philosophy*, last substantive revision October 12, 2017, plato.stanford.edu/entries/scientific-underdetermination/.

75. Jones, *Can Science Explain Religion?*, 75. By the way, Jones is exaggerating

Thus, even if we could convincingly explain religion as a natural phenomenon, it still does not follow that God does not exist or did not reveal himself—or more generally, that religious beliefs are false. As Kelly James Clark and Justin Barrett have claimed: "Showing that natural causes are involved in the production of a belief tells us nothing about the truth or falsity of that belief. . . . Both natural and supernatural explanations may be true" (at the same time).[76]

Consider an analogy. One of the riddles of history concerns the rapid growth of Christianity at the beginning of the common era.[77] The Christian movement started around 30 CE as a small Jewish sect, but despite waves of severe persecution, it ended up as the Roman Empire's dominant religion within a couple of centuries. Christians have always seen God's hand and providential care in this wonderful development. Scholars, however, have pointed to (among other things) the availability of an elaborate infrastructure that was in place throughout the Roman Empire as a major contributing factor. Without the extensive network of roads, bridges, shipping routes, etc., the Christian gospel would never have spread so rapidly across the Middle East and Europe. Although even from a sociological point of view this explanation is grossly incomplete (if only because adherents of other religions at the time could also have used the Roman infrastructure to spread their ideas), let us assume that it is basically correct. Do we now finally know that God was not involved or that the content of the Christian gospel is bogus? That would be an absurd conclusion. For it could well be that, "in the fullness of time" (Gal. 4:4), God put the existing Roman infrastructure to use in order to have the gospel effectively proclaimed throughout the world. Similarly, the cognitive modules and evolutionary mechanisms that allegedly contributed to the emergence of religion—including the Christian faith—can be seen as the roads, shipping routes, and bridges of the Roman Empire. We may actually feel inclined to

---

when he writes that it is "almost universally" contended that CSR shows religion to be a purely (= exclusively) natural phenomenon.

76. Kelly James Clark and Justin L. Barrett, "Reidian Religious Epistemology and the Cognitive Science of Religion," *Journal of the American Academy of Religion* 79 (2011): 655.

77. See on this theme, e.g., Rodney Stark, *The Rise of Christianity: How the Obscure, Marginal Jesus Movement Became the Dominant Religious Force in the Western World in a Few Centuries* (Princeton: Princeton University Press, 1996); of course, Stark discusses more factors than the one I single out for discussion here.

thank God for them, since God sovereignly used them to serve as channels of his revelation![78]

The same argument applies to theories on the evolution of morality as discussed in §8.2. Here, too, we should be wary of the genetic fallacy: whether moral norms exist is not dependent on how moral beliefs and convictions come about. Natural explanations of human morality in terms of evolutionary selection pressure do not, as Ruse and Wilson suggest, exclude any "external grounding." Even if our ethics came into existence as the upshot of evolutionary processes, it does not follow that ethics is "an illusion fobbed off on us by our genes to get us to cooperate."[79] For these same processes might just as well be used by God to write on our hearts what his law requires (cf. Rom. 2:15).[80] That is, rather than being a substitute for revelation, they may be a means of revelation.

What Alvin Plantinga argues with regard to religion applies just as well to morality: "To show that there are natural processes that produce religious belief does nothing to discredit it; perhaps God designed us in such a way that it is by virtue of those processes that we come to have knowledge of him."[81] Similarly, the fact (assuming that it is a fact) that our moral values emerged through processes of natural selection because of their fitness-enhancing value does not mean that we cannot act in ways that are really altruistic, as is sometimes suggested.[82] For although in that case on a subconscious level Mother Teresa's decision to help the poor—to give an example—would have stood (in complex ways) in the service of the survival of

---

78. Cf. Barrett, "Cognitive Science, Religion, and Theology," 76: "I see much promise in the cognitive sciences to enrich our understanding of how humans might be 'fearfully and wonderfully made' (Ps. 139:14) to readily (though not inevitably) . . . enjoy a relationship with Him."

79. Ruse and Wilson, "The Evolution of Ethics," 52 (cf. n. 31 above).

80. This argument presupposes that evolutionary randomness and divine guidance are not incompatible, as I have argued in the previous chapter; cf. Barrett, "Cognitive Science, Religion, and Theology," 97.

81. Alvin Plantinga, *Warranted Christian Belief* (Oxford: Oxford University Press, 2000), 145. The conclusion that religious belief is discredited in this way only follows when it is tacitly presupposed that God does not exist (or does not reveal himself). As to the parallel with morality, cf. Erik J. Wielenberg, "On the Evolutionary Debunking of Morality," *Ethics* 120 (2010): 441-64; after having argued against evolutionary debunkings of morality, Wielenberg suggests that "some evolutionary debunkings gain an illegitimate air of plausibility by exploiting many people's moral skepticism" (463).

82. E.g., by Ruse and Wilson, "The Evolution of Ethics."

her genes or group, on the conscious level she still took this self-sacrificing decision in full awareness of the consequences. Apparently, cultural evolution can dissociate itself from its roots in biological evolution and turn against its preference for the fittest (or most adapted) organisms. From a Christian theological point of view, we may add that the gospel indeed turns the principle of natural selection upside down, for example, where the good shepherd leaves the ninety-nine safe sheep alone in order to search for the one that went astray (Matt. 18:12-14), or where it summons us to love our enemies (Luke 6:27), or where it says that many who are first will be last and the last will be first (Mark 10:31; etc.), or where we are called upon to hate our own lives instead of loving them (John 12:25).

It is sometimes argued that, although perhaps not logically impossible, somehow it is still *epistemically inferior* to attribute the rise of morality and religion to God when a completely satisfying natural explanation has become available.[83] Indeed, if one does not believe in God, it will seem rationally untenable or unwarranted or otherwise epistemically substandard to see the evolutionary processes operating here as channels of God's revelation. But in that case, evidently one cannot make an atheistic argument out of these evolutionary processes, since one has already presupposed the nonexistence of God. Moreover, it might just as well be argued, quite to the contrary, that the evolutionary backgrounds of morality and religion point to the general reliability of moral and religious claims. For the development of our minds through evolutionary mechanisms has also produced in us the beliefs that living human bodies are persons with minds like our own mind, that our senses are generally trustworthy, that usually our memory does not cheat us, and that science can discover the truth. Now if these beliefs are true, why would belief in objective moral values and belief in God, which come to us through the very same sort of natural processes, be false? As De Cruz and De Smedt argue, those who hold religion to be illusory because of evolutionary explanations "will have to specify why natural selection would lead us to adopt false beliefs in the religious domain, whereas it allows us to adopt true beliefs under many other conditions."[84] As long as they do not succeed in doing so in

---

83. Cf. Clark and Barrett, "Reidian Religious Epistemology," 656; see also Michael J. Murray, "Belief in God: A Trick of Our Brain?," in *Contending with Christianity's Critics: Answering New Atheists and Other Objectors*, ed. Paul Copan and William Lane Craig (Nashville: B&H, 2009), 47-57 (esp. 55).

84. De Cruz and De Smedt, *Natural History*, 188. De Cruz and De Smedt point to the fact that from an evolutionary perspective "there are severe costs involved in

a convincing way, Reformed Christians and others may see CSR explanations as providing confirmation for the existence of a *sensus divinitatis* as professed by John Calvin.

Apparently, belief in God (or a supernatural agency) comes to us quite naturally. But when the cognitive mechanisms with which we are equipped are normally aimed at truth tracking, we may assume that they are so as well when they install in us the belief that (some) God exists.[85] As Wentzel van Huyssteen contends, "the naturalness of religious ideas actually supports religious claims rather than undermines them: if religious beliefs are largely produced by normal human cognitive systems and if we generally trust these systems, then we should not suspect them in the case of religious beliefs."[86]

In line with this, as a commonsense belief—just like the belief that other minds exist next to my own mind and the belief that the world exists (instead of me dreaming so)—belief in God is seen by contemporary ("Reformed") epistemologists as "properly basic," which roughly means that it is not in need of evidence since it is rationally justified unless proven otherwise.[87]

This may also hold if religion should be considered a by-product of evolution rather than an adaptation, which is the case when mechanisms like HADD and ToM turn out to be its main natural causes. Even when religion arose as "an incidental by-product of cognitive functioning gone awry,"[88] it may still be conducive in finding important truths. For cognitive systems that did not emerge as a result of selection pressure may still turn out to be tremendously important for truth finding. A striking example here is science. Clearly, the cognitive capacities that helped us survive as a human species

---

brain growth," which makes it "extremely unlikely" that many of our natural beliefs are entirely unrelated to what is going on in the world (187).

85. See Kelly James Clark and Justin L. Barrett, "Reformed Epistemology and the Cognitive Science of Religion," *Faith and Philosophy* 27 (2010): 174–89.

86. J. Wentzel van Huyssteen, "From Empathy to Embodied Faith? Interdisciplinary Perspectives on the Evolution of Religion," in *Evolution, Religion, and Cognitive Science: Critical and Constructive Essays*, ed. Fraser Watts and Léon Turner (Oxford: Oxford University Press, 2012), 149. See also Aku Visala, *Naturalism, Theism, and the Cognitive Science of Religion: Religion Explained?* (Farnham, UK: Ashgate, 2011), 185; Barrett, *Why Would Anyone Believe in God?*, 95–105.

87. On properly basic beliefs, cf., e.g., Michael Peterson et al., *Reason and Religious Belief: An Introduction in the Philosophy of Religion*, 5th ed. (New York: Oxford University Press, 2013), 118–24, or the relevant chapter in any other introduction to contemporary philosophy of religion or epistemology.

88. Paul Bloom, "Is God an Accident?," *Atlantic* 126 (December 2005): 105.

throughout the ages did not evolve in order to provide us with current science—for example, to help us discover the theory of relativity. Rather, science split off as recently as, say, four centuries ago as an accidental by-product of the cognitive capacities that helped us deal successfully with the practical challenges of everyday life. If we nevertheless accept the results of science as generally reliable, the theory that *religion* is a by-product does not provide us with a reason to distrust its data. If, alternatively, we should discredit the epistemic outcomes of every by-product of evolution, we should also discredit the results of science (including, if this belongs to the results of science, the theory that religion is an evolutionary by-product).[89]

Of course, the naturalist may go on to look for relevant differences between these two cases. She might, for example, stipulate that the cognitive mechanisms that evolved over time to help us survive, including their by-products, generally speaking work well when applied to the visible world but become utterly unreliable when we apply them to a conjectured extrasensory realm. In that case, science is in and religion is out. The problem, however, is that there is no independent way to establish why we should go in this direction here rather than in another. De Cruz and De Smedt refer to this as the "generality problem":[90] it is not clear on what level of generality we should evaluate the workings of our evolved cognitive faculties—do they just apply to the world of ordinary objects; are they also to be trusted with regard to, for example, quantum phenomena, or perhaps even with regard to metaphysical issues such as the origin of the universe? It is worth quoting the conclusion they draw in this connection:

> Nontheists start from the assumption that the natural world is all there is.... In this view, the inference to a supernatural entity is a mistake, a misapplication of intuitions that have evolved for a different context.... While these cognitive processes are reliable in general, they are unreliable when applied to metaphysical questions. By contrast, theists begin with the supposition that God is responsible for the design of reality, including human minds. From

---

89. See Clark and Barrett, "Reidian Religious Epistemology," 662–63. In this rich essay, Clark and Barrett also explain why Occam's razor (or the principle of simplicity) cannot be deployed to argue for the superfluity of a supernatural explanation of religion next to a natural one (660–61). See also Kelly James Clark, *Religion and the Sciences of Origin: Historical and Contemporary Discussions* (New York: Palgrave Macmillan, 2014), 115–36.

90. De Cruz and De Smedt, *Natural History*, 196–97.

this perspective, it is reasonable to state that our natural cognitive processes are working properly when they generate the intuitions that underlie natural theological arguments. . . . In this view, it is likely that these intuitions are also on the mark when they point us to God's existence."[91]

We finally find ourselves at the level of our most basic assumptions here, and science does not seem fit to help us any further at this level. Rather, the famous "principle of antithesis" posited by Abraham Kuyper between (Christian) theism and naturalism seems still in place: "Not faith and science therefore, but *two scientific systems* or if you choose, two scientific elaborations, are opposed to each other, *each having its own faith.*"[92]

In line with this, in his book-length critique of debunking arguments that take their cue from CSR, James Jones argues that "many disputes about cognitive science and religion are driven by different judgments regarding basic assumptions—judgments that reflect "our most fundamental ways of experiencing our self and the world."[93] Being convinced that CSR is only about how human cognition works and not about what exists in reality, Jones wonders what makes the debunkers' rhetorical deployment of its religiously neutral scientific findings work so well. In response, he points to the *background* or context in which the conviction flourishes that "religious claims are always mistaken and that virtually every religious practice is cognitively pathological." He suspects that the context in which such beliefs appear intuitively obvious "is one in which it is assumed that science is the only arbiter of knowledge and that the only reality is what is disclosed to us through natural science."[94] This twofold assumption is called *scientism* in present-day scholarly literature. So, in one word: in Jones's view, it is scientism that plays a key role here—and he may be quite right on this.[95]

---

91. De Cruz and De Smedt, *Natural History*, 198–99.

92. Abraham Kuyper, *Lectures on Calvinism* (Grand Rapids: Eerdmans, 1931), 133; cf. 132: "This, and no other, is the principal antithesis, which separates the thinking minds in the domain of Science into two opposite battle-arrays."

93. Jones, *Can Science Explain Religion?*, 56, 58.

94. Jones, *Can Science Explain Religion?*, 83.

95. See, on the pervasive influence of scientism in current popular science writing, Jeroen de Ridder, "Science and Scientism in Popular Science Writing," *Social Epistemology Review and Reply Collective* 12, no. 3 (2014): 23–39; cf. on scientism in general, Jeroen de Ridder, Rik Peels, and René van Woudenberg, eds., *Scientism: Prospects and Problems* (Oxford: Oxford University Press, 2018), and Mikael Stenmark's earlier but still instructive book, *Scientism: Science, Ethics, and Religion* (Aldershot, UK: Ashgate, 2001).

## Morality, the Cognitive Science of Religion, and Revelation

In this chapter, we have seen that the data of the cognitive science of religion do not force us to adopt such a scientistic framework.[96] Rather, as Jones and others suggest, the scientific (or some other naturalist) framework is presupposed in the debunkers' attack on religion. For Christians, evolution-based natural explanations of moral and religious belief in general and CSR theories in particular do not form a threat to their faith. Because they have independent reasons to believe in God and God's revelation, they can, as I hope to have shown, very well interpret the data of CSR within this framework. They may, for instance, appeal to the classic dogmatic distinction between a divine first cause and immanent secondary causes, both of which can be at play at the same time.[97] Evolutionary mechanisms, no doubt along with countless cultural factors, may serve as secondary causes that generate faith in God; God himself may be praised for being the primary source of his gracious self-revelation, sovereignly putting to his use any means he in his wisdom has selected. Indeed, theological selection precedes natural selection here.[98]

In brief, CSR explanations, as well as other natural explanations of morality and religion, undoubtedly raise important theological questions.[99] They do not, however, show that morality and religion are illusory phenomena, missing real referents. That is an illusion fobbed off on us by our scientism-

---

96. See for this latter point Clark and Barrett, "Reidian Religious Epistemology," 668-70.

97. Though this distinction goes back to Thomas Aquinas and continues to be more influential in Roman Catholic theology, it is also part and parcel of Reformed accounts of divine concurrence (*concursus*); see Heppe, *Reformed Dogmatics*, 258-61, and cf. Benjamin W. Farley, *The Providence of God* (Grand Rapids: Baker Books, 1988), 37-42. See also n. 47 on p. 218, above.

98. Of course, one can point to other problems in this connection, such as that these evolutionary mechanisms also have outcomes that do not seem to fit God's purposes very well. Here, however, I refer back to earlier chapters in this book (esp. chap. 4) where such problems were discussed.

99. Cf. Barrett, "Cognitive Science, Religion, and Theology," 97; Barrett, for example, thinks of such questions as why God chose such a long and complicated route to make himself known to us; why, instead of equipping us with a true, accurate, and fully formed god-concept, our evolved cognitive faculties provide us with a much more general belief in supernatural agency; etc. Such questions do indeed demand theological reflection, though they do not seem structurally different from the well-known "problem of divine hiddenness." This problem falls beyond the scope of this book, but see, e.g., Adam Green and Eleonore Stump, eds., *Hidden Divinity and Religious Belief: New Perspectives* (Cambridge: Cambridge University Press, 2015).

imbued culture. Thus, there is no need for Christians to keep evolutionary theory at bay out of fear that, in the very end, it will leave them empty-handed with regard to their faith, explaining away their deepest commitments: belief in God's revelation and perhaps even his very existence. For there is actually no way in which science can "explain God away." Therefore, Christians don't have to be afraid of science, not even evolutionary science.

## 8.6 Conclusion

In this chapter we addressed some of the most recent attempts to use the theory of natural selection for discrediting (or "debunking") not only the Christian faith but religion and morality in general. In particular, the doctrine of divine revelation is targeted by these attempts. According to this doctrine, God has "written in the hearts" of all humans both a basic knowledge of good and evil and a basic awareness of himself. God has equipped us with a faculty by which we may come to know him and his will. And although we humans usually suppress or twist the products of this faculty as a result of our sinfulness, God is able to restore these by revealing them to us much more explicitly through his law and gospel. Thus, religion and morality derive from God's gracious revelation. In the wake of attempts to extend the scope of natural selection from nature to human culture, however, scientists have specified ways in which morality and religion might have come to us quite naturally, because they brought us evolutionary advantages or were "unintended" by-products of other traits that did so. Indeed, especially in the case of morality, some of these theories have found some empirical corroboration. In the case of religion, theories that have been put forward in the so-called cognitive science of religion are more contested, but this may change in the future.

Quite a number of naturalistic scholars and philosophers—often adherents of scientism—have argued that belief in God is finally explained away or debunked by these theories. Indeed, don't these theories show that both religion and morality are entirely natural phenomena, which have nothing to do with God or divine revelation? The answer to this question is a clear no: this is not what these theories show. If we grant for the sake of argument that these theories (or some of them) are correct, what they do demonstrate is that natural, human processes are at work in religion. But that is not at all startling. Natural processes are always involved, even in our deepest or most committed religious moments. It does not follow, however, that religion should be *reduced* to such processes. It may just as well be the case that in his revelation God sovereignly makes use of the way in which our brains are hardwired through endless processes of natural selection. Wouldn't that, in fact, enrich our view of the "how" of revelation? In any case, the role of natural explanations does not rule out that God's agency is involved. Reformed theology, along with Catholic theology, has always made a distinction between God as the first cause of things, working on a transcendent level, and natural phenomena as the secondary causes that God puts to use. Reformed Christians, therefore, have no reason to reject evolutionary theory out of fear of losing both morality and true religion.

CHAPTER 9

# Any Other Business?

> It takes theological discernment to judge what should be insisted upon for theological reasons and what can be adjusted in the interests of scientific coherence.
>
> —Celia Deane-Drummond[1]

We are now in a position to take stock of our findings. In the course of the previous chapters, it has become clear that, although evolutionary theory does not leave unaffected our ways of thinking about the doctrine of Scripture, the goodness of God, theological anthropology, the history of redemption, divine providence, and the doctrine of revelation, there is no reason to think that all these classical loci fall apart as soon as one starts to take evolutionary theory seriously. This is an important outcome of the present study, since many observers, both Christians and non-Christians, intuitively think that the very foundations of the Christian religion are at stake in the evolution debate. If my argument has been by and large convincing, it is clear that that is not the case. On the contrary, we even end up with an enriched view of the workings of, for example, God's providence and revelation.

Still, the idea of having to revise particular doctrinal notions in light of scientific developments deters many Christians (even though the church started to do so when it adapted its exegesis of a couple of biblical passages to the newly adopted heliocentric worldview). For where will this, in the end,

---

1. Celia Deane-Drummond, "In Adam All Die? Questions at the Boundary of Niche Construction, Community Evolution, and Original Sin," in *Evolution and the Fall*, ed. William T. Cavanaugh and James K. A. Smith (Grand Rapids: Eerdmans, 2017), 41.

leave us? Especially among Reformed and evangelical Christians, the "domino theory" is quite popular: as soon as one starts to make revisions in one particular doctrine to accommodate some implication of evolutionary theory, it is thought, this domino will hit a second one, knocking it down as it falls, with the second domino in turn toppling over the next one, and so forth, until the entire "grand narrative" of Christianity has collapsed. Remember the famous dictum of Seventh-day Adventist George McCready Price: "No Adam, no Fall; no Fall, no atonement; no atonement, no Savior" (cf. §6.7). Hopefully, by disentangling the various issues and examining them on a one-by-one basis, showing how each of them is related to a particular aspect of evolutionary theory and how none of them is debunked by it, the preceding discussions have made clear that there is no reason to fear such a slippery slope. More than one and a half centuries after Charles Darwin, we have a fairly good overview of what evolutionary theory amounts to and where it causes tensions with orthodox Christianity. I have tried to elucidate that none of these tensions amounts to an incompatibility.

Nevertheless, some readers may wonder whether more doctrines than the ones discussed in this book may be affected by evolutionary theory. Clearly, it is important to avoid the risk of ice floe hopping that we observed in chapter 3. Thus, the question that comes up at this stage is: Could it be that more doctrines than the ones discussed collide with aspects or corollaries of evolutionary theory, so that classical Reformed accounts of them can no longer be upheld? It is hard to allay this concern. I will suggest, however, that whereas other parts of Christian doctrine *could* be transformed in the light of evolutionary theory and, in fact, have been transformed by various theologians, and though it may be tempting or attractive to do so, there is *no need* for further adaptations. I will unpack this claim by briefly surveying several parts of Christian doctrines that are regularly mentioned in this connection.

To begin with, how about *eschatology*? Indeed, eschatology is often mentioned as a doctrine that might have to be reconceived as a result of evolutionary theory. It is sometimes argued that if we have to change our views about protology (the doctrine of the first things), we probably also have to change our eschatological expectations. For if God worked incrementally in creating the universe, should we not expect God to do so as well when bringing the universe, including life on this earth, to its completion? And doesn't such a scenario fit ill with the apocalyptic expectations of a sudden ending that are so pervasive in the New Testament and that (despite the so-called *Parusieverzögerung*, i.e., the delay of the second coming of Christ) have always been upheld in the Christian tradition? The answer to these questions, however, is in the negative. Whereas we can infer from the empirical data that God gradually created the world, we have no such empirical data for the ending of the universe. Nor could we just extrapolate from our current observations to future scenarios such as a "big crunch," a "big freeze," or a "big rip," since such extrapolations presuppose that the universe will still exist for a long time. For all we know, God might put an end to history before any of these scenarios on the ultimate fate of our universe will materialize.[2] Evolutionary theory does not at all forbid that "the Lord will come like a thief in the night" (1 Thess. 5:2), since it does not imply a "closed" worldview in which no divine action is possible.

For the same reason, it does not follow from evolutionary theory that *miracles* (such as those related in the Bible—but also later ones) are impossible, and that as a result one could no longer believe in, for example, the resurrection of Jesus Christ. In that case, it would indeed become impossible to be an orthodox Reformed Christian and an orthodox Darwinian (in the three-layered sense specified in chapter 2) at the same time. To be sure, process theology and related forms of panentheism—according to which God is luring the world from within toward its future rather than acting in the world from the outside—have proven attractive to many thinkers in the field of science and religion (Ian Barbour and Arthur Peacocke are well-known examples).[3] Such a framework clearly puts pressure, to say the least, on our recognition of the great divine acts of salvation as narrated in the gospel. However, evolutionary theory does not force us to adopt a process view on divine action. On the contrary, as Reformed theologian David Fergusson explains, our view of divine

---

2. For an excellent study of the relationship between theological and scientific explorations of the future of the universe, see David Wilkinson, *Christian Eschatology and the Physical Universe* (London: T&T Clark, 2010). See also Mark Harris, *The Nature of Creation: Examining the Bible and Science* (Durham, UK: Acumen, 2013), 161-84, who points out that the big bang model's "ability to predict the long-term future of the universe is seriously hampered by a number of factors," such as the presumed existence of dark matter (163). Harris also, however, rightly points to the hermeneutical difficulties in interpreting the New Testament apocalyptic texts (170-75).

3. Cf. Philip Clayton and Arthur Peacocke, eds., *In Whom We Live and Move and Have Our Being: Panentheistic Reflections on God's Presence in a Scientific World* (Grand Rapids: Eerdmans, 2004). John Polkinghorne is an important exception here; see his *Scientists as Theologians* (London: SPCK, 1996) and cf. Christopher Knight, "John Polkinghorne," in *The Blackwell Companion to Science and Christianity*, ed. J. B. Stump and Alan G. Padgett (Oxford: Blackwell, 2012), 622-31.

action can unduly be "narrowed by an exclusive preoccupation with the laws of natural science." Therefore, we need a more Trinitarian view:

> The danger here is that theology simply becomes a gloss on the prevailing scientific worldview. By contrast, the different types of divine action that characterize the central themes of Scripture and tradition require a differentiated approach that employs a variety of models to understand divine involvement in creation, incarnation, and eschatological fulfillment. The appropriation of particular actions to the different persons of the Trinity is important in pointing to the different forms of divine activity.[4]

Indeed, the Trinitarian nature of Christian theology should remind us that God's activity is not exhausted by the works of creation and evolution but extends to his saving and perfecting work in Jesus Christ and the Spirit.

The reader may have noticed that in the preceding chapters explicit references to the *doctrine of creation* are conspicuous by their absence. One reason is that the entire evolution debate is situated in the quarters of creation theology. A Christian appropriation of evolutionary theory will consider this theory as a specification of the "how" of God's creative and sustaining action. As far as we have been speaking (especially in chapter 6) about redemption and consummation—the other grand parts of the triptych of Christian theology—this reflected the close connections we already find between these three in the Bible.[5] Another reason we have abstained from bringing up the doctrine of creation is that the black-and-white contrast with evolution that is often suggested (as in the typical debates on "creation *or* evolution") is largely illusory. As Benjamin Warfield already observed, since creation is about the beginning of life and evolution about its subsequent development, the two can't conflict.[6] Moreover, the doctrine of creation traditionally deals first and foremost with much more important questions than the "how question," such as: Who is the God that created the world, and to what end did God do so? What does

---

4. David Fergusson, *Creation* (Grand Rapids: Eerdmans, 2014), 86.

5. Cf. on this point, e.g., Fergusson, *Creation*, 1-9; Cornelis van der Kooi and Gijsbert van den Brink, *Christian Dogmatics: An Introduction* (Grand Rapids: Eerdmans, 2017), 218-24; Mike Higton, *Christian Doctrine* (London: SCM, 2008), 169-77; Hendrikus Berkhof, *Christian Faith: An Introduction to the Study of the Faith*, rev. ed. (Grand Rapids: Eerdmans, 1986), 172-80.

6. This is a recurring emphasis in Warfield's writings on creation. See Fred Zaspel, *The Theology of B. B. Warfield: A Systematic Summary* (Wheaton, IL: Crossway, 2010), 372.

creation imply for the nature of the created world—in terms of its contingency, its reliability, its nondivine character, etc.? From a theological point of view, the question *how* God created the world is less exciting and covers only a tiny part of the ground of creation theology.

One specific concern some Christians have about evolutionary theory in the context of the doctrine of creation is whether the human species is indeed the culmination point of God's creative intentions. Doesn't evolutionary theory imply that we are still evolving into ever more complex species? And if so, how does this square with our unique position as the apex of creation? We must realize that from a biblical perspective the emergence of humans was not God's only or even God's highest creative goal. Creation texts like Genesis 1 and Psalm 8 have to be balanced here with others, such as Psalm 104 and Job 38–41 (even with regard to Genesis 1, it can be argued that not the human species but the *Sabbath* serves as creation's pinnacle). Also, if at some places in the world humans evolved into another species, that would not necessarily mean the end of the human race (although that might then become a troubling possibility); due to the rich variety of natural environments, the history of evolution is a history of branching, not one of substitution of one species by another.[7] However, as Denis Alexander explains, even though our genomes are indeed evolving all the time in response to new challenges, it is highly improbable that we are moving toward a new speciation event.[8] A more feasible scenario is that at some point in the (near?) future, through genetic engineering, humans themselves will come to create a new race of beings that can no longer produce offspring with "normal" humans. This is an unsettling scenario indeed. If at some moment, however, this materializes, it will do so irrespective of whether or not evolutionary theory is true. Therefore, in that case we should not blame evolution but the human beings who will put these new techniques in practice.

Finally, how about Christology, pneumatology, and ecclesiology? It is intriguing to ponder what these doctrines would look like if approached from an evolutionary point of view. We could, for example, follow the lead of Hendrikus Berkhof by portraying Jesus—in what in fact is a remarkably concordistic move—as the man in whom (evolutionary) history "makes its most decisive

---

7. This is also the answer to the oft-heard question of how there can still be other primates if we humans evolved from them.

8. Denis Alexander, *Creation or Evolution: Do We Have to Choose?* 2nd rev. ed. (Oxford: Monarch Books, 2014), 280–82. One of his arguments is that speciation usually occurs in isolated populations, whereas the human world has largely become a global village without any isolation.

jump forward."⁹ In Jesus, for the first time in history we find someone who is able not only to preach love for one's enemies (which was exceptional at the time) but also to practice it till the very end of his life. Here, evolution has finally reached its culmination point, breaking out of the cycle of self-preservation, violence, and in-group thinking. Similarly, one might revisit pneumatology in light of evolutionary theory. As Veli-Matti Kärkkäinen argues, talk about the Spirit as "the presence of God in the cosmos . . . makes an important connection with contemporary natural sciences."¹⁰ Viewing the Spirit as God's life-giving presence within the world (cf. Ps. 104:30) easily links up with the scientific picture of a dynamic, ever-evolving world. For example, could *emergence*—the self-transcendence of biological phenomena into ever more complex levels of existence—be seen as a hidden working of the Spirit in nature?¹¹ Could we perhaps even reconceptualize classical notions of sanctification and holiness in evolutionary categories?¹² In line with this, we might attempt to spell out ecclesiology from an evolutionary perspective, envisioning the church as the new community that, enlivened by the Spirit of God, starts to mirror the fully altruistic messianic way of life inaugurated by Jesus.¹³

Indeed, Christian theology may have to move beyond "postevoluntary apologetics"¹⁴ toward a more constructive engagement with evolution, aimed

---

9. Berkhof, *Christian Faith*, 291; it is no coincidence that Berkhof rejected the doctrine of the virginal conception of Jesus (which places Jesus outside evolutionary history). For more recent reflection on the relationship between Christology and evolutionary theory, see Celia Deane-Drummond, *Christ and Evolution: Wonder and Wisdom* (London: SCM, 2009).

10. Veli-Matti Kärkkäinen, *Creation and Humanity* (Grand Rapids: Eerdmans, 2015), 65. Cf. for a broad elaboration of this point, Amos Yong, *The Spirit of Creation: Modern Science and Divine Action* (Grand Rapids: Eerdmans, 2011).

11. That would be in line with John Calvin's comment on Gen. 1:2 (*Joannis Calvini opera quae supersunt omnia* 23:16) about the "secret inspiration" of the natural world by the Spirit of God. Cf. Yong, *Spirit of Creation*, 133-72.

12. Cf. Matthew Nelson Hill, *Evolution and Holiness: Sociobiology, Altruism, and the Quest for Wesleyan Perfection* (Downers Grove, IL: InterVarsity Press, 2016).

13. See, e.g. (though not written from an evolutionary perspective), Jürgen Moltmann, *The Church in the Power of the Spirit: A Contribution to Messianic Ecclesiology* (New York: Harper & Row, 1977). Cf. from a Roman Catholic point of view, Jack Mahoney, *Christianity in Evolution* (Washington, DC: Georgetown University Press, 2011), 111-41, on the church as an "evolutionary community" (137).

14. Mahoney, *Christianity in Evolution*, x, uses this term for contemporary attempts at defending the providence of God, the uniqueness of humanity, etc., in the

at enriching its resources and strengthening its plausibility in the face of critics who dismiss the church's message because of its hopeless entanglement in an outmoded worldview. One could even rewrite the entire biblical faith from an evolutionary perspective, as the German theologian and New Testament scholar Gerd Theissen has attempted.[15] Nevertheless, there are two important caveats here. First, proposals for a more wide-ranging incorporation of evolutionary theory in Christian theology could easily lose continuity with historic orthodox Christianity. For example, Berkhof's preoccupation with Jesus Christ as "the new man" put him in Arian waters, preventing him from acknowledging Christ's divinity next to his humanity. And reconceptualizing pneumatology from an evolutionary perspective easily "pushes theology toward pantheism," thus blurring the distinction between Creator and creation that has been so vital to the Reformed theological stance.[16] Second, we should be cautious not to "baptize" evolutionary theory. Christian theology should take evolution seriously because of its current plausibility, but at the same time it is just one more picture of the world, and we know from history that world pictures change over time. Moreover, thoroughly incorporating the evolutionary process in one's theological reflection might easily identify it too closely with God's purposes.[17] We then lose sight of its unsettling dimensions, such as the enormous amount of suffering it generates. In this connection, Ernst Conradie insightfully speaks of Darwin's "ambiguous gift" to Reformed theology.[18] Whereas in some ways evolutionary theory comes as a blessing, it may also turn into a risk. Conradie not only mentions the problem of natural evil in this connection. He also holds that "the danger of Darwin's gift is that it may direct theological interests toward interesting intellectual inquiries

---

light of evolution (the present book would probably be a perfect example of what he has in mind).

15. Gerd Theissen, *Biblical Faith: An Evolutionary Approach* (Minneapolis: Fortress, 1984). For an attempt to do so from a more broadly scientific perspective, see Klaus Nürnberger, *Informed by Science—Involved by Christ: How Science Can Update, Enrich, and Empower the Christian Faith* (Pietermaritzburg, South Africa: Cluster Publications, 2013).

16. Kärkkäinen, *Creation and Humanity*, 63 (with critical reference to Moltmann).

17. Fergusson, *Creation*, 85.

18. Ernst Conradie, "Darwin's Ambiguous Gift to Reformed Theology," in *Restoration through Redemption: John Calvin Revisited*, ed. Henk van den Belt (Leiden: Brill, 2013), 95-112.

... but away from the primary problem, namely the destructive impact of human sin."[19]

Conradie voices a deep-seated Reformed concern here, namely, the recognition that the world's predicament is not first of all caused by the way it is structured but by human sin. Sin is never caused by natural conditions—though such conditions may precipitate it—but goes back to a flawed human will. As we have seen, Reformed theology (following Augustine here) has always insisted on the seriousness of sin and highlighted the primary importance of the knowledge of our own misery. It can continue to do so even when we may now see our capacity to understand and acknowledge our failures (instead of positing the sources of evil only outside ourselves) as one of our highest evolved moral capacities. Fortunately, this thoroughgoing awareness of our moral fallibility is correlated with an equally thoroughgoing faith in God as the Wholly Other in his holiness and grace.[20] Thus, the basic thrust of Reformed theology is about human frailty over against divine glory and exaltedness, and about human sin over against divine grace, forgiveness, and transformation. This heartbeat of Reformed theology—its attempt to echo the heartbeat of the gospel—is not at all made obsolete by evolutionary theory. To the contrary, the unspeakable glory and majesty of God are underlined by the evolutionary history of the world.[21]

Thus, while it is important to explore further possibilities of connecting evolutionary theory with Christian doctrinal notions, we would be ill advised to revise our entire theology in its light. Although—following the lead of Pierre Teilhard de Chardin (1881-1955)—many have developed proposals in this direction, I see no compelling reason for such an encompassing overhaul.

In this book, in a sense we set out to investigate to what extent classical (Reformed) theology can remain unaltered when evolutionary theory is accepted as a major and enduring scientific development. It turned out that at three places adjustments are needed: we can no longer uphold a concordist hermeneutics (chapter 3), the theory of the cosmic fall (chapter 4), and the idea that human history started with a single couple (chapter 6).[22] Since these

---

19. Conradie, "Darwin's Ambiguous Gift," 111. Particularly, but not exclusively, Conradie has in mind the sins of ecological destruction.

20. See, for this correlation, John Calvin, *Institutes of the Christian Religion*, ed. John T. McNeill, trans. Ford L. Battles (Philadelphia: Westminster, 1960), 1.1.1.

21. Cf., e.g., Christopher Southgate, "Divine Glory in a Darwinian World," *Zygon* 49 (2014): 784-807.

22. In response to the Dutch version of this book, some reviewers pointed out

issues hinge on matters of biblical *interpretation*, in none of these cases, it seems to me, is biblical authority or any other Reformed doctrinal tenet necessarily at stake. We can therefore conclude that each of the specific commitments and concerns that characterize the Reformed theological stance can be articulated against the background of an evolutionary framework.

Therefore, Christian believers do not have to resist evolutionary theory because of their faith commitments; and non-Christians don't have to think that in order to become a Christian they should do the impossible, that is, renounce something that is so evidently true to them as Darwinian evolution.

---

that they found its argument "unconvincing"; unfortunately, however, they did not expound which other doctrinal notions are in need of revision as a result of evolution and for what reasons.

# Acknowledgments

Previous drafts of most chapters in this book were published in various venues during the past couple of years. All of them have been thoroughly reworked, in light of later feedback and (hopefully) growing insight as well as in an attempt to make them fit the overall purpose of this monograph. An earlier version of chapter 1 was coauthored with Johan Smits, MA, and published as "The Reformed Stance: Distinctive Commitments and Concerns," *Journal of Reformed Theology* 9 (2015): 325-47. Parts of "As a Beautiful Book: The Natural World in Article 2 of the Belgic Confession," *Westminster Theological Journal* 73 (2011): 273-91, were also incorporated in this chapter. Materials from the Dutch paper "Als orgels en stofzuigers?," *Theologia Reformata* 53 (2010): 351-68 found their way into chapter 3. Chapter 4 was preceded by "God and the Suffering of Animals," in *Playing with Leviathan*, ed. Koert van Bekkum et al. (Leiden: Brill, 2017), 179-200. An earlier draft of chapter 5 appeared as "Are We Still Special? Evolution and Human Dignity," *Neue Zeitschrift für Systematische Theologie und Religionsphilosophie* 53 (2011): 318-32. Chapters 7 and 8, finally, were preceded by contributions to Dutch *Festschriften*: "God gaat toch veel geheimnisvoller met ons om? Voorzienigheid en toeval in verband met de evolutie," in *Om het Godsgeheim. De theologie van Martien E. Brinkman*, ed. C. van der Kooi (Amsterdam: VU University Press, 2015), 25-38, and "Openbaring overbodig? Een verkenning van enkele hedendaagse CSR-verklaringen van het ontstaan van godsdienstigheid," in *Weergaloze kennis. Opstellen over Jezus Christus, Openbaring en Schrift, Katholiciteit en Kerk aangeboden aan prof. dr. B. Kamphuis*, ed. Ad de Bruijne et al. (Zoetermeer, the Netherlands: Boekencentrum, 2015), 115-25. Bible quotations in this book have usually been taken from the New Revised Standard Version (NRSV).

An earlier version of this book was published in Dutch as *En de aarde bracht voort. Christelijk geloof en evolutie* (Utrecht: Boekencentrum, 2017).

# Bibliography

Adams, Marilyn McCord. *Christ and Horrors: The Coherence of Christology.* Cambridge: Cambridge University Press, 2006.
Alexander, Denis. *Creation or Evolution: Do We Have to Choose?* 2nd rev. ed. Oxford: Monarch Books, 2014.
Allen, R. Michael. *Reformed Theology.* London: T&T Clark, 2010.
Alper, Matthew. *The "God" Part of the Brain: A Scientific Interpretation of Human Spirituality and God.* Naperville, IL: Sourcebooks, 2006.
Anderson, David. "Creation, Redemption and Eschatology." In *Should Christians Embrace Evolution? Biblical and Scientific Responses*, edited by Norman C. Nevin, 73–92. Nottingham: Inter-Varsity Press, 2009.
Arbesman, Samuel. *The Half-Life of Facts: Why Everything We Know Has an Expiration Date.* New York: Current, 2012.
Asselt, Willem J. van. "Calvinism as a Problematic Concept in Historiography." *International Journal of Philosophy and Theology* 74 (2013): 144–50.
———. *The Federal Theology of Johannes Cocceius.* Leiden: Brill, 2001.
———. *Introduction to Reformed Scholasticism.* Grand Rapids: Reformation Heritage Books, 2011.
Asselt, Willem J. van, and Eef Dekker, eds. *Reformation and Scholasticism: An Ecumenical Enterprise.* Grand Rapids: Baker Academic, 2001.
Astley, Jeff. "Evolution and Evil: The Difference Darwinism Makes in Theology and Spirituality." In *Reading Genesis after Darwin*, edited by Stephen C. Barton and David Wilkinson, 163–80. Oxford: Oxford University Press, 2009.
Atran, Scott. *In Gods We Trust: The Evolutionary Landscape of Religion.* Oxford: Oxford University Press, 2002.
Atran, Scott, and Joseph Henrich. "The Evolution of Religion: How Cognitive By-Products, Adaptive Learning Heuristics, Ritual Displays, and Group Competition Generate Deep Commitments to Prosocial Religions." *Biological Theory* 5, no. 1 (2010): 18–30.

Axelrod, Robert. *The Evolution of Cooperation*. London: Penguin Books, 1984.
Ayala, Francisco J. *Darwin's Gift to Science and Religion*. Washington, DC: Joseph Henry, 2007.
———. "The Evolution of Life: An Overview." In *God and Evolution: A Reader*, edited by Mary Kathleen Cunningham, 63-67. London: Routledge, 2007.
Baker, Catherine. *The Evolution Dialogues: Science, Christianity, and the Quest for Understanding*. Washington, DC: American Association for the Advancement of Science, 2006.
Baker, J. Wayne. *Heinrich Bullinger and the Covenant: The Other Reformed Tradition*. Athens: Ohio University Press, 1980.
Balserak, Jon. *Divinity Compromised: A Study of Divine Accommodation in the Thought of John Calvin*. Dordrecht: Springer, 2006.
Barbour, Ian G. *Religion and Science: Historical and Contemporary Issues*. New York: HarperCollins, 1997.
Barr, James. *Biblical Faith and Natural Theology*. Oxford: Clarendon, 1993.
———. *The Garden of Eden and the Hope of Immortality*. London: SCM, 1992.
Barr, Stephen M. "The Concept of Randomness in Science and Divine Providence." In *Divine Action and Natural Selection: Science, Faith, and Evolution*, edited by Joseph Seckbach and Richard Gordon, 464-78. Hackensack, NJ: World Scientific, 2009.
Barrett, Justin L. *Born Believers: The Science of Children's Religious Belief*. New York: Free Press, 2012.
———. "Cognitive Science of Religion: Looking Back, Looking Forward." *Journal for the Scientific Study of Religion* 50 (2011): 229-39.
———. *Cognitive Science of Religion and Theology: From Human Minds to Divine Minds*. West Conshohocken, PA: Templeton, 2011.
———. "Cognitive Science, Religion, and Theology." In *The Believing Primate: Scientific, Philosophical, and Theological Reflections on the Origin of Religion*, edited by Jeffrey Schloss and Michael J. Murray, 76-99. Oxford: Oxford University Press, 2009.
———. "Exploring the Natural Foundations of Religion." *Trends in Cognitive Sciences* 4, no. 1 (2000): 29-34.
———. *Why Would Anyone Believe in God?* Walnut Creek, CA: AltaMira, 2004.
Barth, Karl. *Church Dogmatics*. II/1. Edinburgh: T&T Clark, 1957.
———. *Church Dogmatics*. III/1. Edinburgh: T&T Clark, 1958.
———. *Church Dogmatics*. III/2. Edinburgh: T&T Clark, 1960.
———. *Letters, 1961-1968*. Edited by Jürgen Fangmeier and Hinrich Stoevesandt. Translated by Geoffrey Bromiley. Grand Rapids: Eerdmans, 1981.

Bartholomew, David J. "God and Probability." In *Creation: Law and Probability*, edited by Fraser Watts, 139–53. Minneapolis: Fortress, 2008.

Bauckham, Richard. "Jesus and Animals I: What Did He Teach?" In *Animals on the Agenda: Questions about Animals for Theology and Ethics*, edited by Andrew Linzey and Dorothy Yamamoto, 33–48. Champaign: University of Illinois Press, 1998.

———. *Living with Other Creatures: Green Exegesis and Theology*. Waco, TX: Baylor University Press, 2011.

Bauswein, Jean-Jacques, and Lukas Visscher. *The Reformed Family Worldwide: A Survey of Reformed Churches, Theological Schools, and International Organizations*. Grand Rapids: Eerdmans, 1999.

Bavinck, Herman. "Evolution" (1907). In *Essays on Religion, Science, and Society*, edited by John Bolt, 105–18. Grand Rapids: Baker Academic, 2008.

———. *Reformed Dogmatics*. Vol. 1, *Prolegomena*. Grand Rapids: Baker Academic, 2003.

———. *Reformed Dogmatics*. Vol. 2, *God and Creation*. Grand Rapids: Baker Academic, 2004.

Bechtel, Lyn M. "Genesis 2.4b–3.24: A Myth about Human Maturation." *Journal for the Study of the Old Testament* 67 (1995): 3–26.

Beck, Naomi. "Social Darwinism." In *The Cambridge Encyclopedia to Darwin and Evolutionary Thought*, edited by Michael Ruse, 195–201. Cambridge: Cambridge University Press, 2013.

Beek, Abraham van de. "Evolution, Original Sin, and Death." *Journal of Reformed Theology* 5 (2011): 206–20.

Behe, Michael. *Darwin's Black Box: The Biochemical Challenge to Evolution*. New York: Free Press, 1996.

Bekkum, Koert van, Jaap Dekker, Henk R. van de Kamp, and Eric Peels, eds. *Playing with Leviathan: Interpretation and Reception of Monsters from the Biblical World*. Leiden: Brill, 2017.

Bekoff, Marc. *The Emotional Life of Animals: A Leading Scientist Explores Animal Joy, Sorrow, and Empathy—and Why They Matter*. Novato, CA: New World Library, 2007.

———. "The Evolution of Animal Play, Emotions and Social Morality: On Science, Theology, Spirituality, Personhood, and Love." *Zygon* 26 (2001): 615–55.

———. "Reflections on Animal Emotions and Beastly Virtues: Appreciating, Honoring and Respecting the Public Passions of Animals." *Journal for the Study of Religion, Nature and Culture* 1, no. 1 (2007): 68–80.

———. *Why Dogs Hump and Bees Get Depressed: The Fascinating Science of Animal*

*Intelligence, Emotions, Friendship, and Conservation.* Novato, CA: New World Library, 2013.

Benedict, Philip. *Christ's Churches Purely Reformed: A Social History of Calvinism.* New Haven: Yale University Press, 2002.

Bering, Jesse. "The Cognitive Psychology of Belief in the Supernatural." *American Scientist* 94 (2006): 142–49.

———. "The Existential Theory of Mind." *Review of General Psychology* 6 (2002): 3–24.

———. *The God Instinct: The Psychology of Souls, Destiny, and the Meaning of Life.* London: Nicholas Brealey, 2011.

Berkhof, Hendrikus. *Christian Faith: An Introduction to the Study of the Faith.* Rev. ed. Grand Rapids: Eerdmans, 1986.

Berkhof, Louis. *Systematic Theology: New Combined Edition.* Grand Rapids: Eerdmans, 1996.

Berkouwer, Gerrit C. *Holy Scripture.* Translated by Jack B. Rogers. Grand Rapids: Eerdmans, 1975.

———. *Man: The Image of God.* Grand Rapids: Eerdmans, 1962.

Berry, Robert James. "Did Darwin Dethrone Humankind?" In *Darwin, Creation, and the Fall: Theological Challenges,* edited by Robert James Berry and Thomas A. Noble, 30–74. Nottingham: Apollos, 2009.

———. "This Cursed Earth: Is 'the Fall' Credible?" *Science and Christian Belief* 11 (1999): 29–49.

Bimson, John J. "Doctrines of the Fall and Sin after Darwin." In *Theology after Darwin,* edited by Michael S. Northcott and R. J. Berry, 106–22. Milton Keynes, UK: Paternoster, 2009.

———. "Reconsidering a 'Cosmic Fall.'" *Science and Christian Belief* 18 (2006): 63–81.

Birch, Charles, and Lukas Visscher. *Living with Animals: The Community of God's Creatures.* Geneva: WCC Publications, 1997.

Blancke, Stefaan. "Catholic Responses to Evolution, 1859–2009: Local Influences and Mid-Scale Patterns." *Journal of Religious History* 37 (2013): 353–68.

Blocher, Henri. *In the Beginning: The Opening Chapters of Genesis.* Leicester: IVP, 1984.

———. "The Theology of the Fall and the Origins of Evil." In *Darwin, Creation, and the Fall: Theological Challenges,* edited by Robert James Berry and Thomas A. Noble, 149–72. Nottingham: Apollos, 2009.

Bloom, Paul. "Religious Belief as an Evolutionary Accident." In *The Believing Primate: Scientific, Philosophical, and Theological Reflections on the Origin of Reli-*

gion, edited by Jeffrey Schloss and Michael J. Murray, 118–27. Oxford: Oxford University Press, 2009.

Bonhoeffer, Dietrich. *Letters and Papers from Prison*. Edited by Eberhard Bethge. New York: Simon & Schuster, 1997.

Boston, Thomas. *Human Nature in Its Fourfold State*. Edinburgh: Banner of Truth Trust, 1964.

Bowler, Peter J. *The Eclipse of Darwinism: Anti-Darwinian Evolution Theories in the Decades around 1900*. Baltimore: Johns Hopkins University Press, 1983.

Boyd, Gregory. *Satan and the Problem of Evil*. Downers Grove, IL: InterVarsity Press, 2001.

Boyer, Pascal. "Explaining Religious Ideas: Outline of a Cognitive Approach." *Numen* 39 (1992): 27–57.

———. *Religion Explained: The Evolutionary Origins of Religious Thought*. New York: Basic Books, 2001.

Bratt, James D. *Abraham Kuyper: A Centennial Reader*. Grand Rapids: Eerdmans, 1998.

———. *Abraham Kuyper: Modern Calvinist, Christian Democrat*. Grand Rapids: Eerdmans, 2013.

Breuer, George. *Sociobiology and the Human Dimension*. Cambridge: Cambridge University Press, 1982.

Brink, Gijsbert van den. *Almighty God: A Study of the Doctrine of Divine Omnipotence*. Kampen: Kok Pharos, 1993.

———. "As a Beautiful Book: The Natural World in Article 2 of the Belgic Confession." *Westminster Theological Journal* 73 (2011): 273–91.

———. *Dordt in context. Gereformeerde accenten in katholieke theologie* [The Canons of Dordt in context: Reformed accents in Catholic theology]. Heerenveen: Groen, 2018.

———. "Evolution as a Bone of Contention between Church and Academy: How Abraham Kuyper Can Help Us Bridge the Gap." *Kuyper Center Review* 5 (2015): 92–103.

———. *Philosophy of Science for Theologians: An Introduction*. Frankfurt am Main: Lang, 2009.

———. "Questions, Challenges, and Concerns for Original Sin." In *Finding Ourselves after Darwin: Conversations on the Image of God, Original Sin, and the Problem of Evil*, edited by Stanley P. Rosenberg, 117–29. Grand Rapids: Baker Academic, 2018.

———. "The Reformation, Rationality and the Rise of Modern Science." In *Reformation und Rationalität*, edited by Herman J. Selderhuis and Ernst-Joachim Waschke, 193–205. Göttingen: Vandenhoeck & Ruprecht, 2015.

Brink, Gijsbert van den, and Harro M. Höpfl, eds. *Calvinism and the Making of the European Mind.* Leiden: Brill, 2014.

Brink, Gijsbert van den, and Cornelis van der Kooi. *Christian Dogmatics: An Introduction.* Grand Rapids: Eerdmans, 2017.

Brink, Gijsbert van den, Jeroen de Ridder, and René van Woudenberg. "The Epistemic Status of Evolutionary Theory." *Theology and Science* 15 (2017): 454-72.

Brooke, J. H. *Science and Religion: Some Historical Perspectives.* Cambridge: Cambridge University Press, 1991.

Brosnan, S. F., C. Freeman, and F. B. M. de Waal. "Equitable Behavior, Not Reward Distributions, Affect Capuchin Monkey's (*Cebus apella*) Reactions in a Cooperative Task." *American Journal of Primatology* 68 (2006): 713-24.

Bruce, F. F. *The Epistle of Paul to the Romans.* Tyndale New Testament Commentaries. London: Tyndale Press, 1963.

Brümmer, Vincent. "Introduction: A Dialogue of Language Games." In *Interpreting the Universe as Creation: A Dialogue of Science and Religion*, edited by Vincent Brümmer, 1-17. Kampen: Kok Pharos, 1991.

———. *The Model of Love: A Study in Philosophical Theology.* Cambridge: Cambridge University Press, 1993.

Brunner, Emil. *The Christian Doctrine of Creation and Redemption.* Philadelphia: Westminster, 1952.

Buckland, William. *An Enquiry Whether the Sentence of Death Pronounced at the Fall of Man Included the Whole Animal Creation or Was Restricted to the Human Race.* London: J. Murray, 1839.

Bullinger, Heinrich. *De testamento seu foedere Dei unico et aeterno . . . brevis expositio.* Zürich: Froschauer, 1534.

Bultmann, Rudolf. *The New Testament and Mythology and Other Basic Writings.* Minneapolis: Augsburg Fortress, 1984.

Burdett, Michael. "The Image of God and Human Uniqueness: Challenges from the Biological and Information Sciences." *Expository Times* 127 (2015): 3-10.

Busch, Eberhard. *Reformiert: Profil einer Konfession.* Zürich: TVZ, 2009.

Calvin, John. *Catechism of Geneva.* 1545.

———. *Commentaries on the First Book of Moses called Genesis.* Translated by John King. Grand Rapids: Eerdmans, 1948.

———. *Institutes of the Christian Religion.* Edited by John T. McNeill. Translated by Ford L. Battles. Philadelphia: Westminster, 1960.

Cameron, Nigel M. de S. *Evolution and the Authority of the Bible.* Exeter, UK: Attic, 1983.

Caruana, Louis. *Darwin and Catholicism: The Past and Present Dynamics of a Cultural Encounter.* London: Bloomsbury T&T Clark, 2009.

Cavalli-Sforza, L. L., and M. W. Feldman. *Cultural Transmission and Evolution: A Quantitative Approach*. Princeton: Princeton University Press, 1981.

Cavanaugh, William T., and James K. A. Smith, eds. *Evolution and the Fall*. Grand Rapids: Eerdmans, 2017.

Chappel, Jonathan. "Raymund Schwager: Integrating the Fall and Original Sin with Evolutionary Theory." *Theology and Science* 10 (2012): 179–98.

Charles, R. H., ed. *The Apocrypha and Pseudepigrapha of the Old Testament*. Oxford: Clarendon, 1913.

Chesterton, G. K. *Orthodoxy*. London: John Lane, 1909; London: Fontana, 1961.

Clark, Kelly James. *Religion and the Sciences of Origin: Historical and Contemporary Discussions*. New York: Palgrave Macmillan, 2014.

Clark, Kelly James, and Justin L. Barrett. "Reformed Epistemology and the Cognitive Science of Religion." *Faith and Philosophy* 27 (2010): 174–89.

———. "Reidian Religious Epistemology and the Cognitive Science of Religion." *Journal of the American Academy of Religion* 79 (2011): 639–75.

Clayton, Philip, and Arthur Peacocke, eds. *In Whom We Live and Move and Have Our Being: Panentheistic Reflections on God's Presence in a Scientific World*. Grand Rapids: Eerdmans, 2004.

Clough, David. "All God's Creatures: Reading Genesis on Human and Non-Human Animals." In *Reading Genesis after Darwin*, edited by Stephen C. Barton and David Wilkinson, 145–61. Oxford: Oxford University Press, 2009.

———. *On Animals*. Vol. 1, *Systematic Theology*. London: T&T Clark, 2012.

Cobb, John B., ed. *Back to Darwin: A Richer Account of Evolution*. Grand Rapids: Eerdmans, 2008.

Cohn, Norman. *Noah's Flood: The Genesis Story in Western Thought*. New Haven: Yale University Press, 1996.

Collins, C. John. "Adam and Eve in the Old Testament." In *Adam, the Fall, and Original Sin: Theological, Biblical, and Scientific Perspectives*, edited by Hans Madueme and Michael Reeves, 3–32. Grand Rapids: Baker Academic, 2014.

———. *Did Adam and Eve Really Exist? Who They Were and Why You Should Care*. Wheaton, IL: Crossway, 2011.

———. *Genesis 1–4: A Linguistic, Literary, and Theological Commentary*. Phillipsburg, NJ: P&R, 2006.

———. "A Historical Adam: Old-Earth Creation View." In *Four Views on the Historical Adam*, edited by Matthew Barrett and Ardel B. Caneday, 143–75. Grand Rapids: Zondervan, 2013.

———. *Reading Genesis Well: Navigating History, Poetry, Science, and Truth in Genesis 1–11*. Grand Rapids: Zondervan, 2018.

Collins, Francis S. *The Language of God: A Scientist Presents Evidence for Belief.* New York: Free Press, 2006.
Collins, Robin. "Evolution and Original Sin." In *Perspectives on an Evolving Creation*, edited by Keith B. Miller, 469–501. Grand Rapids: Eerdmans, 2003.
Conradie, Ernst M. "Darwin's Ambiguous Gift to Reformed Theology." In *Restoration through Redemption: John Calvin Revisited*, edited by Henk van den Belt, 95–112. Leiden: Brill, 2013.
———. *An Ecological Christian Anthropology: At Home on Earth?* Aldershot, UK: Ashgate, 2005.
———. *Redeeming Sin? Social Diagnostics and Ecological Destruction.* Lanham, MD: Lexington Books, 2017.
Conway Morris, Simon. "Creation and Evolutionary Convergence." In *The Blackwell Companion to Science and Christianity*, edited by J. B. Stump and Alan G. Padgett, 258–69. Oxford: Blackwell, 2012.
———. "Evolution and the Inevitability of Intelligent Life." In *The Cambridge Companion to Science and Religion*, edited by Peter Harrison, 148–72. Cambridge: Cambridge University Press, 2010.
———. *Life's Solution: Inevitable Humans in a Lonely Universe.* Cambridge: Cambridge University Press, 2003.
Cook, Harry. "Burgess Shale and the History of Biology." *Perspectives on Science and the Christian Faith* 47 (1995): 159–63.
———. "Emergence: A Biologist's Look at Complexity in Nature." *Perspectives on Science and Christian Faith* 65 (2013): 233–41.
Cook, L. M., B. S. Grant, I. J. Saccheri, and J. Mallet. "Selective Bird Predation on the Peppered Moth: The Last Experiment of Michael Majerus." *Biology Letters* 8, no. 4 (November 2, 2011). http://dx.doi.org/10.1098/rsbl.2011.1136.
Cortez, Marc. *Theological Anthropology: A Guide for the Perplexed.* London: T&T Clark, 2010.
Creegan, Nicola Hoggard. *Animal Suffering and the Problem of Evil.* Oxford: Oxford University Press, 2013.
Crisp, Oliver D. *Deviant Calvinism: Broadening Reformed Theology.* Minneapolis: Fortress, 2014.
Cunningham, Conor. *Darwin's Pious Idea: Why the Ultra-Darwinists and Creationists Both Get It Wrong.* Grand Rapids: Eerdmans, 2010.
Cunningham, David S. "The Way of All Flesh: Re-thinking the Imago Dei." In *Creaturely Theology: On God, Humans, and Other Animals*, edited by Celia Deane-Drummond and David Clough, 100–117. London: SCM, 2009.
Darwin, Charles. *On the Origin of Species by Means of Natural Selection.* London: John Murray, 1859.

Darwin, Francis, ed. *The Life and Letters of Charles Darwin.* 2 vols. New York: Basic Books, 1959.

Dawkins, Richard. *The Blind Watchmaker: Why the Evidence of Evolution Reveals a Universe without Design.* New York: Norton, 1985.

———. "Darwin Triumphant: Darwinism as Universal Truth." In *A Devil's Chaplain: Reflections on Hope, Lies, Science, and Love,* by Richard Dawkins, 78–90. London: Weidenfeld & Nicolson, 2005.

———. *River out of Eden: A Darwinian View of Life.* New York: HarperCollins, 1995.

———. *The Selfish Gene.* Oxford: Oxford University Press, 1976.

Day, Allan John. "Adam, Anthropology and the Genesis Record: Taking Genesis Seriously in the Light of Contemporary Science." *Science and Christian Belief* 10 (1998): 115–43.

Deane-Drummond, Celia. *Christ and Evolution: Wonder and Wisdom.* London: SCM, 2009.

———. "In Adam All Die? Questions at the Boundary of Niche Construction, Community Evolution, and Original Sin." In *Evolution and the Fall,* edited by William T. Cavanaugh and James K. A. Smith, 23–47. Grand Rapids: Eerdmans, 2017.

———. "Response: *Homo Divinus*—Myth or Reality?" In *Darwinism and Natural Theology: Evolving Perspectives,* edited by Andrew Robinson, 39–46. Newcastle upon Tyne: Cambridge Scholars, 2013.

———. *The Wisdom of the Liminal: Evolution and Other Animals in Human Becoming.* Grand Rapids: Eerdmans, 2014.

De Cruz, Helen, and Johan De Smedt. *A Natural History of Natural Theology: The Cognitive Science of Theology and Philosophy of Religion.* Cambridge, MA: MIT Press, 2015.

Deemter, Kees van. *Not Exactly: In Praise of Vagueness.* Oxford: Oxford University Press, 2010.

Dembski, William B. *The End of Christianity: Finding a Good God in an Evil World.* Nashville: B&H Academic, 2009.

———. *Intelligent Design: The Bridge between Science and Theology.* Downers Grove, IL: InterVarsity Press, 1999.

Dennett, Daniel C. *Darwin's Dangerous Idea: Evolution and the Meanings of Life.* New York: Simon & Schuster, 1995.

Denton, Michael. *Evolution: A Theory in Crisis.* Chevy Chase, MD: Adler & Adler, 1986.

Descartes, René. *Meditations on First Philosophy.* 1641.

Dijksterhuis, Eduard Jan. *The Mechanization of the World Picture: Pythagoras to Newton.* Princeton: Princeton University Press, 1986 [1961].

Dixon, Thomas, Geoffrey Cantor, and Stephen Pumfrey, eds. *Science and Religion: New Historical Perspectives*. Cambridge: Cambridge University Press, 2010.

Dobzhansky, Theodosius. "Nothing Makes Sense in Biology Except in the Light of Evolution." *American Biology Teacher* 35 (1973): 125–29.

Domning, Daryl P., and Monika K. Hellwig. *Original Selfishness: Original Sin and Evil in the Light of Evolution*. Aldershot, UK: Ashgate, 2006.

Dougherty, Trent. *The Problem of Animal Pain: A Theodicy for All Creatures Great and Small*. New York: Palgrave Macmillan, 2014.

Dougherty, Trent, and Justin P. McBrayer, eds. *Skeptical Theism: New Essays*. Oxford: Oxford University Press, 2014.

Dunbar, Robin. *Human Evolution: A Pelican Introduction*. London: Penguin Books, 2014.

Dunn, James D. G. *Romans 1–8*. Dallas: Word, 1988.

Dupree, A. Hunter. *Asa Gray: American Botanist, Friend of Darwin*. Baltimore: Johns Hopkins University Press, 1988.

Edwards, Rem B. *What Caused the Big Bang?* Amsterdam: Rodopi, 2001.

Eldredge, Niles, and Stephen J. Gould. "Punctuated Equilibria: An Alternative to Physical Gradualism." In *Models of Paleobiology*, edited by T. J. M. Schopf, 83–115. San Francisco: Freeman, Cooper, 1972.

Elk, Michiel van. *De gelovige geest: Op zoek naar de biologische en psychologische wortels van religie* [The believing mind: In search of the biological and psychological roots of religion]. Amsterdam: Bert Bakker, 2012.

Elliott, Mark W. *Providence Perceived: Divine Action from a Human Point of View*. Berlin: de Gruyter, 2015.

Engel, Mary Potter. *John Calvin's Perspectival Anthropology*. Atlanta: Scholars Press, 1988.

Enns, Peter. *The Evolution of Adam: What the Bible Does and Doesn't Say about Human Origins*. Grand Rapids: Brazos, 2012.

Etzelmüller, Georg. "The Evolution of Sin." *Religion and Theology* 21 (2014): 107–24.

Falk, Darrel R. *Coming to Peace with Science: Bridging the Worlds between Faith and Biology*. Downers Grove, IL: InterVarsity Press, 2004.

———. "Human Origins: The Scientific Story." In *Evolution and the Fall*, edited by William T. Cavanaugh and James K. A. Smith, 3–22. Grand Rapids: Eerdmans, 2017.

Farley, Benjamin W. *The Providence of God*. Grand Rapids: Baker Books, 1988.

Fergusson, David. "Creation." In *The Oxford Handbook of Systematic Theology*, edited by John Webster, Kathryn Tanner, and Iain Torrance, 72–90. Oxford: Oxford University Press, 2007.

———. *Creation*. Grand Rapids: Eerdmans, 2014.

———. *The Providence of God: A Polyphonic Approach*. Cambridge: Cambridge University Press, 2018.
Fesko, John V. *The Theology of the Westminster Standards: Historical Context and Theological Insights*. Wheaton, IL: Crossway, 2014.
Finlay, Graeme. "*Homo Divinus*: The Ape That Bears God's Image." *Science and Christian Belief* 15, no. 1 (2003): 17–40.
———. *Human Evolution: Genes, Genealogies, and Phylogenies*. Cambridge: Cambridge University Press, 2013.
Flipse, Abraham. "The Origins of Creationism in the Netherlands: The Evolution Debate among Twentieth-Century Dutch Neo-Calvinists." *Church History* 81 (2012): 104–47.
Fodor, Jerry. *In Critical Condition*. Cambridge, MA: MIT Press, 1998.
Fodor, Jerry, and Massimo Piattelli-Palmarini. *What Darwin Got Wrong*. New York: Picador, 2010.
Fowler, Thomas B., and Daniel Kuebler. *The Evolution Controversy: A Survey of Competing Theories*. Grand Rapids: Baker Academic, 2007.
Fraassen, Bas C. van. *The Empirical Stance*. New Haven: Yale University Press, 2002.
Francescotti, Robert. "The Problem of Animal Pain and Suffering." In *The Blackwell Companion to the Problem of Evil*, edited by Justin P. McBrayer and Daniel Howard-Snyder, 117–21. Chichester: Wiley & Sons, 2014.
Fretheim, Terence E. *God and World in the Old Testament: A Relational Theology of Creation*. Nashville: Abingdon, 2005.
Freudenberg, Matthias. *Reformierte Theologie. Eine Einführung*. Neukirchen-Vluyn: Neukirchener Verlag, 2011.
Frost, David. *Billy Graham: Personal Thoughts of a Public Man*. Colorado Springs: Victor Books, 1997.
Gallup, Gordon G., Jr. "Chimpanzees: Self-Recognition." *Science* 167, no. 3914 (1970): 86–87.
Garte, Sy. "New Ideas in Evolutionary Biology: From NDMS to EES." *Perspectives on Science and Christian Faith* 68, no. 1 (2016): 3–11.
Genderen, J. van, and W. H. Velema. *Concise Reformed Dogmatics*. Translated by G. Bilkes and E. M. van der Maas. Phillipsburg, NJ: P&R, 2008.
Gerrish, Brian. "Tradition in the Modern World: The Reformed Habit of Mind." In *Toward the Future of Reformed Theology: Tasks, Topics, Traditions*, edited by David Willis and Michael Welker, 3–20. Grand Rapids: Eerdmans, 1999.
Gillespie, Charles C. *Genesis and Geology: A Study of the Relations of Scientific Thought, Natural Theology, and Social Opinion in Great Britain, 1790–1850*. Cambridge, MA: Harvard University Press, 1996; original 1951.

Gilmour, Michael J. *Eden's Other Residents: The Bible and Animals*. Eugene, OR: Cascade, 2014.
Goldingay, John. *Genesis for Everyone: Part I*. Louisville: Westminster John Knox, 2010.
Goodall, Jane. "Primate Spirituality." In *Encyclopedia of Religion and Nature*, edited by Bron Taylor, 1303–6. London: Continuum, 2005.
Gosse, Edmund. *Father and Son*. London: Heinemann, 1907.
Gosse, Philip H. *Omphalos: An Attempt to Untie the Geological Knot*. London: John Van Voorst, 1857.
Gould, Stephen Jay. *Adam's Navel*. London: Penguin Books, 1995.
———. *Full House: The Spread of Excellence from Plato to Darwin*. New York: Harmony Books, 1996.
———. *Rocks of Ages: Science and Religion in the Fullness of Life*. New York: Ballantine Books, 1996.
———. *Wonderful Life: The Burgess Shale and Natural History*. New York: Norton, 1989.
Gould, Stephen Jay, and Richard C. Lewontin. "The Spandrels of San Marco and the Panglossian Paradigm: A Critique of the Adaptationist Programme." *Proceedings of the Royal Society of London*, ser. B, 205 (1979): 581–98.
Gray, John. *Straw Dogs: Thoughts on Humans and Other Animals*. London: Granta, 2002.
Green, Adam, and Eleonore Stump, eds. *Hidden Divinity and Religious Belief: New Perspectives*. Cambridge: Cambridge University Press, 2015.
Greenblatt, Stephen. *The Rise and Fall of Adam and Eve*. New York: Norton, 2017.
Gregory, Neville G. *Physiology and Behaviour of Animal Suffering*. Oxford: Blackwell Science, 2004.
Grenz, Stanley J. *The Social God and the Relational Self: A Trinitarian Theology of the Imago Dei*. Louisville: Westminster John Knox, 2001.
Gundlach, Bradley J. *Process and Providence: The Evolution Question at Princeton, 1845–1929*. Grand Rapids: Eerdmans, 2013.
Gunton, Colin E. *A Brief Theology of Revelation*. London: T&T Clark, 1995.
Guthrie, Steven. "A Cognitive Theory of Religion." *Current Anthropology* 21 (1980): 181–203.
Haarsma, Deborah B., and Loren D. Haarsma. *Origins: A Reformed Look at Creation, Design, and Evolution*. Grand Rapids: Faith Alive, 2007.
———. *Origins: Christian Perspectives on Creation, Evolution, and Intelligent Design*. Grand Rapids: Faith Alive, 2011.
Haldane, J. B. S. *The Causes of Evolution*. London: Longmans, Green, 1964.

Hamilton, W. D. "The Genetical Evolution of Social Behaviour." *Journal of Theoretical Biology* 7 (1964): 1–52.
Harlow, Dan. "After Adam: Reading Genesis in an Age of Evolutionary Science." *Perspectives on Science and Christian Faith* 62 (2010): 179–95.
Harris, Eugene E. *Ancestors in Our Genome: The New Science of Human Evolution.* New York: Oxford University Press, 2015.
Harris, Mark. *The Nature of Creation: Examining the Bible and Science.* Durham, UK: Acumen, 2013.
Harrison, Peter. *The Bible, Protestantism, and the Rise of Natural Science.* Cambridge: Cambridge University Press, 1998.
———. *The Territories of Science and Religion.* Chicago: University of Chicago Press, 2015.
———. "Theodicy and Animal Pain." *Philosophy* 64 (1989): 72–92.
Haught, John F. "Darwin and Catholicism." In *The Cambridge Encyclopedia of Darwin and Evolutionary Thought*, edited by Michael Ruse, 485–92. Cambridge: Cambridge University Press, 2013.
———. *Deeper Than Darwin: The Prospect of Religion in the Age of Evolution.* Boulder, CO: Westview, 2003.
———. *God after Darwin: A Theology of Evolution.* Boulder, CO: Westview, 2000.
Heidegger, J. H. *Medulla Theologiae Christianae.* Zürich: D. Gessner, 1696.
Heller, Michael. *Creative Tension: Essays on Science and Religion.* West Conshohocken, PA: Templeton Foundation, 2003.
Helm, Paul. *The Providence of God.* Leicester: Inter-Varsity Press, 1993.
Heppe, Heinrich. *Reformed Dogmatics: Set Out and Illustrated from the Sources.* Revised and edited by Ernst Bizer. Grand Rapids: Baker Books, 1978.
Hesselink, I. John. *On Being Reformed: Distinctive Characteristics and Common Misunderstandings.* 2nd ed. New York: Reformed Church Press, 1988.
Hesselink, I. John, and J. Todd Billings, eds. *Calvin's Theology and Its Reception: Disputes, Developments, and New Possibilities.* Louisville: Westminster John Knox, 2012.
Hick, John. *Evil and the God of Love.* Basingstoke, UK: Macmillan, 1966.
Higton, Mike. *Christian Doctrine.* London: SCM, 2008.
Hill, Matthew Nelson. *Evolution and Holiness: Sociobiology, Altruism, and the Quest for Wesleyan Perfection.* Downers Grove, IL: InterVarsity Press, 2016.
Hodge, Charles. "The Bible and Science." *New York Observer*, March 26, 1863, 98–99.
———. *Systematic Theology.* New York: Charles Scribner and Co., 1872.
———. *What Is Darwinism?* New York: Scribner, Armstrong and Co., 1874.
Hoffmeier, James K. "Genesis 1–11 as History and Theology." In *Genesis: History,*

Fiction, or Neither? Three Views on the Bible's Earliest Chapters, edited by Charles Halton, 23-58. Grand Rapids: Zondervan, 2015.

Hooykaas, Reyer. *Christian Faith and the Freedom of Science.* London: Tyndale Press, 1957.

Höpfl, Harro M. "Predestination and Political Liberty." In *Calvinism and the Making of the European Mind,* edited by Gijsbert van den Brink and Harro M. Höpfl, 155-76. Leiden: Brill, 2014.

Howell, Kenneth J. *God's Two Books: Copernican Cosmology and Biblical Interpretation in Early Modern Science.* Notre Dame: University of Notre Dame Press, 2002.

Huijgen, Arnold. *Divine Accommodation in John Calvin's Theology: Analysis and Assessment.* Göttingen: Vandenhoeck & Ruprecht, 2011.

Hurd, James P. "Hominids in the Garden?" In *Perspectives on an Evolving Creation,* edited by Keith B. Miller, 208-33. Grand Rapids: Eerdmans, 2003.

Huxley, Thomas. "Criticisms on 'The Origin of Species'" (1864). In *Darwiniana: Essays,* by Thomas Huxley, 80-106. New York: Appleton and Co., 1896.

———. "On the Reception of the Origin of Species." In *The Life and Letters of Charles Darwin,* edited by Francis Darwin, 2:201. London: John Murray, 1887.

Huyssteen, J. Wentzel van. *Alone in the World? Human Uniqueness in Science and Theology.* Grand Rapids: Eerdmans, 2006.

———. "From Empathy to Embodied Faith? Interdisciplinary Perspectives on the Evolution of Religion." In *Evolution, Religion, and Cognitive Science: Critical and Constructive Essays,* edited by Fraser Watts and Léon Turner, 132-51. Oxford: Oxford University Press, 2012.

———. "When Were We Persons? Why Hominid Evolution Holds the Key to Embodied Personhood." *Neue Zeitschrift für Systematische Theologie und Religionsphilosophie* 52 (2010): 329-49.

Illingworth, J. R. "The Problem of Pain: Its Bearing on Faith in God." In *Lux Mundi: A Series of Studies in the Religion of the Incarnation,* edited by Charles Gore, 113-27. London: John Murray, 1890.

Inwagen, Peter van, ed. *Christian Faith and the Problem of Evil.* Grand Rapids: Eerdmans, 2004.

———. "The Compatibility of Darwinism and Design." In *God and Design: The Teleological Argument and Modern Science,* edited by Neil A. Manson, 347-62. London: Routledge, 2003.

———. "Genesis and Evolution." In *God, Knowledge, and Mystery: Essays in Philosophical Theology,* by Peter van Inwagen, 128-62. Ithaca, NY: Cornell University Press, 1995.

———. "A Kind of Darwinism." In *Science and Religion in Dialogue*, vol. 2, edited by Melville Y. Stewart, 813–24. Malden, MA: Wiley-Blackwell, 2010.

———. *The Problem of Evil*. Oxford: Oxford University Press, 2006.

Irons, Lee, and Meredith G. Kline. "The Framework Interpretation." In *The Genesis Debate: Three Views on the Days of Creation*, edited by David G. Hagopian, 217–304. Mission Viejo, CA: Crux, 2001.

Jacobs, Alan. *Original Sin: A Cultural History*. New York: HarperCollins, 2008.

Jaki, Stanley. *Genesis 1 through the Ages*. New York: Thomas More, 1992.

Jeeves, Malcolm. Afterword to *The Emergence of Personhood: A Quantum Leap?*, edited by Malcolm Jeeves, 220–41. Grand Rapids: Eerdmans, 2015.

———, ed. *The Emergence of Personhood: A Quantum Leap?* Grand Rapids: Eerdmans, 2015.

Jenson, Robert. "The Praying Animal." *Zygon* 18 (1983): 311–25.

———. *Systematic Theology*. Vol. 2. Oxford: Oxford University Press, 1999.

Jersild, Paul. *The Nature of Our Humanity: A Christian Response to Evolution and Biotechnology*. Minneapolis: Fortress, 2009.

Joad, C. E. M., and C. S. Lewis. "The Pains of Animals: A Problem in Theology." *Month* 189 (1950): 95–104.

John Paul II. "Message to the Pontifical Academy of Sciences on Evolution." *Origins* 26, no. 22 (1996): 349–52.

Johnson, Dominic P. *God Is Watching You: How the Fear of God Makes Us Human*. New York: Oxford University Press, 2015.

Johnson, Elizabeth A. *Ask the Beasts: Darwin and the God of Love*. London: Bloomsbury, 2014.

Jones, James W. *Can Science Explain Religion? The Cognitive Science Debate*. Oxford: Oxford University Press, 2016.

Jorink, Eric. *Reading the Book of Nature in the Dutch Golden Age, 1575–1715*. Leiden: Brill, 2010.

Kärkkäinen, Veli-Matti. *Creation and Humanity*. Grand Rapids: Eerdmans, 2015.

Keathley, Kenneth D., and Mark F. Rooker. *40 Questions about Creation and Evolution*. Grand Rapids: Kregel, 2014.

Kelemen, Deborah. "Are Children 'Intuitive Theists'? Reasoning about Purpose and Design in Nature." *Psychological Science* 15, no. 5 (2004): 295–301.

Kelsey, David. *Eccentric Existence: A Theological Anthropology*. Vol. 1. Louisville: Westminster John Knox, 2009.

Kevles, Daniel J. *In the Name of Eugenics: Genetics and the Use of Human Heredity*. Cambridge, MA: Harvard University Press, 1985.

Kidner, Derek. *Genesis*. Leicester: Inter-Varsity Press, 1967.

King, Barbara J. "Are Apes and Elephants Persons?" In *In Search of Self: Interdis-*

ciplinary Perspectives on Personhood, edited by J. Wentzel van Huyssteen and Erik P. Wiebe, 70–82. Grand Rapids: Eerdmans, 2011.

Kingsley, Charles. "The Natural Theology of the Future." In *Westminster Sermons*, v–xxx (= preface). London: Macmillan, 1881.

Kitcher, Philip. *Living with Darwin: Evolution, Design, and the Future of Faith*. Oxford: Oxford University Press, 2007.

Klapwijk, Jacob. *Purpose in the Living World? Creation and Emergent Evolution*. Cambridge: Cambridge University Press, 2008.

Klaver, J. M. I. *Geology and Religious Sentiment: The Effect of Geological Discoveries on English Society and Literature between 1829 and 1859*. Leiden: Brill, 1997.

Knight, Christopher. "John Polkinghorne." In *The Blackwell Companion to Science and Christianity*, edited by J. B. Stump and Alan G. Padgett, 622–31. Oxford: Blackwell, 2012.

Kooi, Cornelis van der. *As in a Mirror: John Calvin and Karl Barth on Knowing God; A Diptych*. Translated by Donald Mader. Leiden: Brill, 2005.

Kooi, Cornelis van der, and Gijsbert van den Brink. *Christian Dogmatics: An Introduction*. Translated by Reinder Bruinsma with James D. Bratt. Grand Rapids: Eerdmans, 2017.

Korpel, Marjo C. A., and Johannes C. de Moor. *Adam, Eve, and the Devil: A New Beginning*. Sheffield: Sheffield Phoenix, 2014.

Kuhn, Thomas S. *The Structure of Scientific Revolutions*. 2nd ed. Chicago: University of Chicago Press, 1970.

Kunz, Bram. *Als een prachtig boek. Nederlandse Geloofsbelijdenis artikel 2 in de context van de vroegreformatorische theologie*. Zoetermeer, the Netherlands: Boekencentrum, 2013.

Kuyper, Abraham. "The Blurring of the Boundaries." In *Abraham Kuyper: A Centennial Reader*, edited by James D. Bratt, 363–402. Grand Rapids: Eerdmans, 1998.

———. "Evolution." *Calvin Theological Journal* 31 (1996): 11–50.

———. *Lectures on Calvinism*. Grand Rapids: Eerdmans, 1931.

———. "Sphere Sovereignty." In *Abraham Kuyper: A Centennial Reader*, edited by James D. Bratt, 461–90. Grand Rapids: Eerdmans, 1998.

Lamoureux, Denis O. "Darwinian Theological Insights: Evolutionary Theodicy and Evolutionary Psychology." *Faith and Thought* 57 (October 2014): 3–20.

———. *Evolution: Scripture and Nature Say Yes!* Grand Rapids: Zondervan, 2016.

———. *Evolutionary Creation: A Christian Approach to Evolution*. Cambridge: Lutterworth, 2008.

———. *I Love Jesus and I Accept Evolution*. Eugene, OR: Wipf & Stock, 2009.

———. "Theological Insights from Darwin." *Perspectives on Science and Christian Faith* 56, no. 1 (2004): 2-12.

———. "Toward an Intellectually Fulfilled Christian Theism." *Faith and Thought* 55 (2013): 2-17.

Largent, Mark A. "Darwinism in the United States, 1859-1930." In *The Cambridge Encyclopedia of Darwin and Evolutionary Thought*, edited by Michael Ruse, 226-34. Cambridge: Cambridge University Press, 2013.

Laudan, Larry. "A Confutation of Convergent Realism." *Philosophy of Science* 48, no. 1 (1981): 19-49.

Lawson, Thomas E., and Robert McCauley. *Rethinking Religion: Connecting Cognition and Culture*. Cambridge: Cambridge University Press, 1990.

Lee, Hoon J. "Accommodation—Orthodox, Socinian, and Contemporary." *Westminster Theological Journal* 75 (2013): 335-48.

Leith, John H. "The Ethos of the Reformed Tradition." In *Major Themes in the Reformed Tradition*, edited by Donald K. McKim, 5-18. Eugene, OR: Wipf & Stock, 1998.

———. *Introduction to the Reformed Tradition: A Way of Being the Christian Community*. Louisville: John Knox, 1981.

Lever, Jan. *Creation and Evolution*. Grand Rapids: Kregel, 1958.

———. *Where Are We Headed? A Biologist Talks about Origins, Evolution, and the Future*. Grand Rapids: Eerdmans, 1970.

Levering, Matthew. *Engaging the Doctrine of Creation: Cosmos, Creatures, and the Wise and Good Creator*. Grand Rapids: Baker Academic, 2017.

———. *Engaging the Doctrine of Revelation: The Mediation of the Gospel through Church and Scripture*. Grand Rapids: Baker Academic, 2014.

Lewis, C. S. *Miracles*. London: HarperCollins, 2002.

———. "The Pains of Animals: A Problem in Theology." In *God in the Dock: Essays on Theology and Ethics*, 161-71. Grand Rapids: Eerdmans, 1972 [1947].

———. *The Problem of Pain*. New York: Macmillan, 1962 [1943].

Lieburg, Fred van. "Dynamics of Dutch Calvinism: Early Modern Programs for Further Reformation." In *Calvinism and the Making of the European Mind*, edited by Gijsbert van den Brink and Harro M. Höpfl, 43-66. Leiden: Brill, 2014.

Linzey, Andrew. *Animal Theology*. London: SCM, 1994.

Linzey, Andrew, and Dorothy Yamamoto, eds. *Animals on the Agenda: Questions about Animals for Theology and Ethics*. Champaign: University of Illinois Press, 1998.

Lipton, Peter. *Inference to the Best Explanation*. London: Routledge, 1991.

Livingstone, David N. *Darwin's Forgotten Defenders: The Encounter between Evangelical Theology and Evolutionary Thought*. Grand Rapids: Eerdmans, 1987.

———. *Dealing with Darwin: Place, Politics, and Rhetoric in Religious Engagements with Evolution*. Baltimore: Johns Hopkins University Press, 2014.

———. "Myth 17: That Huxley Defeated Wilberforce in Their Debate on Evolution and Religion." In *Galileo Goes to Jail and Other Myths about Science and Religion*, edited by Ronald L. Numbers, 152–60. Cambridge, MA: Harvard University Press, 2009.

Lloyd, Michael. "Are Animals Fallen?" In *Animals on the Agenda*, edited by Andrew Linzey and Dorothy Yamamoto, 147–60. Champaign: University of Illinois Press, 1998.

Lucas, J. R. "Wilberforce and Huxley: A Legendary Encounter." *Historical Journal* 22 (1979): 313–30.

MacKay, Donald M. "'Complementarity' in Scientific and Theological Thinking." *Zygon* 9 (1974): 225–44.

Mahlmann, Theodor. "'Ecclesia semper reformanda': Eine historische Aufarbeitung. Neue Bearbeitung." In *Hermeneutica Sacra. Studien zur Auslegung der Heiligen Schrift im 16. und 17. Jahrhundert*, edited by Torbjörn Johansson, Robert Kolb, and Johann Anselm Steiger, 381–442. Berlin: de Gruyter, 2010.

Mahoney, Jack. *Christianity in Evolution*. Washington, DC: Georgetown University Press, 2011.

Mayr, Ernst. *The Growth of Biological Thought: Diversity, Evolution, and Inheritance*. Cambridge, MA: Belknap Press of Harvard University Press, 1982.

———. *One Long Argument: Charles Darwin and the Genesis of Modern Evolutionary Thought*. Cambridge, MA: Harvard University Press, 1991.

———. *Towards a New Philosophy of Biology: Observations of an Evolutionist*. Cambridge, MA: Harvard University Press, 1988.

———. *What Evolution Is*. New York: Basic Books, 2001.

Mays, James Luther. "The Self in the Psalms and the Image of God." In *God and Human Dignity*, edited by R. Kendall Soulen and Linda Woodhead, 27–43. Grand Rapids: Eerdmans, 2005.

McCormack, Bruce L. "Participation in God, Yes; Deification, No: Two Modern Protestant Responses to an Ancient Question." In *Orthodox and Modern: Studies in the Theology of Karl Barth*, by Bruce L. McCormack, 235–60. Grand Rapids: Baker Academic, 2016.

McCosh, James. *Ideas in Nature Overlooked by Dr. Tyndall*. New York: Robert Carter and Brothers, 1875.

———. *The Religious Aspect of Evolution*. New York: Scribner's Sons, 1890.

McDaniel, Jay B. *Of God and Pelicans: A Theology of Reverence for Life*. Louisville: Westminster John Knox, 1989.

McFarland, Ian. "The Fall and Sin." In *The Oxford Handbook of Systematic Theology*, edited by John Webster, Kathryn Tanner, and Iain Torrance, 140-59. Oxford: Oxford University Press, 2007.

———. *In Adam's Sin: A Meditation on the Christian Doctrine of Original Sin*. Oxford: Wiley-Blackwell, 2010.

McGinnis, Andrew M. *The Son of God beyond the Flesh: A Historical and Theological Study of the Extra Calvinisticum*. London: Bloomsbury T&T Clark, 2014.

McGrath, Alister E. *Darwinism and the Divine: Evolutionary Thought and Natural Theology*. Oxford: Wiley-Blackwell, 2011.

———. *The Foundations of Dialogue in Science and Religion*. Oxford: Blackwell, 1998.

———. "Review Conversation [with Willem B. Drees]." *Theology and Science* 8 (2010): 333-41.

McKim, Donald. *Introducing the Reformed Faith*. Louisville: Westminster John Knox, 2001.

McMullin, Ernan. "Could Natural Selection Be Purposive?" In *Divine Action and Natural Selection: Science, Faith, and Evolution*, edited by Joseph Seckbach and Richard Gordon, 114-25. London: World Scientific, 2009.

———. "How Should Cosmology Relate to Theology?" In *The Sciences and Theology in the Twentieth Century*, edited by Arthur Peacocke, 17-57. Notre Dame: University of Notre Dame Press, 1981.

Meer, Jitse M. van der. "European Calvinists and the Study of Nature." In *Calvinism and the Making of the European Mind*, edited by Gijsbert van den Brink and Harro M. Höpfl, 103-30. Leiden: Brill, 2014.

Menzel, Randolf, and Julia Fischer, eds. *Animal Thinking: Contemporary Issues in Comparative Cognition*. Cambridge, MA: MIT Press, 2011.

Messer, Neil. "Natural Evil after Darwin." In *Theology after Darwin*, edited by Michael S. Northcott and R. J. Berry, 139-54. Milton Keynes, UK: Paternoster, 2009.

Mettinger, Tryggve N. D. *The Eden Narrative: A Literary and Religio-Historical Study of Genesis 2-3*. Winona Lake, IN: Eisenbrauns, 2007.

Meyer, Stephen C. *Darwin's Doubt: The Explosive Origin of Animal Life and the Case for Intelligent Design*. New York: HarperOne, 2013.

———. *Signature in the Cell: DNA and Evidence for Intelligent Design*. New York: HarperOne, 2009.

Middleton, Richard J. *The Liberating Image: The Imago Dei in Genesis 1*. Grand Rapids: Brazos, 2005.

Migliore, Daniel L. *Faith Seeking Understanding: An Introduction to Christian Theology*. 2nd ed. Grand Rapids: Eerdmans, 2004.

Miller, Daniel K. "Responsible Relationship: Imago Dei and the Moral Distinction between Humans and Other Animals." *International Journal of Systematic Theology* 13 (2011): 323-39.

Miller, Keith B., ed. *Perspectives on an Evolving Creation*. Grand Rapids: Eerdmans, 2003.

Miller, Kenneth R. *Finding Darwin's God: A Scientist's Search for Common Ground between God and Evolution*. New York: HarperCollins, 1999.

———. "The Flagellum Unspun: The Collapse of 'Irreducible Complexity.'" In *Debating Design: From Darwin to DNA*, edited by William Dembski and Michael Ruse, 81-97. Cambridge: Cambridge University Press, 2004.

Molnar, Paul D. *Thomas F. Torrance: Theologian of the Trinity*. Farnham, UK: Ashgate, 2009.

Moltmann, Jürgen. *The Church in the Power of the Spirit: A Contribution to Messianic Ecclesiology*. New York: Harper & Row, 1977.

Monod, Jacques. *Chance and Necessity: An Essay on the Natural Philosophy of Modern Biology*. New York: Knopf, 1971.

Moore, James R. *The Post-Darwinian Controversies: A Study of the Protestant Struggle to Come to Terms with Darwin in Great Britain and America, 1870-1900*. Cambridge: Cambridge University Press, 1979.

Moreland, J. P., et al., eds. *Theistic Evolution: A Scientific, Philosophical, and Theological Critique*. Wheaton, IL: Crossway, 2017.

Morgan, G. J. "Emile Zuckerkandl, Linus Pauling, and the Molecular Evolutionary Clock, 1959-1965." *Journal of the History of Biology* 31 (1998): 155-78. doi:10.1023/A:1004394418084.

Moritz, Joshua M. "Does Jesus Save the Neanderthals? Theological Perspectives on the Evolutionary Origins and Boundaries of Human Nature." *Dialog* 54, no. 1 (2015): 51-60.

———. "Evolution, the End of Human Uniqueness, and the Election of *Imago Dei*." *Theology and Science* 9 (2011): 307-39.

Morris, Henry M. *Many Infallible Proofs: Practical and Useful Evidences of Christianity*. San Diego: Creation-Life Publishers, 1980.

Mortenson, Terry. *The Great Turning Point: The Church's Catastrophic Mistake on Geology—before Darwin*. Green Forest, AR: Master Books, 2004.

———, ed. *Searching for Adam: Genesis and the Truth about Man's Origin*. Green Forest, AR: Master Books, 2016.

Mortenson, Terry, and Thane H. Ury, eds. *Coming to Grips with Genesis: Biblical Authority and the Age of the Earth*. Green Forest, AR: Master Books, 2008.

Mouw, Richard. *Calvinism at the Las Vegas Airport*. Grand Rapids: Zondervan, 2004.

Mühling, Markus. "Menschen und Tiere—geschaffen im Bild Gottes." In *Geschaf-*

*fen nach ihrer Art. Was unterscheidet Tiere und Menschen?*, edited by Ulrich Beuttler, H. Hemminger, M. Mühling, and M. Rothgangel, 129-43. Frankfurt: Lang, 2017.

———. *Resonances: Neurobiology, Evolution, and Theology; Evolutionary Niche Constructions, the Ecological Brain, and Relational-Narrative Theology.* Göttingen: Vandenhoeck & Ruprecht, 2014.

Muller, Richard A. *Calvin and the Reformed Tradition.* Grand Rapids: Baker Academic, 2012.

———. *Dictionary of Latin and Greek Theological Terms.* Grand Rapids: Baker Books, 1985.

———. "How Many Points?" *Calvin Theological Journal* 28 (1993): 425-33.

———. *Post Reformation Reformed Dogmatics.* Vol. 1, *Prolegomena to Theology.* 2nd ed. Grand Rapids: Baker Academic, 2003.

Murphy, Francesa Aran, and Philip G. Ziegler, eds. *The Providence of God: Deus Habet Consilium.* London: T&T Clark, 2009.

Murphy, George. "Roads to Paradise and Perdition: Christ, Evolution, and Original Sin." *Perspectives on Science and Christian Faith* 58 (2006): 109-18.

Murphy, Nancey. "Cognitive Science and the Evolution of Religion." In *The Believing Primate: Scientific, Philosophical, and Theological Reflections on the Origin of Religion*, edited by Jeffrey Schloss and Michael J. Murray, 265-77. Oxford: Oxford University Press, 2009.

Murray, Michael J. "Belief in God: A Trick of Our Brain?" In *Contending with Christianity's Critics: Answering New Atheists and Other Objectors*, edited by Paul Copan and William Lane Craig, 47-57. Nashville: B&H, 2009.

———. *Nature Red in Tooth and Claw: Theism and the Problem of Animal Suffering.* Oxford: Oxford University Press, 2008.

Murray, Michael J., and Lyn Moore. "Costly Signaling and the Origin of Religion." *Journal of Cognition and Culture* 9 (2009): 225-45.

Nagel, Thomas. *The View from Nowhere.* Oxford: Oxford University Press, 1986.

Nelson, Paul, and John Mark Reynolds. "Young Earth Creationism." In *Three Views on Creation and Evolution*, edited by J. P. Moreland and John Mark Reynolds, 39-75. Grand Rapids: Zondervan, 1999.

Nevin, Norman C., ed. *Should Christians Embrace Evolution? Biblical and Scientific Responses.* Phillipsburg, NJ: P&R, 2011.

Newberg, Andrew B. *Principles of Neurotheology.* Farnham, UK: Ashgate, 2010.

Newberg, Andrew, Eugene d'Aquili, and Vince Rause. *Why God Won't Go Away: Brain Science and the Biology of Belief.* New York: Random House, 2001.

Newman, Robert C. "Progressive Creationism." In *Three Views on Creation and*

*Evolution*, edited by J. P. Moreland and John Mark Reynolds, 103-33. Grand Rapids: Zondervan, 1999.
Niebuhr, Reinhold. *Man's Nature and His Communities*. New York: Scribner's Sons, 1965.
———. *The Nature and Destiny of Man: A Christian Interpretation*. New York: Scribner's Sons, 1941.
Nielsen, Kai. *Naturalism and Religion*. Amherst, NY: Prometheus, 2001.
Noll, Mark. *The Scandal of the Evangelical Mind*. Grand Rapids: Eerdmans, 1995.
Noordmans, O. *Herschepping* [Re-creation]. Zeist, the Netherlands: NCSV, 1934.
Norenzayan, Ara. *Big Gods: How Religion Transformed Cooperation and Conflict*. Princeton: Princeton University Press, 2013.
Nowak, Martin A., and Roger Highfield. *Super Cooperators: Beyond the Survival of the Fittest; Why Cooperation, Not Competition, Is the Key to Life*. New York: Free Press, 2011.
Nowak, Martin A., and Sarah Coakley, eds. *Evolution, Games, and God: The Principle of Cooperation*. Cambridge, MA: Harvard University Press, 2013.
Numbers, Ronald L. *The Creationists: From Scientific Creationism to Intelligent Design*. Expanded ed. Cambridge, MA: Harvard University Press, 2006.
Nürnberger, Klaus. *Informed by Science—Involved by Christ: How Science Can Update, Enrich, and Empower the Christian Faith*. Pietermaritzburg, South Africa: Cluster Publications, 2013.
Odling-Smee, F. John, Kevin N. Laland, and Marcus W. Feldman. *Niche Construction: The Neglected Process in Evolution*. Princeton: Princeton University Press, 2003.
*Our Faith: Ecumenical Creeds, Reformed Confessions, and Other Resources*. Grand Rapids: Faith Alive, 2013.
Pagels, Elaine. *Adam, Eve, and the Serpent*. New York: Random House, 1988.
Pannenberg, Wolfhart. *Systematic Theology*. Vols. 1-3. Grand Rapids: Eerdmans, 1991-1998.
Partee, Charles. "Calvin's Central Dogma Again." *Sixteenth Century Journal* 18 (1987): 191-200.
———. *The Theology of John Calvin*. Louisville: Westminster John Knox, 2008.
Paul, Mart-Jan. *Oorspronkelijk. Overwegingen bij schepping en evolutie* [Originally: Considerations on creation and evolution]. Apeldoorn: Labarum Academic, 2017.
Pauw, Amy Plantinga, and Serene Jones, eds. *Feminist and Womanist Essays in Reformed Dogmatics*. Louisville: Westminster John Knox, 2006.
Peacocke, Arthur. *Theology for a Scientific Age: Being and Becoming—Natural, Divine, and Human*. Enlarged ed. Minneapolis: Fortress, 1993.

Peels, Rik. "Does Evolution Conflict with God's Character?" *Modern Theology* 34 (2018): 544–64.
Peters, Ted, ed. *Science and Theology: The New Consonance*. Boulder, CO: Westview, 1998.
Peterson, Michael, William Hasker, Bruce Reichenbach, and David Basinger. *Reason and Religious Belief: An Introduction in the Philosophy of Religion*. 5th ed. New York: Oxford University Press, 2012.
Philipse, Herman. "The Real Conflict between Science and Religion: Alvin Plantinga's *Ignoratio Elenchi*." *European Journal for Philosophy of Religion* 5 (2013): 87–110.
Pinnock, Clark. *Most Moved Mover: A Theology of God's Openness*. Carlisle, UK: Paternoster, 2001.
Plantinga, Alvin C. *God, Freedom, and Evil*. Grand Rapids: Eerdmans, 1974.
———. "Seeking an Official Definition of 'Randomness': A Reply to Jay Richards." *Evolution News and Science Today*, April 3, 2012. http://www.evolutionnews.org/2012/04/seeking_an_offi058161.htm.
———. "Supralapsarianism, or 'O Felix Culpa.'" In *Christian Faith and the Problem of Evil*, ed. Peter van Inwagen, 1–25. Grand Rapids: Eerdmans, 2004.
———. *Warranted Christian Belief*. Oxford: Oxford University Press, 2000.
———. *Where the Conflict Really Lies: Science, Religion, and Naturalism*. Oxford: Oxford University Press, 2011.
Plantinga, Alvin, and Michael Tooley. *Knowledge of God*. Oxford: Oxford University Press, 2008.
Plantinga, Cornelius, Jr. *Not the Way It's Supposed to Be: A Breviary of Sin*. Grand Rapids: Eerdmans, 1995.
Polkinghorne, John. "Anthropology in an Evolutionary Context." In *God and Human Dignity*, edited by R. Kendall Soulen and Linda Woodhead, 89–103. Grand Rapids: Eerdmans, 2006.
———. *Scientists as Theologians*. London: SPCK, 1996.
———. "Scripture and an Evolving Creation." *Science and Christian Belief* 21 (2009): 163–73.
Pope, Stephen. *Human Evolution and Christian Ethics*. Cambridge: Cambridge University Press, 2007.
Powell, Baden. *Essays on the Spirit of the Inductive Philosophy*. London: Longman, Brown, Green & Longman, 1855.
Poythress, Vern S. *Redeeming Science: A God-Centered Approach*. Wheaton, IL: Crossway, 2006.
Price, George McCready. *Back to the Bible*. Washington: Review & Herald, 1920.

Prothero, Donald R. *Evolution: What the Fossils Say and Why It Matters*. Cambridge: Cambridge University Press, 2007.

Pruss, Alexander. "A New Way to Reconcile Creation with Current Evolutionary Biology." *Proceedings of the American Catholic Philosophical Association* 85 (2011): 213–22.

Rachels, James. *Created from the Animals: The Moral Implications of Darwinism*. New York: Oxford University Press, 1990.

Rahner, Karl. "Evolution and Original Sin." *Concilium: International Review of Theology* 6, no. 3 (1967): 30–35.

———. "Theological Reflexions on Monogenism." In *Theological Investigations*, 1:229–96. London: Darton, Longman & Todd, 1961.

Ramm, Bernard. *The Christian View of Science and Scripture*. Grand Rapids: Eerdmans, 1954.

Rana, Fazale. *The Cell's Design: How Chemistry Reveals the Creator's Artistry*. Grand Rapids: Baker Books, 2008.

Ratzsch, Del. "Design, Chance, and Theistic Evolution." In *Mere Creation: Science, Faith, and Intelligent Design*, edited by William A. Dembski, 289–312. Downers Grove, IL: InterVarsity Press, 1998.

———. "There Is a Place for Intelligent Design in the Philosophy of Biology." In *Contemporary Debates in Philosophy of Biology*, edited by Francisco J. Ayala and Robert Arp, 343–63. Malden, MA: Wiley-Blackwell, 2010.

Rau, Gerald. *Mapping the Origins Debate: Six Models on the Beginning of Everything*. Downers Grove, IL: IVP Academic, 2012.

Renfrew, Colin. "Personhood: Towards a Gradualist Approach." In *Emergence of Personhood: A Quantum Leap?*, edited by Malcolm Jeeves, 51–67. Grand Rapids: Eerdmans, 2015.

Richards, Robert J. "Ernst Haeckel and the Struggles over Evolution and Religion." *Annals of the History and Philosophy of Biology* 10 (2005): 89–115.

Richerson, Peter J., and Robert Boyd. *Not by Genes Alone: How Culture Transformed Human Evolution*. Chicago: University of Chicago Press, 2005.

Ricoeur, Paul. *The Symbolism of Evil*. Boston: Beacon, 1967.

Ridder, Jeroen de. "Science and Scientism in Popular Science Writing." *Social Epistemology Review and Reply Collective* 12, no. 3 (2014): 23–39.

Ridder, Jeroen de, Rik Peels, and René van Woudenberg, eds. *Scientism: Perils and Prospects*. Oxford: Oxford University Press, 2018.

Ridley, Mark. *Evolution*. 3rd ed. Oxford: Blackwell, 2004.

Roberts, Jon H. "Myth 18: That Darwin Destroyed Natural Theology." In *Galileo Goes to Jail and Other Myths about Science and Religion*, edited by Ronald L. Numbers, 161–69. Cambridge, MA: Harvard University Press, 2009.

Roberts, Richard J. "The German Reception of Darwin's Theory, 1860-1945." In *The Cambridge Encyclopedia of Darwin and Evolutionary Thought*, edited by Michael Ruse, 235-42. Cambridge: Cambridge University Press, 2013.

Ross, Hugh. *Creation and Time: A Biblical and Scientific Perspective on the Creation Date Controversy*. Colorado Springs: NavPress, 1994.

———. *The Fingerprint of God: Recent Scientific Discoveries Reveal the Unmistakable Identity of the Creator*. New Kensington, PA: Whitaker House, 1989.

———. *The Genesis Question: Scientific Advances and the Accuracy of Genesis*. Colorado Springs: NavPress, 1998.

Rowe, William. "The Problem of Evil and Some Varieties of Atheism." *American Philosophical Quarterly* 30 (1979): 335-41.

Ruler, Arnold van. "Zonnigheden in de zonde [1965]." In *Van schepping tot Koninkrijk: Teksten (1947-1970) uit het theologisch oeuvre van A. A. van Ruler* [From creation to kingdom], edited by Gijsbert van den Brink and Dirk van Keulen, 121-24. Barneveld: Nederlands Dagblad, 2008.

Rupke, Nicolaas A. "Myth 13: That Darwinian Natural Selection Has Been the 'Only Game in Town.'" In *Newton's Apple and Other Myths about Science*, edited by Ronald L. Numbers and Kostas Kampourakis, 103-11. Cambridge, MA: Harvard University Press, 2015.

———. *Richard Owen: Biology without Darwin*. Chicago: University of Chicago Press, 2009.

Ruse, Michael. *Can a Darwinian Be a Christian? The Relationship between Science and Religion*. Cambridge: Cambridge University Press, 2000.

———. *Darwin and Design: Does Evolution Have a Purpose?* Cambridge, MA: Harvard University Press, 2003.

———. "Evolutionary Ethics: A Phoenix Arisen." In *Issues in Evolutionary Ethics*, edited by Paul Thompson, 225-47. Albany: State University of New York Press, 1995.

Ruse, Michael, and Edward O. Wilson. "The Evolution of Ethics: Is Our Belief in Morality Merely an Adaptation Put in Place to Further Our Reproductive Ends?" *New Scientist* 108 (October 17, 1985): 50-52.

Sanlon, Peter. "Original Sin in Patristic Thought." In *Adam, the Fall, and Original Sin: Theological, Biblical, and Scientific Perspectives*, edited by Hans Madueme and Michael Reeves, 85-107. Grand Rapids: Baker Academic, 2014.

Schloss, Jeffrey P. "Divine Providence and the Question of Evolutionary Directionality." In *Back to Darwin: A Richer Account of Evolution*, edited by John B. Cobb, 330-50. Grand Rapids: Eerdmans, 2008.

Schloss, Jeffrey P., and Michael J. Murray. "Evolutionary Accounts of Belief in

Supernatural Punishment: A Critical Review." *Religion, Brain, and Behavior* 1 (2011): 46-99.

Schneider, John R. "Recent Genetic Science and Christian Theology on Human Origins." *Perspectives on Science and Christian Faith* 62 (2010): 196-212.

Schoock, Martinus. *De scepticismo pars prior.* Groningen: H. Lussinck, 1652.

Schwager, Raymund, SJ. *Banished from Eden: Original Sin and Evolutionary Theory in the Drama of Salvation.* Leominster, UK: Gracewing, 2006.

Schwarz, Hans. *Vying for Truth—Theology and the Natural Sciences from the 17th Century to the Present.* Göttingen: Vandenhoeck & Ruprecht, 2014.

Schweizer, Alexander. *Die protestantischen Centraldogmen in ihrer Entwicklung innerhalb der reformierten Kirche.* 2 vols. Zürich: Orell, Füslli & Co., 1854-1856.

Searle, John. "How to Derive 'Ought' from 'Is.'" *Philosophical Review* 73 (1964): 43-58.

Segal, Robert. *Myth: A Very Short Introduction.* Oxford: Oxford University Press, 2004.

Selderhuis, Herman, ed. *Handbook of Dutch Church History.* Göttingen: Vandenhoeck & Ruprecht, 2015.

Schaik, Carel van, and Kai Michel. *The Good Book of Human Nature: An Evolutionary Reading of the Bible.* New York: Basic Books, 2016.

Shariff, Azim F., and Ara Norenzayan. "Mean Gods Make Good People: Different Views of God Predict Cheating Behavior." *International Journal for the Psychology of Religion* 21 (2011): 85-96.

Shariff, Azim F., Ara Norenzayan, and Joseph Henrich. "The Birth of High Gods: How the Cultural Evolution of Supernatural Policing Agents Influenced the Emergence of Complex, Cooperative Human Societies, Paving the Way for Civilization." In *Evolution, Culture, and the Human Mind*, edited by Mark Schaller, Ara Norenzayan, Steve Heine, Toshi Yamagishi, and Tatsuya Kameda, 119-36. New York: Psychology, 2009.

Shults, F. LeRon. *Reforming Theological Anthropology: After the Philosophical Turn to Relationality.* Grand Rapids: Eerdmans, 2003.

Shuster, Marguerite. *The Fall and Sin: What We Have Become as Sinners.* Grand Rapids: Eerdmans, 2004.

Simpson, George G. *The Meaning of Evolution: A Study of the History of Life and of Its Significance for Man.* Rev. ed. New Haven: Yale University Press, 1967.

Singer, Peter. *Animal Liberation.* 4th ed. New York: HarperCollins, 2009.

Smith, James K. A. "What Stands on the Fall? A Philosophical Exploration." In *Evolution and the Fall*, edited by William T. Cavanaugh and James K. A. Smith, 48-64. Grand Rapids: Eerdmans, 2017.

Snoke, David. *A Biblical Case for an Old Earth.* Grand Rapids: Baker Books, 2006.

———. "Why Were Dangerous Animals Created?" *Perspectives on Science and Christian Faith* 56 (2004): 117–25.

Sober, Elliott. "Evolution without Naturalism." In *Oxford Studies in Philosophy of Religion*, vol. 3, edited by Jonathan L. Kvanvig, 187–221. Oxford: Oxford University Press, 2011.

Sollereder, Bethany N. *God, Evolution, and Animal Suffering: Theodicy without a Fall*. Abingdon: Routledge, 2019.

Sosis, Richard. "Why Aren't We All Hutterites? Costly Signaling Theory and Religious Behavior." *Human Nature* 14 (2003): 91–127.

Southgate, Christopher. "Divine Glory in a Darwinian World." *Zygon* 49 (2014): 784–807.

———. *The Groaning of Creation: God, Evolution, and the Problem of Evil*. Louisville: Westminster John Knox, 2008.

———. *Theology in a Suffering World: Glory and Longing*. Cambridge: Cambridge University Press, 2018.

Spencer, Herbert. *Principles of Biology*. London: Williams & Norgate, 1864.

Sproul, R. C. *Not a Chance: The Myth of Chance in Modern Science and Cosmology*. Grand Rapids: Baker Books, 1994.

Stark, Rodney. *The Rise of Christianity: How the Obscure, Marginal Jesus Movement Became the Dominant Religious Force in the Western World in a Few Centuries*. Princeton: Princeton University Press, 1996.

Starke, Ekkehard. "Animals." In *The Encyclopedia of Christianity*, vol. 1, edited by Erwin Fahlbusch et al. Grand Rapids: Eerdmans, 1999.

Stenmark, Mikael. *Scientism: Science, Ethics, and Religion*. Aldershot, UK: Ashgate, 2001.

Stewart, Kenneth J. "The Points of Calvinism: Retrospect and Prospect." *Scottish Bulletin of Evangelical Theology* 26 (2008): 187–203.

———. *Ten Myths about Calvinism: Recovering the Breadth of the Reformed Tradition*. Downers Grove, IL: InterVarsity Press, 2011.

Stewart-Williams, Steve. *Darwin, God, and the Meaning of Life: How Evolutionary Theory Undermines Everything You Thought You Knew*. Cambridge: Cambridge University Press, 2010.

Stordalen, Terje. *Echoes of Eden: Genesis 2–3 and Symbolism of the Eden Garden in Biblical Hebrew Literature*. Leuven: Peeters, 2000.

Stott, John. *Understanding the Bible*. London: Scripture Union, 1972.

Stove, David. *Darwinian Fairytales: Selfish Genes, Errors of Heredity, and Other Fables of Evolution*. New York: Encounter Books, 1995.

Sudduth, Michael. *The Reformed Objection to Natural Theology*. Burlington, VT: Ashgate, 2009.

Sunshine, Glenn S. "Accommodation Historically Considered." In *The Enduring Authority of the Christian Scriptures*, edited by D. A. Carson, 238-65. Grand Rapids: Eerdmans, 2016.

Swinburne, Richard. *Providence and the Problem of Evil*. Oxford: Clarendon, 1998.

———. *Revelation: From Metaphor to Analogy*. 2nd ed. Oxford: Clarendon, 2007 [1992].

Tattersall, Ian. *Becoming Human: Evolution and Human Uniqueness*. New York: Harcourt Brace, 1998.

———. "Human Evolution: Personhood and Emergence." In *The Emergence of Personhood: A Quantum Leap?*, edited by Malcolm Jeeves, 37-50. Grand Rapids: Eerdmans, 2015.

———. *Paleontology: A Brief History*. West Conshohocken, PA: Templeton, 2010.

Theissen, Gerd. *Biblical Faith: An Evolutionary Approach*. Minneapolis: Fortress, 1984.

Thomkins, Jeffrey P. "Pseudogenes Are Functional, Not Genomic Fossils." Institute for Creation Research, June 28, 2013. https://www.icr.org/article/7532.

Thomson, Keith S. "Marginalia: The Meanings of Evolution." *American Scientist* 70, no. 5 (1982): 529-31.

Thwaite, Ann. *Glimpses of the Wonderful: The Life of Philip Henry Gosse, 1810-1888*. London: Faber & Faber, 2003.

Tomasello, Michael. *A Natural History of Human Morality*. Cambridge, MA: Harvard University Press, 2016.

Toren, Benno van den. "Human Evolution and a Cultural Understanding of Original Sin." *Perspectives on Science and Christian Faith* 68 (2016): 12-21.

———. "Original Sin and the Coevolution of Nature and Culture." In *Finding Ourselves after Darwin: Conversations on the Image of God, Original Sin, and the Problem of Evil*, edited by Stanley P. Rosenberg, 173-86. Grand Rapids: Baker Academic, 2018.

Torrance, Thomas F. *Divine and Contingent Order*. Oxford: Oxford University Press, 1981.

Trivers, Robert. "The Evolution of Reciprocal Altruism." *Quarterly Review of Biology* 46 (1971): 35-57.

Turner, Léon. "Introduction: Pluralism and Complexity in the Evolutionary Cognitive Science of Religion." In *Evolution, Religion, and Cognitive Science: Critical and Constructive Essays*, edited by Fraser Watts and Léon Turner, 1-20. Oxford: Oxford University Press, 2014.

VanDoodewaard, William. *The Quest for the Historical Adam: Genesis, Hermeneutics, and Human Origins*. Grand Rapids: Reformed Heritage Books, 2015.

Van Eyghen, Hans. "Scientific Theories of Religion as Naturalistic Challenges." *Studies in Science and Theology* 15 (2015-2016): 93-109.

Vanhoozer, Kevin. "Human Being, Individual and Social." In *The Cambridge Companion to Christian Doctrine*, edited by Colin E. Gunton, 158-88. Cambridge: Cambridge University Press, 1997.

Velde, Dolf te, ed. *Synopsis Purioris Theologiae / Synopsis of a Purer Theology*. Vol. 1. Leiden: Brill, 2015.

Venema, Dennis R., and Scot McKnight. *Adam and the Genome: Reading Scripture after Genetic Science*. Grand Rapids: Brazos, 2017.

———. "Genesis and the Genome: Genomics Evidence for Human-Ape Common Ancestry and Ancestral Hominid Population Sizes." *Perspectives on Science and Christian Faith* 62 (2010): 166-78.

Vermij, Rienk. *The Calvinist Copernicans: The Reception of the New Astronomy in the Dutch Republic, 1575-1750*. Amsterdam: KNAW, 2002.

———. "The Debate on the Motion of the Earth in the Dutch Republic in the 1650s." In *Nature and Scripture in the Abrahamic Religions: Up to 1700*, vol. 2, edited by Jitse M. van der Meer and Scott Mandelbrote, 605-25. Leiden: Brill, 2008.

Visala, Aku. "*Imago Dei*, Dualism, and Evolution: A Philosophical Defense of the Structural Image of God." *Zygon* 49 (2014): 101-20.

———. *Naturalism, Theism, and the Cognitive Science of Religion: Religion Explained?* Farnham, UK: Ashgate, 2011.

Visscher, Lukas. "Listening to Creation Groaning: A Survey of Main Themes in Creation Theology." In *Listening to Creation Groaning*, edited by Lukas Visscher, 11-31. Geneva: Centre International Réformé John Knox, 2004.

Voetius, Gisbertus. *Thersites heautontimorumenos*. Utrecht, 1635.

Von Eckardt, Barbara. *What Is Cognitive Science?* Cambridge, MA: MIT Press, 1993.

Vorster, Nico. *The Brightest Mirror of God's Works: John Calvin's Theological Anthropology*. Eugene: Wipf & Stock, 2019.

Vriezen, Theodore C. *An Outline of Old Testament Theology*. Oxford: Blackwell, 1962.

Vroom, Hendrik M. "On Being 'Reformed.'" In *Reformed and Ecumenical: On Being Reformed in Ecumenical Encounters*, edited by Christine Lienemann-Perrin, Hendrik M. Vroom, and Michael Weinrich, 153-69. Amsterdam: Rodopi, 2000.

Waal, Frans de. *Good Natured: The Origins of Right and Wrong in Humans and Other Animals*. Cambridge, MA: Harvard University Press, 1996.

———. *Primates and Philosophers: How Morality Evolved*. Edited by Stephen Macedo and Josiah Ober. Princeton: Princeton University Press, 2006.

Waal, Frans de, and Angeline van Roosmalen. "Reconciliation and Consolation among Chimpanzees." *Behavorial Ecology and Sociobiology* 5 (1979): 55-66.

Waal, Frans de, Patricia Smith Churchland, Telmo Pievani, and Stefano Parmigiani, eds. *Evolved Morality: The Biology and Philosophy of Human Conscience.* Leiden: Brill, 2014.

Wahlberg, Mats. *Revelation as Testimony: A Philosophical-Theological Study.* Grand Rapids: Eerdmans, 2014.

Waltke, Bruce. *An Old Testament Theology: An Exegetical, Canonical, and Thematic Approach.* Grand Rapids: Zondervan, 2007.

Walton, John H. "A Historical Adam: Archetypal Creation View." In *Four Views on the Historical Adam*, edited by Matthew Barrett and Ardel B. Caneday, 89-118. Grand Rapids: Zondervan, 2013.

———. *The Lost World of Adam and Eve: Genesis 2-3 and the Human Origins Debate.* Downers Grove, IL: IVP Academic, 2015.

———. *The Lost World of Genesis One: Ancient Cosmology and the Origins Debate.* Downers Grove, IL: IVP Academic, 2009.

Walton, John H., and D. Brent Sandy. *The Lost World of Scripture: Ancient Literary Culture and Biblical Authority.* Downers Grove, IL: InterVarsity Press, 2013.

Ward, Keith. *The Big Questions in Science and Religion.* West Conshohocken, PA: Templeton, 2008.

———. *God, Faith, and the New Millennium.* Oxford: Oneworld, 1998.

Warfield, Benjamin B. "Creation, Evolution, and Mediate Creation." In *Evolution, Scripture, and Science: Selected Writings*, edited by Mark A. Noll and David N. Livingstone, 197-210. Grand Rapids: Baker Books, 2000.

Watson, Francis. "Genesis before Darwin: Why Scripture Needed Liberating from Science." In *Reading Genesis after Darwin*, edited by Stephen C. Barton and David Wilkinson, 23-38. Oxford: Oxford University Press, 2009.

Webb, Stephen. *On God and Dogs: A Christian Theology of Compassion for Animals.* New York: Oxford University Press, 1998.

Weeks, Noel K. "The Ambiguity of Biblical 'Background.'" *Westminster Theological Journal* 72 (2010): 219-36.

———. "The Bible and the 'Universal' Ancient World: A Critique of John Walton." *Westminster Theological Journal* 78 (2016): 1-28.

Wenham, Gordon. *Genesis 1-15.* Word Biblical Commentary 1. Milton Keynes, UK: Word, 1991.

Wennberg, Robert N. *God, Humans, and Animals: An Invitation to Enlarge Our Moral Universe.* Grand Rapids: Eerdmans, 2003.

Westermann, Claus. *Genesis 1-11: A Commentary.* Minneapolis: Fortress, 1994.

Whitcomb, John C., and Henry M. Morris. *The Genesis Flood: The Biblical Record and Its Scientific Implications*. Philadelphia: P&R, 1961.
Whitehead, Alfred North. *Modes of Thought*. New York: Macmillan, 1938.
Whorton, Mark S. *Peril in Paradise: Theology, Science, and the Age of the Earth*. Waynesboro, GA: Authentic Media, 2005.
Wielenberg, Erik J. "On the Evolutionary Debunking of Morality." *Ethics* 120 (2010): 441–64.
Wilkins, John S. "Could God Create Darwinian Accidents?" *Zygon* 47 (2012): 30–42.
Wilkinson, David. *Christian Eschatology and the Physical Universe*. London: T&T Clark, 2010.
Williams, Norman Powell. *The Ideas of the Fall and of Original Sin*. London: Longmans, Green, 1927.
Williams, Patricia. *Doing without Adam and Eve: Sociobiology and Original Sin*. Minneapolis: Fortress, 2001.
Wilson, David Sloan. *Darwin's Cathedral: Evolution, Religion, and the Nature of Society*. Chicago: University of Chicago Press, 2002.
Wilson, E. O. Introduction to *From So Simple a Beginning: The Four Great Books of Charles Darwin*, edited by E. O. Wilson. New York: Norton, 2005.
———. *Sociobiology: The New Synthesis*. Cambridge, MA: Belknap Press of Harvard University Press, 1975.
Winther, Rasmus Grønfeldt. "The Structure of Scientific Theories." *Stanford Encyclopedia of Philosophy*, March 5, 2015. http://plato.stanford.edu/entries/structure-scientific-theories/#SynSemPraVieBas.
Wittgenstein, Ludwig. *Philosophical Investigations*. Translated by G. E. M. Anscombe. Oxford: Basil Blackwell, 1967.
Wolde, Ellen van. *A Semiotic Analysis of Genesis 2–3*. Assen: Van Gorcum, 1989.
Wolterstorff, Nicholas. *Justice: Rights and Wrongs*. Princeton: Princeton University Press, 2008.
———. *Reason within the Bounds of Religion*. Grand Rapids: Eerdmans, 1976.
———. *Understanding Liberal Democracy*. Oxford: Oxford University Press, 2012.
Wood, Charles M. *The Question of Providence*. Louisville: Westminster John Knox, 2008.
Woudenberg, René van. "Chance, Design, Defeat." *European Journal for Philosophy of Religion* 5 (2013): 31–41.
———. "Darwinian and Teleological Explanations." In *Evolution and Ethics: Human Morality in Biological and Religious Perspective*, edited by Philip Clayton and Jeffrey Schloss, 171–86. Grand Rapids: Eerdmans, 2004.

———. "Limits of Science and the Christian Faith." *Perspectives of Science and Christian Faith* 65, no. 1 (2013): 24–36.

Woudenberg, René van, and Joëlle Rothuizen-van der Steen. "Both Random and Guided." *Ratio* 28 (2014): 332–48.

Wright, Robert. *The Moral Animal: Why We Are the Way We Are—the New Science of Evolutionary Psychology.* New York: Vintage Books, 1994.

Xygalatas, Dimitris. "Cognitive Science of Religion." In *Encyclopedia of Psychology and Religion,* edited by David A. Leeming, 343–47. 2nd ed. New York: Springer, 2014.

Yong, Amos. *The Spirit of Creation: Modern Science and Divine Action.* Grand Rapids: Eerdmans, 2011.

Young, Davis A., and Ralph F. Stearley. *The Bible, Rocks, and Time: Geological Evidence for the Age of the Earth.* Downers Grove, IL: InterVarsity Press, 2008.

Zakai, Avihu. *Jonathan Edwards' Philosophy of Nature: The Re-enchantment of the World in the Age of Scientific Reasoning.* London: T&T Clark, 2010.

Zaspel, Fred. *The Theology of B. B. Warfield: A Systematic Summary.* Wheaton, IL: Crossway, 2010.

Ziegler, Philip G., and Francesca Aran Murphy, eds. *The Providence of God: Deus Habet Consilium.* London: T&T Clark, 2009.

Zuckerberg, P. "Atheism: Contemporary Numbers and Patterns." In *The Cambridge Companion to Atheism,* edited by Michael Martin, 47–65. Cambridge: Cambridge University Press, 2007.

# Index of Names

Adams, Marilyn McCord, 132
Alexander, Denis, 37, 52-53, 68, 105, 112, 162-63, 166-67, 169, 170-71, 174, 178, 188, 197, 199, 226-27, 232-33, 270
Allen, R. Michael, 13, 16, 18-19, 23, 160-61, 178
Alper, Matthew, 253
Anderson, David, 200
Anscombe, G. E. M., 20
Aquili, Eugene d', 242
Aquinas, Thomas, 217-18, 263
Arbesman, Samuel, 44
Aristotle, 147, 217
Asselt, W. J. van, 12, 14-15, 96
Astley, Jeff, 165
Atran, Scott, 251, 255
Augustine, 28, 176-77, 180-81, 186-87, 200, 203, 273
Austin, J. L., 90
Axelrod, Robert, 238
Ayala, Francisco J., 51, 65, 88, 216

Bach, Johann Sebastian, 89
Baker, Catherine, 34, 49-50, 61
Baker, J. Wayne, 15
Balserak, Jon, 90
Barbour, Ian G., 8, 81, 206, 268
Barr, James, 30, 198

Barr, Stephen M., 217-18
Barrett, Justin L., 243, 245-46, 248, 255, 257-61, 263
Barrett, Matthew, 173, 190-91
Barth, Karl, 17-19, 22, 24-25, 31, 57, 73, 81, 89, 92, 125-27, 139, 148, 152, 170-71, 254
Bartholomew, David J., 228
Barton, Stephen C., 82, 150, 165
Bauckham, Richard, 104
Bauswein, Jean-Jacques, 19-20
Bavinck, Herman, 22, 69, 94-95, 118, 192, 256
Bechtel, Lyn M., 167, 182
Beck, Naomi, 231
Beek, Abraham van de, 186
Behe, Michael, 66
Bekkum, Koert van, 153, 275
Bekoff, Marc, 109, 144, 155
Belt, Henk van den, 272
Benedict, Philip, 13-14, 26
Bering, Jesse, 246, 248-50
Berkhof, Hendrikus, 23, 86, 93, 184, 269-72
Berkhof, Louis, 139
Berkouwer, Gerrit C., 94, 117, 152, 179
Berry, Robert J., 160, 163, 169, 178
Bethge, Eberhard, 67
Billings, J. Todd, 15

## INDEX OF NAMES

Bimson, John J., 114, 115, 172, 178, 182-83, 198
Birch, Charles, 102
Blancke, Stefaan, 2
Blocher, Henri, 89, 92, 106-7, 183, 184
Bloom, Paul, 247, 260
Boersema, Jan J., 153
Bonhoeffer, Dietrich, 67
Boston, Thomas, 203
Bowler, Peter, 231
Boyd, Gregory, 124
Boyd, Robert, 234
Boyer, Pascal, 205, 244, 248, 251
Bratt, James D., 15, 25, 211
Brès, Guy de, 28, 31
Breuer, Georg, 238
Brink, Gijsbert van den, 3, 5, 11, 13, 16, 25, 29, 31, 43, 44, 56, 70, 80, 102, 139, 207, 211, 269
Brooke, John Hedley, 27, 29
Brosnan, S. F., 239
Bruce, F. F., 115
Brümmer, Vincent, 87, 154
Brune, Johan de, 32
Brunner, Emil, 118, 152
Bucer, Martin, 12, 14
Buckland, William, 114
Bulgakov, Sergei, 127
Bullinger, Heinrich, 12-15, 23, 161
Bultmann, Rudolf, 91-92
Burdett, Michael, 153, 157
Busch, Eberhard, 18-20

Calvin, John, 2, 10, 12-18, 23-24, 43, 83, 88-90, 117, 155, 199, 243, 245, 253
Cameron, Nigel M. de S., 79
Cantor, Geoffrey, 27
Caruana, Louis, 2
Cavalli-Sforza, L. L., 235
Cavanaugh, William T., 185
Chappel, Jonathan, 194
Chesterton, G. K., 62

Clark, Kelly James, 257, 259-61, 263
Clayton, Philip, 268
Clement of Alexandria, 177
Clough, David, 102, 146, 150, 153
Coakley, Sarah, 126
Cobb, John B., 223
Cocceius, John, 96-97, 161
Cohn, Norman, 113
Collins, C. John, 92-93, 116-17, 173, 178, 185, 193, 197, 201
Collins, Francis S., 51-52, 88, 172
Collins, Robin, 194
Conradie, Ernst, 146, 185, 196, 272-73
Conway Morris, Simon, 223-28
Cook, Harry, 52, 66
Cook, L. M., 59
Cortez, Marc, 151, 158
Creegan, Nicola Hoggard, 33, 126-29
Crisp, Oliver, 10, 12
Cruz, Helen de, 244, 246-47, 251, 259, 261-62
Cunningham, Conor, 60, 62
Cunningham, David, 146-49, 151

Daley, Brian E., 156
Darwin, Charles, 35-37, 39-41, 48, 51, 55, 60, 62, 82, 84, 106, 110, 113-14, 131-32, 138, 147, 204-5, 209-12, 215, 222, 229-32, 256, 267, 272
Darwin, Francis, 62
Dawkins, Richard, 33, 60, 110, 208-9, 216, 235
Day, Allan John, 171
Deane-Drummond, Celia, 61, 127, 132, 146, 157, 189-90, 266, 271
Deemter, Kees van, 20
Dekker, Cees, 5
Dekker, Eef, 14
Dembski, William A., 67, 118
Dennett, Daniel C., 33, 224, 236
Descartes, René, 42, 107, 110, 148
Dijksterhuis, Eduard J., 72

## Index of Names

Dixon, Thomas, 27
Dobzhansky, Theodosius, 60
Domning, Daryl P., 164, 200
Dougherty, Trent, 107-8, 120, 123, 132
Drees, Willem B., 8
Dunbar, Robin, 162
Dunn, James D. G., 177, 181
Dupree, A. Hunter, 213
Durkheim, Émile, 242, 245

Edwards, Jonathan, 29
Edwards, Rem B., 44
Eldredge, Niles, 60, 63
Elk, Michiel van, 252
Elliott, Mark W., 207
Engel, Mary Potter, 139-40
Enns, Peter, 84, 168
Etzelmüller, Georg, 193, 203

Falk, Darrel R., 88, 162-63
Farel, William, 12
Farley, Benjamin W., 207, 263
Feldman, Marcus W., 61, 235
Fergusson, David, 102, 117, 207, 268-69, 272
Fesko, John V., 30
Finlay, Graham, 53, 163, 191-92
Fischer, Julia, 144
Flipse, Abraham, 4
Fodor, Jerry, 61-62
Fowler, Thomas B., 36-37, 60-62
Fraassen, Bas C. van, 21
Francescotti, Robert, 108, 121
Freeman, C., 239
Fretheim, Terence E., 196
Freud, Sigmund, 242, 245
Freudenberg, Matthias, 18, 21
Frost, David, 97

Gallup, Gordon G., Jr., 107
Galton, Francis, 231-32
Garte, Sy, 62

Genderen, J. van, 46
Gerrish, Brian, 19
Gillespie, Charles C., 39
Gilmour, Michael, 102-5
Goldingay, John, 135
Goodall, Jane, 155
Gosse, Edmund, 41
Gosse, Philip H., 41
Gould, Stephen J., 41, 60, 63, 66, 206, 223-26, 249
Graham, Billy, 97
Gray, Asa, 106, 145, 212-13, 215, 219, 228
Gray, John, 145
Green, Adam, 263
Greenblatt, Stephen, 115
Gregory, Neville G., 109
Grenz, Stanley J., 151
Grudem, Wayne, 79
Gundlach, Bradley J., 208
Gunton, Colin E., 254
Guthrie, Stephen, 244

Haarsma, Deborah B., 2, 7, 38-39, 42, 48, 50, 52-53, 69, 163
Haarsma, Loren D., 2, 38-39, 42, 48, 50, 52-53, 69, 163
Haeckel, Ernst, 209, 211-12, 230
Haldane, J. B. S., 237
Ham, Ken, 75
Hamilton, W. D., 237-38
Harlow, Dan, 167
Harris, Eugene E., 163
Harris, Mark, 197, 223, 268
Harrison, Peter, 27-28, 30, 108, 225
Haught, John F., 2, 11, 208, 223
Heidegger, J. H., 218
Heller, Michael, 245
Hellwig, Monica K., 164, 200
Helm, Paul, 207
Henrich, Joseph, 251, 255

## INDEX OF NAMES

Heppe, Heinrich, 141, 218-19, 254, 263
Hesselink, I. John, 10-12, 15
Highfield, Roger, 126, 238
Higton, Mike, 269
Hill, Matthew Nelson, 271
Hitler, Adolf, 232
Hodge, Charles, 69, 71, 210-12, 222
Hoffmeier, James K., 35
Hooff, Jan van, 236
Hooker, Mark F., 40-41, 49, 64, 75, 112, 124, 232
Hoornbeeck, Johannes, 22
Hooykaas, Reyer, 32
Höpfl, Harro M., 11, 13, 15, 27
Howell, Kenneth J., 29-30
Huijgen, Arnold, 90
Hume, David, 20
Hurd, James P., 188-89
Hutton, James, 38
Huxley, Thomas H., 138, 209-10, 212
Huyssteen, J. Wentzel van, 149, 151-52, 155, 157, 176, 189, 260

Illingworth, J. R., 99
Inwagen, Peter van, 61, 63, 108, 122, 124, 177, 220
Irenaeus, 89, 93
Irons, Lee, 89

Jablonka, Eva, 62
Jacobs, Alan, 180
Jaki, Stanley, 74
Jeeves, Malcolm, 136, 175
Jenson, Robert, 128, 154
Jersild, Paul, 58
Joad, C. E. M., 108
John Paul II, 4, 57, 78, 142, 156
Johnson, Dominic, 250
Johnson, Elizabeth A., 102
Jones, James W., 230, 243, 245, 250, 255-56, 262-63

Jones, Serene, 18
Jorink, Eric, 28-29

Kärkkäinen, Veli-Matti, 146, 148, 156, 159, 271-72
Keathley, Kenneth D., 40-41, 49, 64, 75, 112, 124, 232
Kelemen, Deborah, 246
Keller, Tim, 188
Kelsey, David, 148
Kettlewell, Bernard, 59
Kevles, Daniel J., 231
Kidner, Derek, 188
King, Barbara J., 144
Kingsley, Charles, 55
Kitcher, Philip, 214, 239
Klapwijk, Jacob, 175, 227
Klaver, J. M. I., 39
Kline, Meredith, 89
Knight, Christopher, 268
Kooi, Cornelis van der, 25, 56, 80, 89, 102, 129, 139, 151, 207, 269
Koonin, Eugene, 62
Korpel, Marjo C. A., 166
Kuebler, Daniel, 36-37, 60-62
Kuhn, Thomas S., 96
Kunz, Abraham J., 31
Kuyper, Abraham, 11, 15, 22, 25, 95, 210-12, 222, 256, 262

Laland, Kevin N., 61
Lamarck, Jean-Baptiste de, 35, 231
Lamoureux, Dennis O., 4, 84-93, 106, 209
Largent, Mark A., 232
Lasco, John à, 12
Laudan, Larry, 44
Lawson, Thomas E., 244
Lee, Hoon J., 90
Leith, John, 12, 14, 19-20
Lennox, James, 213
Lever, Jan, 4

*Index of Names*

Levering, Matthew, 187, 196, 198, 202, 254
Lewens, Tim, 235
Lewis, C. S., 62, 93, 107-10, 124-25, 128, 172, 237
Lewontin, Richard C., 249
Lieburg, Fred van, 22
Linzey, Andrew, 102, 104, 124
Lipton, Peter, 65
Livingstone, David N., 56, 138, 208, 212
Lloyd, Michael, 124
Lucas, J. R., 138
Luther, Martin, 7, 12
Lyell, Charles, 38-40

MacKay, Donald M., 226
Mahlmann, Theodor, 22
Mahoney, Jack, 271
Majerus, Michael, 59
Marcion, 84, 128, 133
Marx, Karl, 242, 245
Mayr, Ernst, 47, 50, 52, 210, 213, 215, 216
Mays, James L., 149
McBrayer, Justin P., 123, 132
McCauley, Robert, 244
McCormack, Bruce L., 24
McCosh, James, 1, 9, 212
McFarland, Ian, 180-81
McGinnis, Andrew M., 19
McGrath, Alister, 8, 138, 195, 207, 210, 216, 227
McKim, Donald K., 14, 20, 24
McKnight, Scot, 163
McMullin, Ernan, 216, 226
Meer, Jitse M. van der, 73, 192
Menzel, Randolf, 144
Messer, Neil, 125
Mettinger, Tryggve N. D., 178-79, 182, 198
Meyer, Stephen C., 66

Michel, Kai, 166, 169, 194
Middleton, Richard J., 151-52
Migliore, Daniel L., 154-55
Miller, Daniel K., 148, 152
Miller, Keith B., 88, 188
Miller, Kenneth R., 66, 88
Milton, John, 115
Mivart, George, 57
Molnar, Paul D., 125
Moltmann, Jürgen, 22, 271-72
Monod, Jacques, 213
Moor, Johannes C. de, 166
Moore, G. E., 232
Moore, James R., 131
Moore, Lyn, 255
Moreland, J. P., 7, 55, 75-76, 79
Morgan, G. J., 50
Moritz, Joshua M., 146, 156, 176, 188, 192
Morris, Henry M., 43, 48, 80, 113, 174
Mortenson, Terry, 39, 75, 79, 137
Mouw, Richard, ix-xi, 14
Mühling, Markus, 61, 146
Müller, Gerd, 62
Muller, Richard A., 12-13, 16, 25, 141, 253-54
Murphy, Francesca A., 207
Murphy, George, 201
Murphy, Nancey, 242
Murray, Michael J., 108, 114-16, 119-20, 122, 124, 255, 259

Nagel, Thomas, 62
Nelson, Paul, 75
Nevin, Norman C., 4, 79, 111, 200
Newberg, Andrew B., 242
Newman, Robert C., 55, 76
Niebuhr, Reinhold, 155
Nielsen, Kai, 242
Noble, Denis, 62
Noll, Mark, 71
Noordmans, Oepke, 57

## INDEX OF NAMES

Noordtzij, Arie, 89
Norenzayan, Ara, 250-51
Nowak, Martin, 126, 238
Numbers, Richard L., 49, 75
Nürnberger, Klaus, 272

Odling-Smee, F. John, 61
Oecolampadius, Johannes, 12
Olevian, Caspar, 161
Oosterhoff, B. J., 4-5
Oostveen, Daan, 214
Origen, 89, 177
Owen, John, 161
Owen, Richard, 35

Pagels, Elaine, 180-81
Paley, William, 208
Pannenberg, Wolfhart, 30, 156, 159, 183, 199
Partee, Charles, 15
Paul, Mart-Jan, 174
Pauw, Amy Plantinga, 18
Peacocke, Arthur, 114, 165, 268
Peels, R., 33, 133, 262
Peters, Ted, 226
Peterson, Michael, 100, 260
Philipse, Herman, 214-16, 219-21, 255
Philo, 89
Piatelli-Palmarini, Massimo, 61
Pinnock, Clark, 124
Pius XII, 78
Planck, Max, 96
Plantinga, Alvin, 7, 63, 124, 128, 132, 214-16, 220-21, 258
Plantinga, Cornelius, Jr., 92, 181, 184
Polkinghorne, John, 157-58, 193, 268
Pope, Stephen, 142
Powell, Baden, 55, 204
Poythress, Vern S., 43
Price, George McCready, 200, 267
Prothero, Donald R., 50
Pruss, Alexander, 220

Pumfrey, Stephen, 27

Rachels, James, 145, 156
Rahner, Karl, 164, 190
Ramm, Bernard, 54, 74
Rana, Fazale, 66
Ratzsch, Del, 65, 220
Rau, Gerald, 77
Rause, Vince, 242
Renfrew, Colin, 176
Reynolds, John M., 75
Richards, Jay, 215
Richards, Robert J., 209
Richerson, Peter J., 234
Ricoeur, Paul, 186
Ridder, Jeroen de, 3, 33, 70, 184, 217, 262
Ridley, Mark, 221
Roberts, Jon H., 210
Roberts, Richard J., 232
Roosmalen, Angeline van, 239
Rosenberg, Stanley P., 195
Ross, Hugh, 54, 76
Rothuizen-van der Steen, Joëlle, 69, 221
Rowe, William, 121
Ruler, Arnold A., 184
Rupke, Nicholas A., 35, 39
Ruse, Michael, 2, 11, 66, 204-5, 210, 220, 231, 240-41, 258

Sandy, D. Brent, 90
Sanlon, Peter, 180
Schaik, Carel van, 166, 169, 194
Schenderling, Jacques, 109
Schleiermacher, Friedrich, 18
Schloss, Jeffrey P., 220, 223, 242, 255
Schneider, John, 92, 164, 180
Scholten, J. H., 15
Schoock, Martin, 73
Schwager, Raymund, 194, 198
Schwarz, Hans, 1

## Index of Names

Schweizer, Alexander, 14
Searle, John, 232
Segal, Robert, 167
Selderhuis, Herman J., 14
Shapiro, James, 62
Shariff, Azim F., 250-51
Shults, F. LeRon, 151
Shuster, Marguerite, 85, 193
Simpson, George G., 213
Singer, Peter, 145, 239
Smedt, Johan de, 244, 246-47, 251, 259, 261-62
Smith, James K. A., 179, 182, 185, 193, 195-96
Smith, John Maynard, 237
Smith, Martin R., 66
Snoke, David, 114, 116, 130-31
Sober, Elliott, 214-15
Socinus, Faustus, 89
Sosis, Richard, 252
Southgate, Christopher, 105, 112-14, 117, 119-20, 122, 125, 129-30, 132, 183, 186, 200, 273
Spencer, Herbert, 59, 230-32
Sproul, R. C., 228
Stanford, Kyle, 256
Stark, Rodney W., 257
Starke, Ekkehard, 103-4
Stearley, Ralph F., 38, 44-45
Stenmark, Mikael, 217, 262
Stewart, Kenneth J., 12, 16
Stordalen, T., 196
Stott, John, 169
Stove, David, 235
Stump, Eleonore, 263
Stump, Jim, 226, 268
Sudduth, Michael, 30
Sunshine, Glenn S., 90
Swinburne, Richard, 254

Tattersall, Ian, 50, 52, 158, 176
Teilhard de Chardin, Pierre, 273

Tennyson, Alfred, 239
Theissen, Gerd, 272
Theophilus of Antioch, 115
Thomkins, Jeffrey P., 54
Thomson, Keith S., 36
Thwaite, Ann, 41
Thysius, Antonius, 140, 158
Tinbergen, Nico, 59
Tomasello, Michael, 206
Tooley, Michael, 63
Toren, Benno van der, 195
Torrance, Thomas F., 125, 128
Trivers, Robert, 238
Turner, Léon, 247
Turretin, Francis, 161

Urk, Eva van, 110, 155
Ury, Thane H., 75

VanDoodewaard, William, 177
Van Eyghen, Hans, 247, 255
Vanhoozer, Kevin, 144
Velde, Dolf te, 112, 140
Velema, W. H., 46
Venema, Dennis R., 162-63
Vermigli, Peter Martyr, 12
Vermij, Rienk, 73, 96-97
Viret, Pierre, 12
Visala, Aku, 150, 157, 260
Visscher, Lukas, 19-20, 101-2
Voetius, Gisbertus, 29, 72-73, 80, 86, 96
Von Eckardt, Barbara, 243
Vorster, Nico, 131, 138
Vriezen, Theodore C., 152
Vroom, Hendrik M., 11

Waal, Frans de, 145, 239-40
Wahlberg, Matts, 254
Waltke, Bruce K., 77, 89
Walton, John H., 80, 86, 89-90, 135, 163, 166, 177, 186, 189-90, 193

INDEX OF NAMES

Ward, Keith, 192, 219
Warfield, Benjamin, 56, 269
Watson, Francis, 82-84
Webb, Stephen, 102
Weeks, Noel K., 86
Weinrich, M., 11
Welker, Michael, 19
Wenham, Gordon, 80, 89
Wennberg, Robert N., 102-3, 108, 111
Westermann, Claus, 135, 171
Whitcomb, John C., 43, 48, 113
Whitehead, Alfred North, 158, 223
Whorton, Mark S., 113
Wielenberg, Eric J., 258
Wilberforce, Samuel, 138
Wilkins, John S., 228
Wilkinson, David, 268
Williams, N. P., 185-86
Williams, Patricia, 181
Williams, Steve Stewart, 143, 145
Wilson, David Sloan, 205, 251

Wilson, Edward O., 205, 213, 233, 240-41, 258
Winther, Rasmus Grønfeldt, 35
Wittgenstein, Ludwig, 20
Wolde, Ellen van, 182
Wolterstorff, Nicholas, 117, 144, 159
Wood, Charles M., 207
Woudenberg, René van, 3, 33-34, 69-70, 148, 213, 220-21, 262
Wright, Robert, 234

Xygalatas, Dimitris, 244

Yamamoto, Dorothy, 102
Yong, Amos, 271
Young, Davis A., 38, 44-45

Zakai, Avihu, 29
Zaspel, Fred, 269
Ziegler, Philip G., 207
Zuckerberg, P., 241
Zwingli, Huldrych, 12, 14

# Index of Subjects

accommodation, principle/doctrine of, 89-90
Adam, 190. *See also* first couple; historical Adam and Eve
Adam's navel, 41
afterlife, 246, 249, 253, 255n72
agency. *See* hyperactive agency detection device (HADD)
age of the earth, 40-43, 45-46, 76
agnosticism: cosmological, 44; origins, 43-45
altruism, 192, 236-40, 258, 271; reciprocal, 238; self-denial, 252; self-sacrifice, 104, 237, 259
Amish, 252n65
angels, fallen, 54, 123, 124n72, 127-28n85, 129
animal rights movement, 146
animals, relationship to God, 148, 153
animal sacrifice: active, 126, 237; passive, 103
animal suffering, 47, 84-85, 100, 105-35; and demonic forces, 123-27; eschatological compensation for, 132; and greater goods, 120-22, 132; and human sin, 110-19; as part of God's plan, 119-23; real?, 106-10
anthropic principle, 229
Anthropocene, the, 111

anthropomonism, 101, 119
anti-being, 125
apologetics, postevolutionary, 271-72
Apostles' Creed, 56
appearance of age theory, 41-42, 44
asceticism, 252
astronomy, 40, 43, 83
atheism, 33, 65, 91, 96, 210, 221, 241, 244-45; Darwinism as, 211, 233; new, 7, 247, 253; old, 7
atonement, 24, 199, 200
autism, 249n56

Barthianism, 17-18
Belgic Confession, 17; article 2 of, 28-31
Bible: allegorical/symbolic interpretation of, 23, 28; ancient Near Eastern context of, 80n16, 166; and animals, 101-6; anthropocentric but not anthropomonistic, 101; authority of, 2, 39, 72-74, 81, 87, 90, 96-97, 274; as book of God, 31, 94-95, 126; inspiration of, 95; interpretation vis-à-vis evolution, 71-98; literal ("plain") reading of, 23, 28, 31, 46, 72, 88-89; message of, 81, 87-88, 91, 94, 200; not a manual

INDEX OF SUBJECTS

for science, 79-80, 97; theological priority of, 11, 16, 22, 26, 98, 149
big bang, 65, 268n2
big crunch/freeze/rip, 268
Burgess Shale, 223

Calvinism, 12-16, 26n50, 228
Cambrian explosion, 66-67, 70, 223
Canons of Dordt (1618-1619), 16-18
Cartesianism, 30, 96
catastrophism, 38, 39n17
*causa prima* and *causae secundae*, 131, 218, 263, 265
cell, 66, 67, 70
Chalcedonian Definition, 6
chance, 129, 227n74, 228-29; in the Bible, 217n46; as part of Darwinian evolution, 32, 64, 68-69, 206-22; statistical/stochastic, 218-20, 227
children as intuitive theists, 246
children's Bible, 96
Christ-event, 32, 58, 93, 128, 135, 161, 199, 200
Christology, 19n32, 132n95, 146n25, 270, 272
cognitive dissonance, 78
cognitive psychology, 243, 246
cognitive science, 243
cognitive science of religion (CSR), 69, 230, 241-65; beginning of, 244; debunking religious belief?, 248, 252-64; definition of, 244; relation to Christian faith, 245
colonialism, 232
common descent, 36, 44, 47-58, 70, 76, 99, 136, 138, 142, 159, 161, 211, 240
common function, theory of, 52-55, 70
complementarity (between science and faith), 226. See also consonance (between science and faith)

concordism, 74-79, 81n20, 93-94, 117, 179, 270, 273
*concursus*, 218, 263n97
*Consensus Tigurinus* (1549), 13, 19
consonance (between science and faith), 194, 203, 222-29
convergence (evolutionary), 222-28
cooperation: as evolutionary value, 126, 134, 164, 194, 238n24, 251-52; of God, 218
Copernicanism, 72, 96. See also heliocentrism
cosmic fall theory, 111-19, 273; in the Bible?, 114-17; in church history, 114-15
cosmological agnosticism, 44. See also origins agnosticism
cosmology, 87, 89n45
costly signaling theory (CST), 251-52, 255
covenant, 2, 15-16, 23-24, 32, 58, 136, 160-61, 169, 178, 199; of grace, 58, 161; and infant baptism, 16, 23; of works, 58, 161
creation, doctrine of, 8, 56, 269-70; Barth on, 73n5; not at odds with common descent, 56
creation, goodness of, 100, 105, 126, 130, 134, 187n76
creation, stewardship of, 158, 169, 194
creation and redemption, 127, 134
cultural big bang/revolution, 176, 189
cultural evolution, 32, 64, 157, 166, 205, 234-35, 246, 250, 259
culture wars, 4, 221

Darwinian "just so stories," 235, 255
Darwinism, 32, 35, 61n69, 63n79, 69, 130, 210-13, 222; as "ambiguous gift," 272; extrapolation to economics and politics, 230-33; as a form of

unity, 211; orthodox, 61, 205; social, 185, 231–34; weak, 63n79
day-age creationism, 54
death, 44, 47, 86–87, 100, 106, 114–17, 125, 132, 164–65, 168, 170, 174; of humans, 32, 85, 112, 116, 173, 196–99; Paul on, 196–98; as spiritual, 85, 173, 196–97; violent, 198n111
deep time/geological timescale, 36–37, 40, 44–46, 55, 65, 70, 72, 99, 176
deism, 218
demonic forces, 42, 104, 123–30, 133
descent with modification, 48
design, 208, 212, 220n52, 223, 226. *See also* intelligent design (ID) movement; providence (of God)
determinism, biological, 209, 233
devil. *See* Satan
dinosaurs, 2, 48–49
divination, 170, 176
DNA, 51, 66
domino theory, 267
dualism between body and soul, 78
dualism between good and evil: metaphysical, 183; modified, 126–28; radical/unmodified, 127–28

earth, age of, 40–43, 45–46, 76
*ecclesia reformata quia semper reformanda* (a Reformed church should always be reforming according to the Word of God), 22–26, 98
ecclesiology, 22, 270–71
ecology, 127n85, 152, 153n48, 224, 273n19
eco-theology, 105
election, 191–92; and selection, 192, 263. *See also* historical Adam and Eve: as elected; predestination, doctrine of
emergence (of life/life-forms), 52, 56n60, 78, 85, 91, 113, 118, 120, 130,
169, 175, 188, 218–19, 224–25, 229, 271
empathy, 126, 134, 164, 239
endosymbiosis, 61
eschatology, 105, 116, 127, 132, 139, 199, 203, 267–68
ethics, 240–41
eugenics, 231, 233
euthanasia program, 231–32
evangelical(ism), x, 2, 13, 51n46, 89n44, 200; and evolution, 3, 11, 72, 267
evil: dualism between good and, 126–28, 183; evolutionary, 47, 64, 101, 110, 119, 129–35, 222n58; natural evil, 100–101, 111–12, 118, 124–28, 133, 135, 272; problem of, 119, 121n66, 214n32; "shadow sophia," 127. *See also* theodicy
"evo-devo" explanations, 61
evolution/evolutionary theory, 33–37, 56, and *passim*; common meaning of, 35–36; in crisis?, 63–64; different levels or layers, 36; directionality?, 223–26; epistemic status, 3; and God's character, 133–34; guided, 220–21; historical, 36; Hodge on, 211–12; Kuyper on, 210–11; Lyell's view of, 39; strong Darwinian, 36, 64n81; weak Darwinian, 63n79
evolutionary biology, 47–48, 74, 185, 199, 215, 217n44
evolutionary creationism, 85, 87
evolutionary evil, 47, 64, 101, 110, 119, 129–35, 222n58. *See also* natural evil
evolutionary psychology, 233–35
evolutionism (as distinct from evolutionary theory), 7, 33, 231
exaptation, 61
existential theory of mind (EToM), 249

319

INDEX OF SUBJECTS

extended evolutionary synthesis (EES), 62
extinction, 47, 100, 103, 111, 113, 116
*extra calvinisticum*, 19n32

Fall (into sin), 32, 44, 58, 85, 93, 112–15, 137, 139, 141, 161, 164, 175, 179–95; angelic, 124n72; Augustine on, 180–81, 186; as coming of age, 167, 182; as gradual process, 193; as historical, in some way, 182–87, 193; Paul on, 181, 186; retroactive, 118
fallenness of the world, 123, 127–28; as created, 165
federal (or covenantal) theology, 23, 160
fine-tuning of the universe, 65
first and second causes. See *causa prima* and *causae secundae*
first couple, 58, 162–63. See also historical Adam and Eve
fitness, 47, 69, 205, 220n51; direct and indirect, 238; inclusive, 216, 238
flood. See Noah's flood
flood geology, 39, 42–44. See also Noah's flood
"Forms of Unity," 17, 28; Darwinism as a, 211
fossil record, 36, 41–43, 47–50, 54–55, 76, 99, 113, 161, 176n37; transitional fossils, 48–49, 55, 76
framework interpretation (of first chapter of Genesis), 88–89
freedom, human, 120, 134, 187n76, 194
freeriders, 238, 251–52

gap theory, 54, 75, 123–24
genealogies: ancient Near Eastern, 177; in Genesis, 46, 76, 172
generality problem, 261
Genesis, second and third chapters of, 165–80; ahistorical or paradigmatic approach, 167–68; ancient Near Eastern context/parallels, 166, 167n20; Augustine on, 176–77; etiological function, 177, 183; as linked to first chapter of Genesis, 173–74; as myth, 167n20, 168, 186; old-historical approach, 172–74; as paradigmatic and event depicting, 178; as pointing to Neolithic period, 169, 188; prehistorical approach, 168–70, 187; primordial or proto-historical approach, 170–71; as saga, 171; theological reading of, 168; as *Urgeschichte*, 171; young-historical approach, 174–76
genetic drift, 61
genetic engineering, 270
genetic fallacy, 143, 258
genetic mutations (random), 36, 58, 69, 205, 212, 219, 222; caused?, 214, 216, 219; unguided?, 216, 221n56
genetic revolution, 51–52
genetics, 40, 48, 52, 191n90; classical Mendelian, 36, 51
gene transfer, horizontal, 36n10, 62n75
geological timescale. See deep time/geological timescale
geology, 38–40, 43, 45, 48, 113, 124
global flood model, 39, 42–44
God: animals' relationship to, 148, 153; and animal suffering, 119–23; character of, and evolution/evolutionary theory, 133–34; cooperation of, 218; grace of, 24, 31–32, 139, 181, 191, 200, 273; omnipotence of, 43n26, 106n16, 111, 121, 128; *potentia absoluta* (unbound power) of, 42; proofs for existence of, 246n50; sovereignty of, 15, 24, 32, 68, 205; transcendence of, 90. See also providence (of God)

God of the gaps, 55, 67, 70, 223, 226, 256
"God spot" in human brain, 242
goodness of creation, 100, 105, 126, 130, 134, 187n76
gospel, 58, 128, 133, 200, 268; communication of, 6; of free grace, 22; heart of, 200; historical nature of, 91
grace (of God), 24, 31–32, 139, 181, 191, 200, 273
gradualism, 36–47, 54, 65, 70, 76, 99, 117
group selection, 60, 233, 238n23

Hamilton's rule, 237
Hebrews, letter to the, 80
Heidelberg Catechism, 17–18, 25n48, 129, 139, 219n49
heliocentrism, 30, 72–73, 95–97. *See also* Copernicanism
hermeneutics, 23, 71–98, 149, 166, 186–87, 268, 273
hierarchical selection, 61
historical Adam and Eve, 32, 84, 86–87, 93, 137, 165–80, 273; as elected, 169, 188; as federal heads, 178, 190; as part of a group, 169, 175, 189
history, 87, 92–93; role in second and third chapters of Genesis, 167–80
history of redemption/salvation, 32, 82, 91–92, 103, 136–37, 161, 166, 195, 199, 268; infralapsarian, 202; supralapsarian, 202–3. *See also* soteriology
Holocaust, the, 232
hominins, 77, 87, 155, 161, 192n92
*Homo divinus*, 169, 178, 190–91
homologies, 50
*Homo sapiens*, 52, 158, 162, 166, 170–71, 188–89; appearance of, 87, 114
horizontal gene transfer, 36n10, 62n75
human agency. *See* hyperactive agency detection device (HADD)

human being: behaviorally modern, 189, 192n94; as God's vice-regent, 152; as psychosomatic unity, 78; as related to the chimp, 145, 162, 239; similarities and differences with other creatures, 157–59. See also *Homo sapiens*; humankind
human depravity, 16, 139
human dignity, 56–57, 138–46, 154, 156–59
human genome, 51–54
*Humani generis*, 11
humankind: as created in God's image, 32, 93, 137–59; fuzzy boundaries, 155; still evolving?, 270; unity of, 136, 163–64
human origins, 161–65; monophyletic account, 163; multiregional hypothesis of, 163; out of Africa hypothesis, 163, 190
human rights, 144, 159
human uniqueness, 56–57, 77, 136, 142–43, 270; as special uniqueness, 143–45, 149–50, 159; as species specificity, 143, 153; theological account of, 150–58
hyperactive agency detection device (HADD), 248–49, 260

image of God (*imago Dei*), 56–57, 136–59, 189–93; as central canonical idea, 149; christological view of, 146n25; eschatological view of, 152; as ethical challenge, 153; functional view of, 151–54; as including animals, 146–50; intrinsic and extrinsic part, 141n12; relational view of, 152, 154–55; structural/substantive view of, 150n37, 151, 157n63
immortality, 198–99
inclusive fitness, 216, 238
independence model (of science and religion), 8, 81n20, 206

inference to the best explanation, 65
in-group thinking, 240, 271
intelligent design (ID) movement, 64–68, 220n51, 226
intervention(ism) (divine), 47, 55, 70, 113, 116, 172, 220
irreducibly complex systems, 66
Isaiah, book of, 103–4, 105n15
Israel, 23–24, 72, 150, 168

Job, book of, 103, 122, 147n30, 203
Jonah, book of, 104
justification, doctrine of, 14, 22, 25

kin selection, 237

language, 147, 150; different in animals and humans, 157
law: and gospel, 17; three uses of, 25
Luke, Gospel of, 95
Lutheranism, 14, 22–23, 26, 27, 92

macroevolution, 59–60, 73, 76
magisterial Reformation, 13, 27
Manichaeism, 185
Marxism, 185n70, 232
materialism, 126, 209–10, 222, 237n18
mature creation. *See* appearance of age theory
meaning of life, 81, 217n44
memes, 235
meta-Darwinism, 60n68, 61, 70
microevolution, 59–60
mind, 62, 243, 248–49
miracles, 199, 268
mirror test, 107
missing links, 49, 55
mission, 6
modern evolutionary synthesis. *See* neo-Darwinian synthesis
molecular clock, 50
monism, 183, 209
monogenism, 163–64n10

monophyletism, 163, 164n10, 190
morality/moral awareness, 32, 64, 69, 144, 194, 206, 210, 232, 238; in animals?, 144–45n22, 157, 239; evolution of, 236–41, 254; illusory?, 240–41, 253, 258
moral skepticism, 258n81
mutations (random genetic), 36, 58, 69, 205, 212, 219, 222; caused?, 214, 216, 219; unguided?, 216, 221n56

natural evil, 100–101, 111–12, 118, 124–28, 133, 135, 272. *See also* evolutionary evil
naturalism, 63n78, 256, 261–63; metaphysical, 7; methodological, 242
naturalistic fallacy, 232
natural religion, 212
natural selection, 35–37, 39, 46, 55n58, 58–70, 99, 114, 164, 204, 233, 246–48; blind?, 65, 206–22, 226–28; external goal?, 217; harshness of, 126; immanent goal of, 216; turned upside down, 259; universal, 60; universal acid, 236
natural theology, 31, 195n103, 213n30, 262
nature: as a book, 28–30, 42, 46, 67, 74, 126; cruciformity of, 132; red in tooth and claw?, 125–26, 239; study of, 27–29
Neanderthals, 156, 162, 176n37
neo-Cartesianism, 106–10, 131
neo-Darwinian synthesis, 35, 37, 58–65, 68, 70, 145, 215, 220–21, 227; alternatives to, 61–63
neolythic revolution, 194n101
Neoplatonism, 57
neurosciences, 242
Nicene Creed, 6
niche construction, theory of, 61, 62n75

## Index of Subjects

Noah's flood, 38, 43, 76, 103, 113. See also global flood model
NOMA (non-overlapping magisteria) principle, 206n4
nothingness (*das Nichtige*), 125–26, 127n84

observer's perspective interpretation, 73
Occam's razor, 261n89
old-earth creationism, 54–55, 74n6, 75, 78
omnipotence of God, 43n26, 106n16, 111, 121, 128
omphalism (appearance of age theory), 41–43
only way arguments, 119, 122, 129
original sin, 32, 58, 141, 161, 164, 180–81, 195. See also sin
origins agnosticism, 43–45

pain, 99, 106–9, 112, 125, 164, 201n119
paleontology, 48, 113, 155
panentheism, 268
pantheism, 272
paradise/garden of Eden, 85, 167, 170, 172, 174, 187
peppered moth (as example of natural selection), 58–59
personhood, 144, 157, 178; rise of, 175
perspectivism, 74, 81–87, 117n58
pessimistic meta-induction argument, 44n31
philosophical anthropology, 78, 104
philosophy of mind, 243
physico-theology, 29
pneumatology, ix, 25, 270–72; Reformed emphasis on, 25
polygenism, 93n58, 190
polyphyletism, 163, 190n89
population bottleneck, 163, 191n90
post-Reformation Protestant/Reformed scholasticism, 140–41, 218

*potentia absoluta* (unbound power) of God, 42
prayer, effect of, 246n46
predestination, doctrine of, 14, 15n20, 17, 24
Presbyterianism, 10n2
problem of divine hiddenness, 263n99
problem of evil, 119, 121n66, 214n32. See also theodicy
process theism, 223n59, 223n60, 268
progressive creation, 36
progressive creationism, 54
proofs for God's existence, 246n50
pro-social behavior, 239, 251
protomorality, 239
providence (of God), 56, 68, 205, 225–26, 257; doctrine of, 64, 69, 112n34, 129, 131, 206–29
pseudogenes, 53
punctuated equilibrium, 60n67, 61, 63

Qur'an, 40, 60

racism, 232
randomness. See chance
rationality, 147, 150–51, 157; evolution of, 62–63
reason, 144, 151; as opposed to faith, 29–30
redemption, doctrine of, 202; history of, 32, 82, 91–92, 103, 136–37, 161, 166, 195, 199, 268; infralapsarian, 202; supralapsarian, 202–3. See also soteriology
Reformed epistemology, 260
Reformed theology, 10–32, 57, 70–71; appreciation of Old Testament, 23–24; as catholic, 19; central-dogma approach to, 14–16; and chance, 228; covenantal structure, 32, 160; defining characteristics, 19–20; as distinct from Calvinism, 12–14; dynamic character, 26, 32, 95; and

## INDEX OF SUBJECTS

experimental research, 28; as "habit of mind," 19; heartbeat of, 22, 273; on the human being, 57, 138–43; iconoclastic tendency, 23; as intensification of Christian heritage, 21, 180; and natural knowledge of God, 30; and natural world, 26–31; plurality of, 18; and political theory, 15n20, 25n50; as a stance, x, 21–26, 81, 129, 274; theocentric focus, 24, 31, 129, 207; as Wittgensteinian family, 20

religion/religiosity, 32, 64, 69, 206; as by-product, 247–49, 260–61; distinctively human, 155; as evolutionary adaptation, 247, 250–52; evolution of, 176, 241–65; illusory?, 253, 256–63; innateness/naturalness of, 246, 249, 260

reprobation, eternal, 17

Revelation, book of, 105, 167

revelation, doctrine of, 32, 69, 242–43, 245, 253–65; Reformed, 10–11, 25, 30, 241, 254

Roman Catholicism, 26n50, 263; and evolution, 2, 11; on the Fall, 187n76; and human dignity, 142; on the human soul, 57, 77–78; and natural knowledge of God, 30; on sin and atonement, 201–2n120; and the study of nature, 29n58

Romans, letter to the, 186

sanctification, 25, 271

Satan, 54, 104, 123–25, 127–28n85, 141, 183

science, 157; ancient Near Eastern, 85–86; appreciation of, 4, 26–27; and Bible, 71–98; as evolutionary by-product, 261; fallibility of, 44, 85; and historical scholarship, 4n4; limits of, 34n3, 221; as occasion to reconsider biblical exegesis, 94; rise of, 29

science and religion: conflict model, 27, 29; independence model, 8, 81n20

science skepticism, 44

scientific theory, x, 7–8, 26, 34–35, 44; underdetermination of, 70, 256n74

scientism, 7, 33–34, 262–64

Scopes trial, x

Scripture: doctrine of, 22, 45–46, 71–98; inspiration of, 95. *See also* Bible

selection: group, 60, 233, 238n23; hierarchical, 61; kin, 237; sexual, 61, 252. *See also* natural selection

selection, unit of, 60, 164, 237

self-assertion, 119, 194–95

self-consciousness, 144, 157, 225; in animals?, 107–9

self-denial, 252

self-hate, 24n45

self-interest, 164

selfish genes, 60

self-organization, 61

self-sacrifice, 104, 237, 259

self-transcendence, 155, 271

*sensus divinitatis*, 260; Calvin on, 243n38, 245, 253

sexuality, 187, 193, 232–33, 235, 252

sexual selection, 61, 252

"shadow sophia," 127

sin, 24, 27, 32, 57, 84, 139, 141, 180–87, 264, 273; and animal suffering, 111–19; Augustine on, 180–81, 186, 203; consequences of, 195–96; as contingent, 184–86; definition of, 180n49; as exclusively human, 193; not inherent in creation, 92, 183, 201. *See also* original sin

skeptical theism, 122, 132, 134

## Index of Subjects

social Darwinism, 185, 231–34
socially strategic information (SSI), 251
sociobiology, 233, 242
*sola Scriptura*, 23
soteriology, 16, 58, 161, 199–203; synergism in, 24
soul, 77–78, 141–42, 151, 156
sovereignty of God, 15, 24, 32, 68, 205
spandrel (accidental by-product), 249
special creation, 52; of humans, 76–77, 177
species, 143n17
specified complexity, 66
spiritual battle, 7
spiritual body, 202
spirituality, 155; among animals?, 155n54
stance, 21
stasis, 60
*status integritatis* (state of righteousness), 139–41, 193n97
stewardship of creation, 158, 169, 194. See also image of God (*imago Dei*): functional view of
supernatural punishment theory (SPT), 250, 255
survival of the fittest, 59, 122, 133, 230–31
symbiosis, 126
symbolic behavior/thought, capacity for, 157; evolution of, 175–76
Synod of Dordt, 16, 18, 28
*Synopsis Purioris Theologiae*, 140–41

teleology, 68, 208–9, 212, 215–16, 220, 222; two types of, 217n43
theism, 68–69, 209n11, 215, 220–21, 262; intuitive, 246
theistic evolution, 7n12, 51, 76, 78, 82, 85, 220
theodicy, 101, 131–33

theological anthropology, 56, 104, 136–59
theology: distinctive character of, 187; task of, 8, 45; Trinitarian, 129, 269
theory. See scientific theory
theory of mind (ToM), 248–49, 260
*theōsis*, 24
thermodynamics, second law of, 125
*tiktaalik*, 49n38
*tota Scriptura*, 23, 100
tragic dimension/view of life, 123, 201
transcendence of God, 90
transitional fossils, 48–49, 55, 76
tree of life, 36, 50–52
TULIP, 16
two-books metaphor/paradigm, 29–30, 74, 226n70

underdetermination. See scientific theory
uniformitarianism, 38, 40
union with Christ (*unio cum Christo*), 15
unit of selection, 60, 164, 237
universe, fine-tuning of the, 65

vegetarianism: prelapsarian?, 117–18n58
violence, 86n36, 124, 128, 152, 153n47, 194, 271

weak Darwinism, 63n79
Westminster Confession/Standards, 17, 30
world picture, 86, 90–92, 96, 272; of three-layered universe, 86–87
worldview, 72, 82, 195n103, 208–10, 217n44, 268, 272

young-earth creationism, 4, 40, 43, 46, 59, 75, 78–80, 113, 137; latent, 4; in the Netherlands, 4–5

# Index of Scripture References

## OLD TESTAMENT

**Genesis**
| | |
|---|---|
| 1 | 42, 54, 57, 72–76, 85, 88, 89, 103, 105, 134, 166, 173–76, 270 |
| 1–2 | 171, 174 |
| 1–3 | 4, 80, 94, 167 |
| 1–11 | 171 |
| 1:1 | 75, 124, 135, 152 |
| 1:2 | 54, 75, 124, 271 |
| 1:6–7 | 86 |
| 1:16 | 82–83, 88 |
| 1:21 | 114 |
| 1:24 | 100 |
| 1:25 | 100, 105 |
| 1:26 | 140, 152, 153 |
| 1:26–28 | 148 |
| 1:28 | 101, 151, 152, 154 |
| 1:29–30 | 117 |
| 1:31 | 134 |
| 2 | 85, 117, 140, 173–75 |
| 2–3 | 5, 92, 116, 165–74, 176, 178–82, 186–89, 193, 196–99 |
| 2–4 | 190 |
| 2:7 | 77, 85, 172, 179 |
| 2:16–17 | 178 |
| 2:17 | 170, 173, 174, 197, 199 |
| 2:19 | 101, 174 |
| 2:45 | 202 |
| 3 | 85, 115, 117, 165, 168, 169, 172, 173, 175, 182, 185, 186, 196, 197, 202 |
| 3:5 | 167 |
| 3:14 | 116 |
| 3:14–19 | 182 |
| 3:15 | 168, 195 |
| 3:17 | 115 |
| 3:17–19 | 116 |
| 3:19 | 170 |
| 3:20 | 170, 177 |
| 3:22 | 167, 198 |
| 4 | 181, 198 |
| 4–11 | 181 |
| 4:14 | 189, 190 |
| 4:17 | 169, 189, 190 |
| 4:21 | 177 |
| 5 | 177 |
| 5:1 | 148 |
| 5:29 | 115 |
| 6–8 | 103 |
| 6:5 | 182 |
| 9 | 148, 154 |
| 9:3 | 148 |
| 9:6 | 144, 148 |
| 12:1–3 | 191 |

**Exodus**
| | |
|---|---|
| 20:4 | 86 |
| 20:10 | 103 |
| 23:5 | 103 |
| 23:11 | 103 |
| 23:12 | 103 |

**Leviticus**
| | |
|---|---|
| 17:10–16 | 103 |
| 25:7 | 103 |

**Deuteronomy**
| | |
|---|---|
| 5:14 | 103 |
| 22:10 | 103 |
| 25:4 | 105 |
| 26:5 | 191 |

**Joshua**
| | |
|---|---|
| 10:12 | 73 |

**1 Kings**
| | |
|---|---|
| 22:34 | 217 |

# Index of Scripture References

## Job
| | |
|---|---|
| 7:12 | 114 |
| 19:25 | 6 |
| 38–40 | 114 |
| 38–41 | 270 |
| 39–41 | 103 |
| 39:3 | 130 |
| 39:16–17 | 130 |
| 39:30 | 130 |
| 41:14 | 130 |

## Psalms
| | |
|---|---|
| 8 | 101, 270 |
| 8:6–8 | 152 |
| 19 | 30, 147 |
| 19:6 | 73 |
| 36:6 | 103, 128 |
| 74:13 | 114 |
| 104 | 102, 103, 270 |
| 104:5 | 73 |
| 104:11 | 102 |
| 104:12 | 102 |
| 104:14 | 102 |
| 104:17 | 102 |
| 104:18 | 102 |
| 104:19–21 | 114 |
| 104:20 | 102 |
| 104:21 | 102 |
| 104:23 | 102 |
| 104:24 | 102 |
| 104:25 | 102 |
| 104:26 | 102 |
| 104:30 | 271 |
| 139 | 219 |
| 139:14 | 258 |
| 145:16 | 103 |
| 147:9 | 103 |

## Proverbs
| | |
|---|---|
| 12:10 | 104 |

## Ecclesiastes
| | |
|---|---|
| 1:5 | 73 |
| 7:29 | 193 |
| 9:11 | 217 |

## Isaiah
| | |
|---|---|
| 1:3 | 104 |
| 11:6–9 | 104, 116 |
| 14:12–15 | 124 |
| 24 | 115 |
| 24:4–6 | 115, 116 |
| 35:9 | 105 |
| 65:25 | 104, 105, 115, 116 |

## Jeremiah
| | |
|---|---|
| 8:7 | 104 |

## Ezekiel
| | |
|---|---|
| 28:12–19 | 124 |

## Hosea
| | |
|---|---|
| 2:18 | 104 |

## Jonah
| | |
|---|---|
| 4:11 | 104 |

## NEW TESTAMENT

## Matthew
| | |
|---|---|
| 6:26 | 104 |
| 10:16 | 104 |
| 10:29 | 104 |
| 10:30 | 219, 228 |
| 13:24–30 | 126 |
| 13:31–32 | 86 |
| 13:32 | 81 |
| 13:36–43 | 126 |
| 18:12–14 | 259 |

## Mark
| | |
|---|---|
| 1:13 | 104 |
| 5:1–20 | 125 |
| 5:11–13 | 104 |
| 10:31 | 259 |
| 10:45 | 201 |
| 13:32 | 82 |

## Luke
| | |
|---|---|
| 2 | 179 |
| 6:27 | 259 |
| 6:32–33 | 240 |
| 10:31 | 217 |
| 13:16 | 127 |
| 14:28 | 98 |
| 15:32 | 197 |

## John
| | |
|---|---|
| 3:3 | 197 |
| 3:16 | 207 |
| 5:17 | 55 |
| 12:24 | 82 |
| 12:25 | 259 |

## Acts
| | |
|---|---|
| 2:36 | 191 |
| 14:17 | 253 |
| 17:26 | 163 |

## Romans
| | |
|---|---|
| 1 | 30 |
| 1:19–20 | 253 |
| 1:20 | 28, 147 |
| 2:14–15 | 241 |
| 2:15 | 258 |
| 5 | 85, 116, 167, 177, 178, 181, 190, 197 |
| 5:8–10 | 197 |
| 5:12 | 116, 181, 196 |
| 5:13 | 193 |
| 5:14 | 190 |
| 5:18 | 190 |
| 6:2–11 | 197 |

| | | | | | |
|---|---|---|---|---|---|
| 6:4 | 197 | **2 Corinthians** | | **1 Thessalonians** | |
| 6:21 | 197 | 3:18 | 152 | 5:2 | 268 |
| 6:23 | 112, 196, 199 | 4:4 | 152 | | |
| 8:10 | 197 | | | **1 Timothy** | |
| 8:19–20 | 127 | **Galatians** | | 5:17–18 | 105 |
| 8:19–22 | 105 | 4:4 | 257 | | |
| 8:21–22 | 105 | 5:16–21 | 203 | **Hebrews** | |
| 8:22 | 85, 202 | | | 1:1 | 253 |
| | | **Ephesians** | | 11:3 | 80 |
| **1 Corinthians** | | 2:1 | 197 | | |
| 1:27–38 | 191 | 2:5 | 197 | **2 Peter** | |
| 9:9 | 107 | 4:24 | 139 | 2:4 | 124 |
| 9:9–10 | 105 | | | | |
| 9:19–22 | 6 | **Philippians** | | **1 John** | |
| 15 | 8, 178, 202 | 2:10 | 86 | 4:9 | 134 |
| 15:20–26 | 197 | | | | |
| 15:21 | 116, 196 | **Colossians** | | **Jude** | |
| 15:21–22 | 198 | 1:15 | 152 | 6 | 124 |
| 15:36 | 82 | 1:20 | 105, 203 | | |
| 15:42 | 199 | 2:13 | 197 | **Revelation** | |
| 15:42–44 | 117, 202 | 2:15 | 128 | 2:11 | 197 |
| 15:45 | 117 | 3:10 | 139 | 20:14 | 197 |
| 15:46 | 117, 203 | | | 21:1–8 | 201 |
| | | | | 22 | 115 |